Quantum Plasma

Punit Kumar

Quantum Plasma

Springer

Punit Kumar
Department of Physics
University of Lucknow
Lucknow, Uttar Pradesh, India

ISBN 978-981-95-0480-0 ISBN 978-981-95-0481-7 (eBook)
https://doi.org/10.1007/978-981-95-0481-7

This Springer imprint is published by the registered company Springer Nature Singapore Pte Ltd.
The registered company address is: 152 Beach Road, #21-01/04 Gateway East, Singapore 189721, Singapore

Preface

The domain of quantum plasmas is rapidly emerging as a cornerstone of modern physics, where the principles of quantum mechanics and plasma dynamics intersect to explain a wide spectrum of natural and experimental phenomena. From warm dense matter in laser-compressed targets to degenerate electron fluids in white dwarfs, the relevance of quantum plasmas has grown steadily across fields such as astrophysics, nanotechnology, and high-energy-density physics. This book has been written to serve as a detailed and accessible guide for graduate students, researchers, and educators who wish to explore the fundamentals, theoretical frameworks, and recent advancements in this rapidly evolving discipline.

As with any scholarly endeavor, the journey from idea to publication has been made possible with the generous support of many individuals. I express my heartfelt gratitude to Mr. Prabhat Kumar Singh for his dedicated assistance in conducting an extensive literature survey and for his meticulous efforts in the typesetting of this manuscript.

I am deeply thankful to my wife, Dr. Priti Saxena, for her constant encouragement, patience, and emotional support throughout the preparation of this work. Her presence has been a source of strength and inspiration at every step.

I hope this guide proves useful to students, researchers, and professionals seeking to explore and contribute to the fascinating world of quantum plasmas.

Lucknow, India
May 2025

Dr. Punit Kumar

Contents

1 Introduction 1
1.1 Classical to Quantum 2
1.2 Milestones 2
1.3 The Domain of Quantum Plasma 3
1.4 Key Features of Quantum Plasmas 3
1.5 Theoretical Frameworks 5
1.6 Structure and Organization of This Book 5
1.7 Importance and Relevance 6
1.8 Who Should Read This Book? 6

2 Plasma Physics 9
2.1 Historical Aspects 10
2.2 Importance of Plasma in Astrophysics, Fusion Energy, and Industry 12
2.3 Criteria for a Charged Gas to Be Considered Plasma 15
2.3.1 Debye Length and Electrostatic Shielding 15
2.3.2 Plasma Frequency and Collective Oscillations 16
2.3.3 Number of Particles in a Debye Sphere 17
2.4 Collective Behaviour in Plasma 17
2.4.1 Non-local Transport and Instabilities 19
2.4.2 Electrostatic Plasma Oscillations (Langmuir Waves) 19
2.4.3 Ion-Acoustic Waves 20
2.4.4 Electromagnetic Waves in Plasma 21
2.5 Plasma Confinement and Stability 22
2.5.1 Plasma Confinement Mechanisms 22
2.5.2 Stability Considerations 23
2.6 Charged Particle Trajectories in Electric and Magnetic Fields 25

2.7 Classification of Plasma ... 28
2.7.1 Thermal Versus Non-thermal Plasma ... 28
2.7.2 Fully Ionized Versus Partially Ionized Plasma ... 29
2.7.3 Magnetized Versus Unmagnetized Plasma ... 30
2.7.4 Natural Versus Laboratory Plasmas ... 31
2.7.5 Relativistic Versus Non-relativistic Plasma ... 32
References ... 32

3 Electrodynamics in Plasmas ... 35
3.1 Maxwell's Equations in Vacuum and Media ... 35
3.1.1 Wave Propagation in Linear and Nonlinear Media ... 37
3.1.2 Energy–Momentum Conservation and Poynting's Theorem in Electrodynamics ... 40
3.2 Gauge Choices in Electrodynamics ... 43
3.2.1 Lorenz Gauge ... 43
3.2.2 Coulomb Gauge ... 44
3.2.3 Axial Gauge ... 44
3.2.4 Gauge Redundancy ... 45
3.3 Limitations of Maxwell's Equations in Media ... 48
3.3.1 Non-locality and Dispersion Relations ... 48
3.3.2 Breakdown of Classical Maxwellian Theory at High Frequencies ... 49
3.3.3 Classical Versus Quantum Corrections ... 50
3.3.4 Linear Response Theory (LRT) ... 52
References ... 54

4 Introduction to Quantum Plasma ... 57
4.1 Defining Features ... 59
4.1.1 Thermal De Broglie Wavelength ... 59
4.1.2 Fermi Energy and Fermi Temperature ... 60
4.1.3 Quantum Coupling ... 61
4.1.4 Quantum Degeneracy ... 61
4.1.5 Zero-Temperature Limit ... 62
4.2 Physical Characteristics of Quantum Plasma ... 62
4.2.1 Wave–Particle Duality and Quantum Coherence ... 62
4.2.2 Landau Damping and Phase Space Description ... 64
4.3 Dimensionless Parameters in Quantum Plasma ... 66
4.3.1 Quantum Degeneracy Parameter ... 67
4.3.2 Coupling Parameter ... 67
4.3.3 Temperature–Density Phase Diagrams ... 68
4.4 Regimes in Plasma Physics ... 70
4.4.1 Classical Regime ... 70
4.4.2 Semiclassical Regime ... 71
4.4.3 Fully Quantum Regime ... 71

4.5 Quantum Effects in Plasma Dynamics 72
4.5.1 Bohm Potential 72
4.5.2 Quantum Tunneling 74
4.5.3 Wigner Function Formalism 75
4.5.4 Spin Effects and Spin-Polarized Plasmas 77
4.5.5 Role of Exchange–Correlation Potential 79
4.6 Multi-species and Component-Specific Quantum Plasmas 79
4.7 Theoretical Models of Quantum Plasma 81
4.7.1 Fluid Models in Quantum Plasmas 81
4.7.2 Kinetic Models in Quantum Plasma Theory 84
4.7.3 Field-Theoretical Approaches to Quantum Plasma 87
4.8 Applications and Emerging Domains 90
References 94

5 Theoretical Models 97
5.1 Kinetic Theory Approaches to Quantum Plasma 98
5.1.1 Quantum Generalization of Vlasov Equation 98
5.1.2 Wigner Distribution Function: Derivation from the von Neumann Equation 99
5.1.3 Linearized Wigner–Poisson and Wigner–Maxwell Systems 102
5.1.4 Pauli Equation-Based Kinetic Models for Spin-½ Plasmas 105
5.1.5 Landau Damping in Quantum Plasmas 106
5.1.6 Kinetic Models in Ultra-Cold and Dense Plasma Environments 109
5.2 Quantum Hydrodynamic Model 111
5.2.1 Bohm Potential and Quantum Pressure 114
5.2.2 Spin Quantum Hydrodynamics and Magnetization Current Dynamics 117
5.3 Density Functional Theory in Quantum Plasmas 121
5.3.1 The Kohn–Sham Construction 123
5.3.2 Thomas–Fermi and Thomas–Fermi–Dirac Approximations 124
5.3.3 Orbital-Free Density Functional Theory 127
5.3.4 Exchange–Correlation Functionals 130
5.3.5 Time-Dependent Density Functional Theory 133
5.4 Quantum Field Theory 136
5.4.1 Coherent States of Plasma Modes 138
5.4.2 Canonical and Path-Integral Quantization 141
5.4.3 Derivation of QED Plasmas via Functional Integrals 144

5.5 Comparative Analysis 147
5.5.1 Hydrodynamic Models 147
5.5.2 Kinetic Models 147
5.5.3 Field-Theoretic Models 148
References 150

6 Magnetized Quantum Plasma 153
6.1 Overview of Regimes 154
6.1.1 Regime Map and Classification 155
6.1.2 Quantum Versus Classical Magnetization Responses (Pauli, Landau, Diamagnetic Effects) 157
6.1.3 Role of External Versus Self-generated Magnetic Fields 160
6.2 Quantum Effects in Magnetized Plasma: Theoretical Preliminaries 163
6.2.1 Landau Quantization 163
6.2.2 Spin Magnetization and Spin–Orbit Coupling 166
6.2.3 Relativistic Landau Levels and Quantum Hall Features 169
6.2.4 Quantum-Statistical Pressure and Degeneracy Pressure 172
6.2.5 Effective Quantum Potential in the Presence of Magnetic Fields 175
6.2.6 Physical Interpretation and Regime Classification 177
6.3 Quantum Magnetohydrodynamics (QMHD) 178
6.3.1 Single-Fluid and Multi-species QMHD Extensions 181
6.4 Spin-1/2 and Spin-3/2 Fluid Models Under Magnetization 184
6.5 Quantum Instabilities in Magnetized Plasmas 187
6.5.1 The Weibel Instability: Quantum Kinetic Anisotropy 187
6.5.2 Firehose Instability: Quantum Anisotropic Pressure Balance 188
6.5.3 Magnetoacoustic Instabilities: Quantum Magnetosonic Modes 189
6.5.4 Streaming Instabilities: Quantum Counterstreaming and Anisotropic Drift 189
6.5.5 Beam–Plasma Instabilities: Quantum Langmuir–Cherenkov Dynamics 190
6.6 Quantum Turbulence in Magnetized Regimes 190
References 193

7 Spin Polarization in Quantum Plasmas 195
7.1 What is Spin Polarization? 196
7.1.1 Plasma Environments Where Spin is Relevant 197
7.1.2 Challenges: Incorporating Spin Consistently into Collective Plasma Models 199
7.2 Methods: From Quantum Kinetic Theory to Quantum Hydrodynamics 200
7.3 Quantum Hydrodynamic Framework with Spin Degrees of Freedom 201
7.3.1 Wigner–Pauli Approach and Madelung Decomposition 202
7.3.2 Ensemble-Averaged Dirac (Pauli) Equation Leading to Spin QHD 203
7.4 Governing Variables in Spin Quantum Hydrodynamics 203
7.5 Spin Transport Equations in Quantum Hydrodynamics 205
7.5.1 Conservation of Spin Density in the Absence of Spin Relaxation 205
7.5.2 Derivation from the Pauli Equation 206
7.5.3 Role of Spin Torque and Precession 207
7.6 Spin Momentum 208
7.6.1 Spin Force Contributions: Magnetic Torque and Gradient of Spin Pressure 208
7.6.2 Coupling to Magnetization Current and External Magnetic Field 210
7.7 Spin Diffusion Equation 210
7.8 Outlook and Future Directions 212
7.8.1 Extensions to Relativistic Spin Quantum Hydrodynamics (Spin-QHD) 212
7.8.2 Inclusion of Spin–Spin and Exchange Correlations 213
7.8.3 Coupling to Quantum Kinetic and 2PI Spinor Field Theory Approaches 214
7.8.4 Opportunities in Quantum Simulation Platforms and Spintronic Plasmas 215
References 215

8 Quantum Plasma in Astrophysics 217
8.1 White Dwarfs 217
8.2 Neutron Stars 220
8.3 Quark-Gluon Plasma 223
8.4 Brown and Red Dwarfs 225
8.5 Quantum Plasma in Accretion Disks and Magnetospheres 227
8.6 Quantum Plasma in Nebulae and Stellar Birth Regions 230
8.7 Quantum Plasma During Star Formation and Protostellar Collapse 231
8.8 Quantum Effects in Dusty Astrophysical Plasma 232

8.9 Instabilities in Quantum Astrophysical Plasmas ... 234
8.9.1 Quantum-Modified Weibel, Jeans, and Rayleigh–Taylor Instabilities in Astrophysical Plasmas ... 234
8.9.2 Quantum Exchange-Driven and Spin-Induced Instabilities in Astrophysical Quantum Plasmas ... 236
8.9.3 Quantum Hall Drift Instabilities and Turbulence Under Strong Confinement ... 239
References ... 241

9 Quantum Plasma in Fusion Research ... 243
9.1 Overview of Confinement Strategies ... 244
9.1.1 Magnetic Confinement ... 244
9.1.2 Inertial Confinement ... 245
9.1.3 Gravitational Confinement ... 246
9.2 Magnetic Confinement Fusion in the Quantum Regime ... 247
9.2.1 Quantum-Modified Drift Kinetics and Gyrokinetics ... 250
9.2.2 Quantum Hall-MHD Effects in Strong-Field Environments ... 253
9.2.3 Spin Alignment and Quantum Vorticity in Magnetized Plasma Flows ... 256
9.2.4 Quantum Landau Damping ... 259
9.3 Inertial Confinement Fusion (ICF) and Quantum Effects ... 262
9.3.1 Quantum Kinetic Models in Shock and Compression Dynamics ... 264
9.3.2 Laser–Plasma Interaction Under Quantum Dispersion Regimes ... 267
9.4 Gravitational and Relativistic Confinement Considerations ... 270
9.4.1 Quantum Plasma Dynamics in Compact Object Analogs ... 273
9.4.2 Application of Relativistic Quantum Plasma Theory to Extreme Confinement ... 276
9.4.3 Coupling of Curvature Effects with Quantum Corrections in Plasma Flows ... 279
References ... 282

10 Quantum Plasmas in High Energy Contexts ... 285
10.1 Thermal Quantum Plasmas Versus Strong-Field Relativistic Plasmas ... 286
10.1.1 Kinetic Descriptions and Field-Particle Interactions ... 287
10.1.2 Role of Temperature Versus Field Strength ... 288
10.1.3 Quantum Coherence and Radiation Effects ... 288
10.1.4 Overview of Energy Scales: From keV–MeV (Laser-Induced) to GeV–TeV ... 289

10.1.5 Relevance to High-Intensity Laser Physics, Particle Colliders, and Astrophysical Observations 291
10.2 Relativistic Quantum Plasma 294
10.2.1 Dirac–Maxwell and Klein–Gordon–Maxwell Systems: Relativistic Field-Theoretic Descriptions of Quantum Plasmas 296
10.2.2 Wave Propagation and Collective Excitations in High-Energy Quantum Regimes 299
10.2.3 Applications—Gamma-Ray Bursts, Relativistic Shocks, and Quark-Gluon Plasma Precursors 303
10.3 Laser–Plasma Interaction in the Quantum Regime 305
10.3.1 Quantum Corrections to Ponderomotive Force 308
10.3.2 Nonlinear Compton Scattering and Electron Acceleration 310
10.3.3 Importance in Laser Wakefield Acceleration and Next-Generation Accelerators 312
10.3.4 Applications in Attosecond Pulse Generation and Extreme Ultraviolet (EUV) Sources 314
10.4 Critical Fields and the Schwinger Limit 315
10.4.1 Quantum Plasma Behaviour in Sub-critical, Critical, and Supercritical Regimes 316
10.5 Inverse Compton Scattering 318
10.6 Quantum Recoil, Nonlinear Compton Shift, and Spectral Broadening 320
10.7 Application in Compact Gamma-Ray Sources and Diagnostics 323
References 325

11 Experimental Techniques and Simulations 327
11.1 Laboratory Platforms 327
11.1.1 Ultracold Plasma Experiments 327
11.1.2 Laser-Produced Dense Plasmas 328
11.1.3 High-Pressure Shock Tubes and Diamond Anvil Cells 329
11.1.4 Inertial Confinement Fusion (ICF) Experiments 329
11.1.5 Quantum Plasmas in Condensed Systems 330
11.2 Diagnostics and Measurement Tools 331
11.2.1 Thomson Scattering: Probing Electron Density, Temperature, and Quantum Effects 331
11.2.2 X-ray Absorption and Emission Spectroscopy: Detecting Ionization Potential Depression and Quantum Screening 332
11.2.3 Interferometry and Reflectometry: Reconstructing Quantum-Scale Density Profiles 333

11.2.4 Collective Oscillation Probes: Langmuir Frequency Shifts and Quantum Dispersion Relations ... 334
11.3 Simulation Techniques for Quantum Plasma ... 334
11.3.1 Particle-In-Cell (PIC) with Quantum Corrections ... 335
11.3.2 Monte Carlo Simulations in Quantum Plasma ... 336
11.3.3 Molecular Dynamics Simulations in Quantum Plasma ... 338
11.3.4 Lattice Gauge and Dirac Simulations ... 341
11.3.5 Time-Dependent Density Functional Theory (TDDFT) ... 342
11.4 Real-Time and Non-equilibrium Simulations ... 343
11.4.1 Schwinger–Keldysh Techniques: Quantum Statistical Ensembles and Decoherence ... 344
11.4.2 Quantum Quench Simulations: Prethermalization and Hydrodynamization ... 344
11.4.3 A2PI Effective Action Methods: Quantum Turbulence and Coherent Mode Dynamics ... 345
11.5 Simulation Benchmarks and Validation ... 346
11.6 Hybrid Approaches ... 349
11.6.1 Quantum–Classical Hybrid Modeling ... 349
11.6.2 Data-Driven Simulations ... 351
11.6.3 Reinforcement Learning in Control of Experimental Quantum Plasmas ... 352
11.6.4 Bayesian Inference for Extracting Quantum Parameters from Spectral Data ... 354
11.7 Challenges, Limitations, and Future Directions ... 355
11.7.1 Experimental Limits ... 355
11.7.2 Simulation Challenges ... 356
11.7.3 Bridging Laboratory and Astrophysical Plasmas ... 357
11.7.4 Role of Quantum Computing in Modeling Many-Body Quantum Plasmas ... 358
11.7.5 Synergies Between Experiments, Theory, and Simulations ... 360
References ... 361

12 Future Directions and Challenges ... 365
12.1 Open Theoretical Problems in Quantum Plasma Physics ... 365
12.1.1 Nonperturbative and Strong Coupling Regimes ... 365
12.1.2 Gauge Symmetry, Renormalization, and Higher-Order QED Corrections ... 367
12.1.3 Quantum Turbulence and Topological Effects ... 367
12.1.4 Unified Frameworks Across Scales ... 369

12.2 Frontier Computational and Experimental Challenges 369
12.2.1 Quantum Simulation and Emulation 369
12.2.2 High-performance and Quantum Computing for Plasma Simulation 371
12.2.3 Diagnostic Gaps and Instrumentation 372
12.3 Cross-Disciplinary and Technological Applications 373
12.3.1 Quantum Plasma in Quantum Information Science 373
12.3.2 Nanoplasma and Quantum Photonics 375
12.3.3 Biomedical and Environmental Prospects 376
12.3.4 Compact Plasma-Based Accelerators and Fusion Reactors 377
12.4 Interfacing Quantum Plasma with Emerging Paradigms 379
12.4.1 Quantum Thermodynamics and Entropy Production 379
12.4.2 Quantum Gravity, Cosmology, and Early Universe Plasmas 381
12.4.3 Machine Learning and Data-Driven Models 382
12.5 Synthesis and Future Vision 384
12.5.1 Recapitulation of Theoretical, Computational, and Experimental Themes 384
12.5.2 Defining Central Questions for the Next Decades 384
12.5.3 Toward Unified Frameworks: QED, QCD, Optics, and Plasma 385
12.5.4 Quantum Matter Under Extreme Conditions: Toward an Integrated Theory 386
References 387

Index 389

Chapter 1
Introduction

Plasma, often referred to as the fourth state of matter, is an electrified gas of charged particles that exhibits collective behavior. From the fiery corona of the Sun to the ionized trails of comets, from fusion reactors on Earth to the interstellar medium, plasma is the most abundant form of visible matter in the universe. Classical plasma physics, developed throughout the twentieth century, has provided a solid theoretical and experimental foundation for understanding such systems. However, as our scientific capabilities advanced, we began probing environments that lie at the very edge of physical understanding conditions of extremely high densities, ultra-low temperatures, and intense electromagnetic fields. These frontier conditions demand a new framework of thinking, one where the laws of quantum mechanics become essential. This is the realm of quantum plasma.

Quantum plasmas are not just a theoretical curiosity. They are central to understanding the physics of white dwarfs and neutron stars, designing future nanoelectronic devices, and exploring the behavior of ultra-intense laser-solid interactions. With increasing interest in quantum technologies, high-energy–density physics, and astrophysical modeling, the study of quantum plasma has become one of the most vibrant interdisciplinary domains in contemporary physics.

This book, *A Quantum Plasma*, is a rigorous attempt to bring together the vast and diverse developments in this field into a structured, unified, and pedagogically sound format. Covering theory, applications, modeling, and experiments, the aim is to make this work a definitive reference for students, researchers, and professionals involved in plasma physics, quantum mechanics, astrophysics, material science, and applied physics.

P. Kumar, *Quantum Plasma*, https://doi.org/10.1007/978-981-95-0481-7_1

1.1 Classical to Quantum

The evolution from classical to quantum plasma physics arises from a fundamental realization, classical descriptions break down when quantum effects become significant. In classical models, plasma particles are treated as point-like entities with well-defined positions and velocities. However, when conditions involve extremely high particle densities, or very low temperatures, the quantum nature of matter comes to the forefront. Specifically, the de Broglie wavelength of electrons becomes comparable to the average inter-particle spacing, and the assumption of distinguishable, classical particles no longer holds. Under such circumstances, electrons obey Fermi–Dirac statistics and the system transitions into a quantum degenerate state. In this regime, thermal energy becomes secondary, and pressure arises predominantly from the Pauli's exclusion principle, which prevents electrons from occupying identical quantum states leading to what is known as degeneracy pressure.

Several uniquely quantum phenomena emerge in this environment. Quantum diffraction effects cause wave-like interference, tunneling allows particles to pass through potential barriers, and zero-point energy persists even at absolute zero. Additionally, spin polarization and exchange–correlation interactions, rooted in the indistinguishability of fermions begin to influence plasma dynamics. Classical kinetic and fluid theories are inadequate for describing such phenomena; instead, specialized quantum kinetic equations or quantum hydrodynamic models become necessary.

Typical conditions where quantum plasma modeling becomes essential include electron number densities exceeding 10^{26} m^{-3}, temperatures below the Fermi temperature, magnetic field strengths approaching the Schwinger critical field (~10^{13} Gauss), and systems operating at ultrafast (attosecond) time scales, or nanoscale dimensions. The transition from classical to quantum plasma physics is not merely a change in mathematical treatment, it reflects a deep shift in our understanding of plasma behavior, moving from the realm of statistical disorder to one of coherent, collective quantum effects that govern the dynamics of matter under extreme conditions.

1.2 Milestones

Although the concept of 'quantum plasma' has gained prominence in recent decades, its theoretical origins stretch back to the foundational years of quantum mechanics. In 1926, Erwin Schrödinger introduced wave mechanics, marking the birth of quantum theory and establishing the principle of wave-particle duality, which would later prove crucial in understanding the collective behavior of particles in a plasma. During the 1930s–1950s, significant progress was made when Enrico Fermi and Paul Dirac developed quantum statistical mechanics, leading to the formulation of Fermi–Dirac statistics. Around the same time, Lev Landau introduced kinetic theory and the concept of Landau damping, which laid the groundwork for quantum kinetic modeling. In the 1960s, pioneering work by David Pines and Landau on quantum many-body systems led to the description of collective excitations, such as plasmons, within degenerate electron gases, core elements of quantum plasma theory. The 1980s saw renewed interest as the fields of semiconductor physics and laser-plasma interactions expanded, exposing the relevance of quantum effects in technologically significant regimes. Entering the twenty-first century, rapid developments in high-energy–density physics, attosecond laser science, spintronics, and nanoplasmonics elevated quantum plasma research to a critical interdisciplinary field, merging insights from atomic, condensed matter, plasma, and quantum field physics into a unified and dynamic discipline (Fig. 1.1).

1.3 The Domain of Quantum Plasma

Quantum plasma encompasses a remarkably diverse range of physical environments, extending far beyond the typical realms explored in classical plasma physics, such as fusion devices, or interstellar media. In astrophysics, quantum plasma plays a pivotal role in the interiors of white dwarfs, neutron stars, and magnetars. These celestial bodies exhibit extreme conditions, enormous densities and intense magnetic fields that give rise to degenerate electron fluids, Landau quantization, and in many cases, relativistic quantum phenomena. In the domain of high-energy–density physics, inertial confinement fusion (ICF) experiments and interactions involving petawatt-class lasers generate transient quantum plasmas characterized by highly non-linear and non-classical behavior. In condensed matter physics, semiconductors, quantum wells, metallic nanostructures, and nanowires display quantum plasma features through the collective oscillations of conduction electrons, known as plasmons. These systems are inherently quantum mechanical due to their reduced dimensionality and confinement effects. Laboratory-created cold plasmas, where ultracold electrons are trapped or cooled to near absolute zero, show behaviors that are accurately captured only through quantum hydrodynamics. Even in the emerging realm of quantum information science, superconducting circuits and qubit arrays sometimes exhibit plasma-like collective excitations. Each of these domains presents unique challenges, necessitating specialized models that integrate quantum statistics, wave mechanics, and field-theoretical approaches to accurately capture the underlying physics.

1.4 Key Features of Quantum Plasmas

Quantum plasmas display a range of distinct characteristics that set them apart from classical plasmas, arising primarily from the quantum nature of their constituent particles. One of the most fundamental features is quantum statistical pressure, which emerges under degenerate conditions where electrons fill up energy states up to the Fermi level. Unlike classical pressure, this degeneracy pressure is independent of

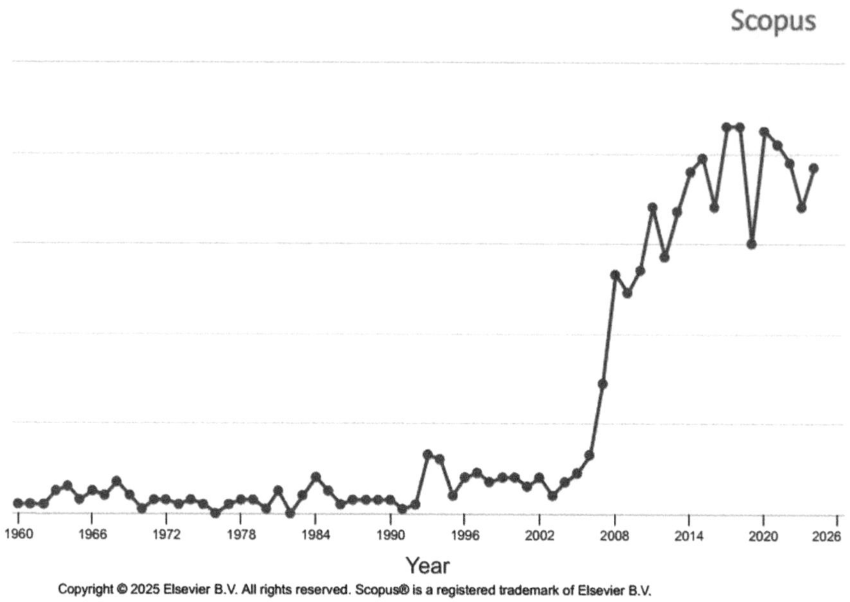

Fig. 1.1 Annual number of publications on quantum plasma in Scopus-indexed journals, highlighting the growing research interest and advancements in the field over recent years (*Source* https://www.scopus.com)

temperature and plays a crucial role in supporting dense astrophysical objects like white dwarfs against gravitational collapse. Another hallmark is the Bohm potential, a term in quantum fluid models that captures the effects of quantum diffraction and interference. This term enables phenomena such as quantum tunneling, which has no counterpart in classical theory. Additionally, spin and magnetic effects become significant in strongly magnetized plasmas, where electron spin gives rise to spin waves, magnetization, and Pauli paramagnetism, particularly in spin-polarized, or ultra-cold environments. Exchange and correlation effects, stemming from the indistinguishability of fermions and the antisymmetry of their wavefunctions, introduce additional complexity, especially in dense, low-temperature plasmas. Moreover, quantum corrections alter the nonlinear and dispersive dynamics of plasma waves, modifying the dispersion relations for Langmuir and ion-acoustic modes and enabling the formation of quantum solitons, vortices, and shock waves. These features collectively render quantum plasma a uniquely rich and challenging field, where classical approximations often break down, and quantum mechanics must guide our understanding of collective behavior.

1.5 Theoretical Frameworks

The study of quantum plasmas requires a layered and flexible theoretical approach, as no single model can capture the wide range of scales, densities, and quantum effects involved. At the most fundamental level, quantum kinetic models such as the Wigner–Moyal formalism and the quantum Vlasov equation provide a statistical description of plasma behavior, incorporating phase-space quantum effects and non-local dynamics. For situations where a fluid-like approximation is valid, quantum hydrodynamics (QHD) offers a more tractable framework by extending classical fluid equations to include quantum-specific terms like Fermi pressure, the Bohm potential, and spin contributions. When magnetic fields play a dominant role, quantum magnetohydrodynamics (QMHD) becomes essential, incorporating quantum corrections into the magnetohydrodynamic equations to describe plasmas under strong magnetic confinement or in magnetized astrophysical settings. Density Functional Theory (DFT), although traditionally a tool of condensed matter physics, has proven useful for describing strongly correlated electron systems and can be adapted to model quantum plasmas in specific regimes, especially at equilibrium. For extreme conditions, such as those involving relativistic speeds or ultra-strong magnetic fields, relativistic and spin-resolved models like the Dirac–Maxwell framework are necessary. The choice of model depends on key plasma parameters including particle density, temperature, magnetic field strength, and system size, ensuring the theoretical description accurately captures the relevant physical phenomena.

1.6 Structure and Organization of This Book

The structure and organization of this book have been carefully designed to guide the reader progressively from foundational principles to advanced applications in the field of quantum plasma physics. Chapter 2 transitions into the basics of plasma physics, offering a comprehensive overview of plasma states, the criteria for a gas to become plasma, essential plasma parameters, and the collective behavior of charged particles. It further categorizes different types of plasmas and their respective domains.

Chapter 3 delves into the electrodynamics of uncharged media and plasmas, highlighting Maxwell's equations, gauge choices, and the limitations of classical formulations. Chapter 4 explores the distinctive nature of quantum plasma, focusing on quantum effects, electron degeneracy, and relevant theoretical models. Chapter 5 elaborates on various theoretical approaches, including kinetic theory (Vlasov and quantum kinetic equations), quantum hydrodynamics (QHD), density functional theory (DFT), and quantum field theory (QFT), emphasizing QED applications and challenges.

Chapter 6 focuses on magnetized quantum plasma, exploring magnetic field effects, magnetohydrodynamics (MHD), and quantum MHD models. Chapter 7 examines spin polarization in quantum plasmas, emphasizing spin transport equations, spin-wave phenomena, and QHD-based simulations. Chapter 8 connects quantum plasma to astrophysical contexts such as white dwarfs, neutron stars, and accretion disks. Chapter 9 discusses quantum plasma in fusion research, including confinement techniques and plasma diagnostics. Chapter 10 explores high-energy physics applications, including laser-plasma interactions and particle generation. Chapter 11 covers experimental methods and simulations, while Chap. 12 discusses future directions, unresolved challenges, and interdisciplinary potentials, concluding with a summary of key insights and the path ahead in quantum plasma research.

1.7 Importance and Relevance

Quantum plasma has evolved beyond its traditional association with exotic astrophysical phenomena and now plays a crucial role in various advanced technologies and scientific fields. In the realm of compact object physics, it helps explain the behavior of stellar remnants, equilibrium states, and the spectral emissions from these objects. In plasmonics, quantum plasma principles enable the development of miniaturized photonic circuits and advanced quantum sensors. In fusion research, it contributes to modeling dense plasmas, essential for the advancement of energy generation through fusion. Further, quantum plasma forms the foundation of quantum devices, influencing the physics of semiconductors and low-dimensional systems that are central to modern electronics. In space missions, the study of quantum plasmas enhances radiation shielding, propulsion mechanisms, and communication technologies, improving the efficiency and safety of space exploration. As science ventures into extreme conditions of temperature, density, and scale, understanding quantum plasmas becomes indispensable, highlighting their increasing relevance and necessity in both research and practical applications.

1.8 Who Should Read This Book?

This book is designed for a diverse audience, including graduate students in physics, astrophysics, and applied sciences who wish to establish a strong foundation in quantum plasma. It serves as a valuable reference for researchers seeking theoretical formulations, equations, and models related to quantum plasmas. Engineers and technologists working with nanomaterials, quantum optics, or high-energy systems will find practical insights to enhance their work. Additionally, educators and academicians involved in curriculum design or teaching advanced plasma physics will benefit from its comprehensive content. The book assumes that readers have a working knowledge of quantum mechanics, statistical mechanics, and electromagnetism. For those in need of a refresher, an appendix is included to quickly review prerequisite concepts. Whether you're a student, researcher, or professional, this book offers essential knowledge for understanding and exploring the field of quantum plasma.

Quantum plasma physics represents one of the most intellectually stimulating and practically significant frontiers of modern science. It challenges our classical understanding of matter and forces us to reconsider the behavior of collective systems at quantum scales. As we step into an era dominated by quantum technologies, space exploration, and ultra-fast computing, the insights offered by quantum plasma research will continue to grow in importance.

This book is an invitation to explore this fascinating domain with depth, precision, and curiosity. Welcome to the world of Quantum Plasma.

Chapter 2
Plasma Physics

Plasma physics is the branch of physics that studies the behaviour, properties, and governing principles of plasma, a state of matter distinct from solids, liquids, and gases, characterized by the presence of a significant number of free electrons and ions. It emerges when a gas is sufficiently ionized, meaning that a substantial fraction of its constituent atoms or molecules have been stripped of their electrons due to thermal energy, electromagnetic fields, or other ionizing mechanisms. This ionization endows plasma with unique properties, such as collective behaviour, long-range electromagnetic interactions, and the ability to sustain wave-like phenomena and instabilities. Unlike neutral gases, where interactions are typically governed by short-range collisions, plasmas are dominated by long-range Coulomb forces, resulting in complex and often nonlinear dynamics that require advanced mathematical and computational frameworks for their analysis.

The significance of plasma physics extends across both fundamental and applied domains. Fundamentally, plasma physics provides insights into the behaviour of astrophysical and laboratory plasmas, helping to elucidate phenomena such as stellar formation, magnetohydrodynamic (MHD) processes in the solar corona, and cosmic ray propagation. It plays a critical role in space physics, explaining the dynamics of planetary magnetospheres, the solar wind, and interstellar plasmas. In applied contexts, plasma physics underpins the development of controlled nuclear fusion, which aims to harness thermonuclear reactions for energy production through devices such as Tokamaks and Stellarators. Industrial applications include plasma processing in semiconductor manufacturing, plasma propulsion for spacecraft, and plasma-assisted material fabrication. Additionally, high-energy density plasmas are central to inertial confinement fusion, laser-plasma interactions, and particle acceleration technologies.

Mathematically, plasma behaviour is described using a hierarchy of models, ranging from kinetic theory, which employs the Boltzmann and Vlasov equations to represent phase-space distributions of charged particles, to fluid models such as magnetohydrodynamics (MHD), which treat plasma as a conducting fluid governed

P. Kumar, *Quantum Plasma*, https://doi.org/10.1007/978-981-95-0481-7_2

by the coupled Maxwell and Navier–Stokes equations. The kinetic approach captures microscopic dynamics, including Landau damping and particle trapping, while the fluid approach provides macroscopic descriptions of equilibrium, stability, and wave propagation. The fundamental equations governing plasma dynamics include the continuity equation, the momentum equation incorporating the Lorentz force, and the energy equation accounting for dissipative and non-ideal effects. These are supplemented by Maxwell's equations, which determine the self-consistent electromagnetic fields that mediate plasma interactions. In many cases, simplifying assumptions such as the quasi-neutrality condition, wherein charge densities approximately balance over macroscopic scales allow for reduced models that retain essential physics while facilitating analytical and numerical treatments.

Plasma physics, therefore, represents a cornerstone of modern theoretical and applied physics, bridging microscopic particle dynamics with macroscopic collective effects. Its study necessitates sophisticated mathematical tools, including perturbation theory, spectral analysis, and nonlinear dynamical systems. The interplay between electromagnetic fields and charged particle motion leads to a rich array of wave phenomena, instabilities, and transport processes that define plasma behaviour across vastly different scales, from laboratory experiments to cosmic structures. Given its foundational and practical importance, plasma physics continues to be an area of extensive research, with implications ranging from fundamental physics to advanced technological innovations.

2.1 Historical Aspects

The historical development of plasma physics is deeply intertwined with the broader evolution of electromagnetism, statistical mechanics, and astrophysical studies. While the concept of ionized gases can be traced back to early observations of lightning and aurorae, the formal scientific treatment of plasma as a distinct state of matter emerged in the late nineteenth and early twentieth centuries. The systematic study of plasma dynamics, driven by theoretical advancements and experimental discoveries, has shaped our understanding of both Natural and laboratory plasmas, culminating in modern applications such as controlled nuclear fusion and Space plasma physics.

The foundational principles of plasma physics were laid by James Clerk Maxwell, whose formulation of Maxwell's Equations (1861–1865) provided the fundamental framework for understanding electromagnetic interactions. Maxwell's work established the relationship between electric and magnetic fields, forming the basis for later studies of charged particle motion in plasmas. In parallel, Ludwig Boltzmann and J. Willard Gibbs developed statistical mechanics in the late nineteenth century, providing a rigorous mathematical foundation for describing the collective behaviour of particles, which later became essential for kinetic plasma theory.

The experimental discovery of cathode rays in the nineteenth century provided early evidence for the existence of free electrons, a defining feature of plasma. The pioneering work of Sir William Crookes in 1879 on discharge tubes led him to introduce the term "radiant matter" to describe ionized gas-like phenomena. Later, J. J. Thomson's discovery of the electron in 1897, along with his measurements of charge-to-mass ratios in ionized gases, reinforced the notion that ionized matter exhibits distinct electrical properties. Irving Langmuir, in the 1920s, formalized the term 'plasma' to describe ionized gases in discharge tubes, noting their quasi-neutrality and collective behaviour. His contributions extended to the development of Langmuir probes, which enabled direct measurements of electron temperature and density in laboratory plasmas.

The mid-twentieth century saw significant advances in both theoretical and applied plasma physics. Hannes Alfvén's introduction of magnetohydrodynamics (MHD) in 1942 revolutionized the study of plasma behaviour in astrophysical and laboratory contexts. His work on Alfvén waves, a fundamental mode of wave propagation in magnetized plasmas provided insights into solar and space plasmas, earning him the Nobel Prize in 1970. Simultaneously, Lev Landau's 1946 discovery of Landau damping elucidated the mechanism by which plasma waves dissipate energy without particle collisions, a phenomenon critical to wave-particle interactions and plasma stability.

The post-war era witnessed a surge in plasma research, primarily motivated by nuclear fusion. The development of controlled fusion concepts, such as the Tokamak design by Igor Tamm and Andrei Sakharov in the 1950s, demonstrated the feasibility of confining high-temperature plasmas using strong magnetic fields. Lyman Spitzer's theoretical work on Stellarators and plasma confinement in the same period laid the groundwork for ongoing efforts in magnetic confinement fusion. Additionally, the study of shock waves, turbulence, and instabilities, as pioneered by Lev Artsimovich and MarsHall Rosenbluth, became integral to understanding plasma transport and confinement.

Space plasma physics also gained prominence with the advent of satellite missions in the latter half of the twentieth century. The discovery of the Van Allen radiation belts by James Van Allen in 1958 and subsequent studies of the solar wind by Eugene Parker expanded our knowledge of magnetospheric dynamics and planetary plasmas. Parker's theoretical prediction of the continuous solar wind, later confirmed by spacecraft observations, underscored the universality of plasma behaviour across cosmic scales.

In contemporary plasma physics, research has expanded into high-energy-density plasmas, laser-plasma interactions, and quantum plasmas. The development of ultra-intense lasers has enabled experimental studies of relativistic plasma dynamics, while advancements in computational plasma physics, such as particle-in-cell (PIC) simulations, have provided unprecedented insights into nonlinear plasma phenomena. Plasma physics continues to evolve as a multidisciplinary field, integrating aspects of fluid dynamics, statistical mechanics, and field theory to address fundamental questions in astrophysics, fusion energy, and space science.

2.2 Importance of Plasma in Astrophysics, Fusion Energy, and Industry

Plasma, as the most prevalent state of matter in the observable universe, plays an indispensable role in a vast range of physical phenomena, spanning astrophysical systems, controlled fusion energy, and numerous industrial applications. Its ubiquity in cosmic structures, its potential as a limitless energy source, and its technological utility in modern engineering and manufacturing underscore its fundamental importance. A rigorous examination of plasma's significance necessitates a systematic exploration of its role in astrophysics, its application in nuclear fusion, and its diverse industrial implementations, each governed by distinct yet interconnected physical principles (Fig. 2.1).

Plasma in Astrophysics

Astrophysical systems are inherently dominated by plasma processes, where charged particles interact with electromagnetic fields to govern large-scale cosmic structures and small-scale energetic phenomena. The interstellar medium (ISM), composed predominantly of ionized hydrogen, is a vast plasma environment where star formation, shock waves, and magnetohydrodynamic (MHD) turbulence dictate the evolution of galaxies. Theoretical and observational studies of MHD plasmas elucidate the role of magnetic reconnection, a fundamental plasma process responsible for explosive energy release in astrophysical settings such as solar flares, coronal mass ejections (CMEs), and accretion disk dynamics.

The Sun itself is an archetypal plasma system, exhibiting a highly dynamic structure ranging from the convective plasma region in its interior to the outer corona, where plasma temperatures exceed several million Kelvin. The solar wind, a continuous outflow of ionized particles from the Sun's corona, interacts with planetary magnetospheres, shaping space weather phenomena and geomagnetic storms. The Earth's magnetosphere, another critical plasma environment, consists of trapped charged particles in the Van Allen radiation belts, where kinetic and fluid plasma theories describe particle dynamics, wave-particle interactions, and magnetic substorms.

Beyond the solar system, plasma physics underpins our understanding of exotic astrophysical environments, including neutron star magnetospheres, pulsar winds, and active galactic nuclei (AGN). Relativistic plasmas, where particle velocities approach the speed of light, exhibit unique phenomena such as synchrotron radiation and pair plasma dynamics, essential for interpreting observations of quasars and gamma-ray bursts. The study of plasma turbulence in astrophysical jets and cosmic ray propagation relies on non-linear plasma equations and kinetic simulations, providing deeper insights into the mechanisms governing high-energy astrophysical processes.

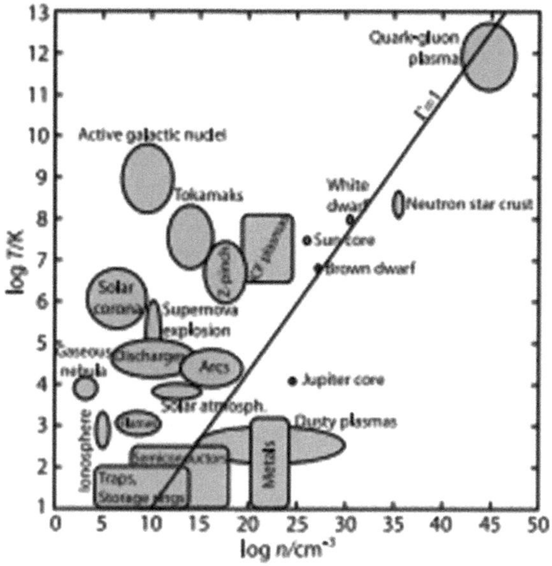

Fig. 2.1 Various plasma types mapped across the density–temperature spectrum (*Source* Donkó, Z., Hartmann, P., & Kalman, G. J. (2007). Strongly coupled plasma liquids. arXiv preprint arXiv:0710.5229).

Plasma in Controlled Fusion Energy

The pursuit of controlled nuclear fusion as a viable energy source is a cornerstone of plasma physics, driven by the need for a sustainable and nearly limitless supply of energy. Fusion occurs in high-temperature plasmas where thermal kinetic energies enable the overcoming of Coulombic repulsion between nuclei, facilitating nuclear reactions that release vast amounts of energy, analogous to stellar fusion processes.

Magnetic confinement fusion (MCF) and inertial confinement fusion (ICF) represent the two principal approaches for achieving fusion conditions in laboratory settings. In MCF, as exemplified by the Tokamak and Stellarator configurations, plasma is confined by strong toroidal and poloidal magnetic fields, mitigating energy losses while sustaining temperatures exceeding 10^7 K, necessary for deuterium–tritium fusion. The physics of magnetic reconnection, turbulence-induced transport, and gyrokinetic instabilities are central to optimizing plasma confinement and reducing anomalous energy losses. Advanced plasma diagnostics, including Thomson scattering and laser interferometry, enable precise measurement of density and temperature profiles, guiding theoretical and computational models aimed at achieving net energy gain.

In ICF, high-energy lasers, or ion beams compress a small pellet of fusion fuel to extreme densities, initiating thermonuclear burn via shock-driven implosion. The governing equations of laser-plasma interaction involve nonlinear wave coupling, plasma instabilities (such as Rayleigh–Taylor and Richtmyer-Meshkov instabilities), and relativistic electron transport. The high-energy laser experiments have demonstrated significant progress in achieving ignition conditions, pushing the boundaries of plasma physics in extreme states of matter.

Beyond energy production, fusion plasmas provide a platform for studying high-energy–density physics (HEDP), relevant to astrophysical conditions found in white dwarfs and supernova explosions. The interplay of MHD turbulence, radiation hydrodynamics, and nuclear reaction kinetics in fusion plasma research extends well beyond terrestrial applications, informing our understanding of stellar and cosmological plasma environments.

Plasma in Industry and Technological Applications

Plasma physics has found extensive applications across multiple industrial and technological domains, where its unique properties enable high-precision manufacturing, material processing, and advanced propulsion technologies. One of the most significant industrial applications is in plasma enhanced semiconductor fabrication, where low-temperature plasmas are employed in etching, deposition, and surface modification processes essential for microelectronics and nanotechnology. Plasma assisted chemical vapour deposition (PECVD) and reactive ion etching (RIE) leverage controlled plasma interactions to fabricate nanoscale structures with unprecedented precision, facilitating advancements in transistor miniaturization and quantum device engineering.

Plasma propulsion systems, particularly in the form of Hall-effect thrusters and magnetoplasmadynamic (MPD) thrusters, provide efficient alternatives to chemical propulsion for space exploration. These systems utilize ionized propellants accelerated by electromagnetic forces, achieving significantly higher specific impulse compared to conventional chemical rockets. The physics of sheath formation, plasma-wall interactions, and electrostatic confinement in these devices exemplifies the intricate control required for optimizing performance in plasma-based propulsion technologies.

Atmospheric pressure plasmas play a crucial role in sterilization, environmental remediation, and biomedical applications. Cold plasma discharges are utilized for bacterial decontamination, wound healing, and surface sterilization, demonstrating a broad spectrum of applications in medical and healthcare technologies. Additionally, plasma based environmental technologies facilitate air purification and wastewater treatment through the generation of reactive oxygen and nitrogen species, contributing to sustainable environmental management.

In materials science, plasma processing techniques such as plasma nitriding and plasma-enhanced surface coatings enhance the mechanical and thermal properties of materials, improving wear resistance and durability in aerospace and automotive industries. Plasma arc welding and plasma torches offer highly controlled and efficient methods for precision machining and metal cutting, leveraging the high-energy density of ionized plasmas to achieve superior performance in industrial fabrication.

The pervasive role of plasma across astrophysical, energy, and industrial applications highlights its fundamental importance in both scientific inquiry and technological innovation. From the vast cosmic plasma structures governing stellar evolution and galactic dynamics to the controlled thermonuclear reactions promising an abundant energy future, and the transformative impact on industrial processes, plasma physics remains at the forefront of modern physics and engineering. The mathematical and physical frameworks describing plasma behavior, encompassing kinetic theory, fluid models, and electromagnetic interactions continue to evolve, driven by both fundamental research and applied advancements. The continued exploration of plasma in extreme conditions, novel confinement regimes, and interdisciplinary applications reaffirms its status as a foundational pillar of contemporary physics and a key enabler of future scientific and technological breakthroughs.

2.3 Criteria for a Charged Gas to Be Considered Plasma

A charged gas is not automatically classified as a plasma, it must satisfy specific criteria that distinguish it from an ordinary ionized gas. The fundamental requirements stem from the necessity of collective behaviour, effective shielding of electric fields, and the presence of characteristic plasma oscillations. Mathematically, these criteria are encapsulated by three essential parameters, the Debye length, plasma frequency, and the number of particles within a Debye sphere. These parameters collectively determine whether the system exhibits the defining characteristics of plasma physics rather than behaving as a simple collection of independent charged particles.

2.3.1 Debye Length and Electrostatic Shielding

The Debye length λ_D is a fundamental scale in plasma physics that quantifies the distance over which electrostatic interactions are screened by the surrounding charge distribution. Unlike in vacuum, where the Coulomb potential of a charged particle extends indefinitely, in a plasma, free charge carriers rearrange themselves in response to an external perturbation, forming a screening cloud that reduces the influence of the external charge beyond a characteristic length. This effect, known as Debye shielding, ensures that long-range electric fields do not dominate the plasma behaviour, a necessary condition for collective interactions to take precedence over binary Coulomb interactions.

The Debye length is derived from the Poisson's equation combined with the Boltzmann distribution for electron density perturbations. Assuming a small electrostatic potential perturbation $\phi(\mathbf{r})$, the charge density $\rho(\mathbf{r})$ follows:

$$\rho = q(n_i - n_e),$$

where n_i and n_e are the ion and electron densities, respectively, and q is the elementary charge. The equilibrium Boltzmann relation for electrons in a weak electrostatic potential is given by:

$$n_e = n_{e0} \exp\left(-\frac{q\phi}{k_B T_e}\right) \approx n_{e0}\left(1 - \frac{q\phi}{k_B T_e}\right),$$

where the approximation holds for small perturbations $\frac{q\phi}{k_B T_e} << 1$. Substituting this into Poisson's equation:

$$\nabla^2 \phi = -\frac{\rho}{\varepsilon_0} = \frac{q}{\varepsilon_0}(n_e - n_i),$$

and assuming quasi-neutrality ($n_{e0} \approx n_{i0}$), we obtain:

$$\nabla^2 \phi = \frac{q^2 n_{e0}}{\varepsilon_0 k_B T_e} \phi.$$

This leads to the definition of the Debye length:

$$\lambda_D = \sqrt{\frac{\varepsilon_0 k_B T_e}{q^2 n_{e0}}},$$

which sets the characteristic scale over which electric potentials are screened. For a system to exhibit plasma-like behaviour, the Debye length must be much smaller than the system size (L), ensuring that local electrostatic perturbations do not extend throughout the entire volume:

$$\lambda_D \ll L.$$

Physically, this means that individual charges do not exert long-range Coulomb forces on all other particles, allowing collective interactions to dominate.

2.3.2 Plasma Frequency and Collective Oscillations

A second fundamental requirement for a gas to be considered a plasma is the ability to sustain collective oscillations. Unlike a neutral gas where particle dynamics are dominated by individual collisions, in a plasma, charge perturbations induce restoring forces that lead to coherent charge oscillations. The natural frequency of these oscillations is called the plasma frequency (ω_p), given by:

$$\omega_p = \sqrt{\frac{n_e q^2}{\varepsilon_0 m_e}}.$$

This frequency represents the rapidity with which electrons oscillate when displaced from equilibrium, while ions, due to their much larger mass, remain relatively immobile on these timescales. The plasma frequency provides an intrinsic timescale for the system, distinguishing regimes of plasma behaviour from those where individual particle motion dominates. For a gas to qualify as a plasma, the characteristic frequency of external perturbations or interactions ν must be lower than the plasma frequency:

$$\nu << \omega_p.$$

This ensures that the system can respond collectively rather than behaving as an assembly of independently moving charges. If ω_p is high, external electromagnetic waves experience dispersion and cutoff phenomena, a defining feature of plasmas that is absent in ordinary ionized gases.

2.3.3 *Number of Particles in a Debye Sphere*

The third crucial criterion is the number of charged particles within a Debye sphere, which is the spherical volume of radius λ_D around any given charge:

$$N_D = \frac{4}{3}\pi\lambda_D^3 n_e.$$

For the system to be considered a plasma, this number must be significantly greater than unity:

$$N_D \gg 1.$$

Physically, this requirement ensures that any individual charged particle is surrounded by a sufficient number of oppositely charged carriers to effectively screen its electric field and participate in collective interactions. If N_D is too small, the system behaves as a correlated ionized gas rather than a true plasma, as screening effects and collective interactions become negligible.

2.4 Collective Behaviour in Plasma

Plasma, as an ionized medium of free charges, exhibits a defining characteristic that distinguishes it from neutral gases, the dominance of collective effects. These effects arise due to the long-range nature of Coulomb interactions, which lead to complex behaviours such as wave propagation, shielding, instabilities, and self-organized structures. Unlike neutral fluids where only short-range collisions govern dynamics, plasmas exhibit interactions over macroscopic scales, where every charged particle influences and is influenced by many others simultaneously. This collective nature profoundly alters transport properties, energy distribution, and stability conditions in laboratory and astrophysical plasmas.

In an ordinary gas, particles interact primarily via short-range collisions, where interactions decay rapidly with distance $e^{-r/\lambda}$, where λ is the mean free path). However, in a plasma, the interaction potential follows Coulomb's law:

$$V(r) = \frac{q_1 q_2}{4\pi \epsilon_0 r},$$

which decays only as $1/r$, implying that every charge exerts an influence on all others, regardless of distance. This non-locality results in significant many-body correlations that do not exist in neutral fluids.

The impact of such long-range forces manifests in several fundamental ways:

Electrostatic Coupling and Self-Consistent Fields: Charges do not move independently, but rather in response to the collective potential generated by all other charges.

Plasma oscillations and Wave Propagation: Due to correlated charge displacements, plasmas exhibit high-frequency oscillations and structured wave dynamics.

Debye shielding: Despite long-range forces, the presence of mobile charges ensures that perturbations are screened over the Debye length.

Non-Local transport properties: Unlike classical gases where diffusion depends on nearest-neighbor collisions, plasma transport is governed by long-range electric fields and collective drifts.

To rigorously establish these properties, one must consider Poisson's equation, kinetic theory, and linear response functions. The fundamental equation governing electrostatic interactions in a plasma is Poisson's equation, given by:

$$\nabla^2 \phi = -\frac{\rho}{\epsilon_0},$$

where ρ is the charge density, given by:

$$\rho = e(Zn_i - n_e).$$

Since every charge contributes to the total potential, the self-consistent electric field follows from:

$$\mathbf{E} = -\nabla \phi.$$

This equation illustrates that the electric field at any point in a plasma results from the superposition of the contributions from all charges, leading to a highly correlated system. Unlike a simple gas where forces act via localized collisions, plasma particles respond to a global electric field that evolves dynamically based on the charge distribution.

The existence of a long-range interaction means that a single charge cannot be considered in isolation. Instead, it polarizes the surrounding plasma, leading to the formation of plasma oscillations, waves, and instabilities, which can be understood through the plasma frequency.

The presence of long-range Coulomb forces enables a rich variety of wave phenomena in plasma, where disturbances propagate as collective excitations rather than as individual particle motions.

2.4.1 Non-local Transport and Instabilities

Beyond wave propagation, long-range interactions profoundly affect plasma transport properties. Unlike neutral fluids, where momentum transfer occurs via local particle collisions, plasma transport is influenced by electric fields that extend across the system, leading to:

Anomalous diffusion: Non-local electric field fluctuations lead to superdiffusive behaviour in magnetized plasmas.

Beam-plasma instabilities: Interaction between drifting electrons and background plasma generates growing oscillations.

Turbulent cascades: Long-range interactions lead to self-sustained turbulence that enhances energy transport. These effects are essential for understanding fusion plasma confinement, space plasmas, and astrophysical accretion flows.

Plasma oscillations and wave propagation arise from the intrinsic collective behaviour of charged particles interacting via long-range Coulomb forces. Unlike neutral gases, where disturbances decay due to collisional damping, plasmas sustain and propagate waves through coherent charge motion driven by electrostatic and electromagnetic fields. These waves represent fundamental excitations in plasma, playing a crucial role in energy transport, instability development, and interactions with external fields. The study of plasma waves provides deep insights into fusion plasma confinement, astrophysical plasma dynamics, and technological applications such as plasma-based accelerators and communication systems [1, 2].

2.4.2 Electrostatic Plasma Oscillations (Langmuir Waves)

One of the most fundamental modes in plasma is the Langmuir wave, an electrostatic oscillation of electrons against a nearly stationary ion background. The origin of this oscillation can be understood through a simple displacement argument: when electrons are displaced from their equilibrium positions, the resulting charge separation creates a restoring electrostatic force that drives coherent oscillations [3].

To rigorously analyze wave propagation, consider a small perturbation to the plasma variables (density, velocity, and electric field) and linearize the governing equations. The fluid description of plasma dynamics is given by:

Continuity equation (Charge Conservation):

$$\frac{\partial n}{\partial t} + \nabla \cdot (n\mathbf{v}) = 0.$$

Momentum equation (Electron Dynamics in an Electric Field):

$$m_e\left(\frac{\partial \mathbf{v}}{\partial t} + \mathbf{v} \cdot \nabla \mathbf{v}\right) = -e\mathbf{E} - \frac{\nabla P}{n}.$$

Poisson's equation (Electrostatic Potential):

$$\nabla^2 \phi = -\frac{e(n_i - n_e)}{\epsilon_0}.$$

For a small perturbation $\tilde{n}, \tilde{v}, \tilde{E} \sim e^{i(kx-\omega t)}$, linearizing these equations leads to the electrostatic wave dispersion relation:

$$\omega^2 = \omega_p^2 + 3k^2 v_{th}^2,$$

where $v_{th} = \sqrt{k_B T_e / m_e}$ is the electron thermal velocity. The second term accounts for finite temperature effects, which introduce a restoring pressure force, modifying the simple cold-plasma Langmuir wave.

This dispersion relation reveals a crucial aspect of plasma wave propagation: For large wavelengths ($\mathbf{k} \to 0$), the oscillation frequency approaches ω_p, meaning the plasma supports stationary charge oscillations. For finite k, wave propagation occurs, with thermal effects influencing phase velocity. Langmuir waves thus do not propagate below the plasma frequency ($\omega < \omega_p$), emphasizing the cutoff behaviour that distinguishes them from ordinary acoustic waves.

2.4.3 *Ion-Acoustic Waves*

In addition to Langmuir waves, plasmas also support low-frequency density waves known as Ion-Acoustic Waves. These are the plasma analog of sound waves in neutral gases, but they arise due to ion inertia and electron pressure balance. For low-frequency perturbations where $\omega << \omega_p$, , electron motion is nearly instantaneous, ensuring quasi-neutrality ($n_e \approx n_i$). The ion equation of motion then governs wave dynamics:

$$m_i \frac{\partial \mathbf{v}_i}{\partial t} = -e\mathbf{E} - \frac{\nabla P}{n}.$$

Linearizing and combining with the continuity equation, the dispersion relation for *Ion-Acoustic Waves* is obtained:

$$\omega^2 = k^2 \frac{\gamma_e k_B T_e}{m_i}.$$

Since $m_i >> m_e$, the phase velocity of Ion-Acoustic Waves is significantly lower than Langmuir waves, leading to low-frequency wave propagation in plasma.

2.4.4 Electromagnetic Waves in Plasma

When considering electromagnetic perturbations, Maxwell's equations must be coupled with plasma dynamics. The general wave equation for the electric field is:

$$\nabla^2 \mathbf{E} - \frac{1}{c^2} \frac{\partial^2 \mathbf{E}}{\partial t^2} = \mu_0 \frac{\partial \mathbf{J}}{\partial t}.$$

For a plane wave solution $\mathbf{E} \sim e^{i(kx-\omega t)}$, the dispersion relation for electromagnetic waves in a cold plasma emerges:

$$\omega^2 = \omega_p^2 + c^2 k^2.$$

This equation reveals two regimes:

For $\omega > \omega_p$, waves propagate as light waves modified by plasma dispersion.

For $\omega < \omega_p$, waves are evanescent, meaning that plasmas exhibit cutoff frequencies where wave propagation ceases.

These properties are crucial for understanding radio wave propagation in the ionosphere, plasma mirrors, and controlled fusion confinement.

Plasma, as a system of charged particles interacting via self-consistently generated electromagnetic fields, supports a vast range of wave phenomena that fundamentally distinguish it from neutral fluids. The classification of waves in plasma broadly falls into electrostatic waves, where the oscillating field is predominantly longitudinal (i.e., parallel to the wave vector), and electromagnetic waves, where transverse oscillations dominate. The behaviour of these waves is governed by Maxwell's equations coupled with plasma fluid or kinetic descriptions, leading to a set of characteristic dispersion relations. Understanding these wave modes is essential for applications in fusion plasma confinement, astrophysical plasmas, and plasma-based accelerators.

2.5 Plasma Confinement and Stability

Plasma confinement and stability are fundamental concerns in both natural and laboratory plasmas, particularly in applications such as controlled nuclear fusion, astrophysical plasmas, and plasma-based devices. The highly dynamic nature of charged particles in a plasma, coupled with their tendency to diffuse due to collisions, collective interactions, and turbulence, necessitates external mechanisms to ensure confinement. Various confinement strategies, such as magnetic, inertial, and electrostatic confinement, are employed to counteract plasma loss. However, confinement alone is insufficient for sustaining a stable plasma, as instabilities can arise due to fluid-like perturbations, kinetic effects, or wave-particle interactions, leading to energy losses and disruption of equilibrium conditions.

2.5.1 Plasma Confinement Mechanisms

Plasma confinement refers to the methods used to prevent charged particles from escaping the system while maintaining a desirable density and temperature for applications such as thermonuclear fusion. The primary confinement methods are:

Magnetic Confinement

In many plasma systems, particularly in fusion devices (Tokamaks, Stellarators) and astrophysical contexts (magnetospheres, accretion disks), magnetic fields play a crucial role in confining plasma by exerting a force on charged particles that restricts their motion. The Lorentz force acting on a charged particle in a magnetic field is given by:

$$\mathbf{F} = q(\mathbf{E} + \mathbf{v} \times \mathbf{B}).$$

In the absence of an electric field, the motion of a charged particle is characterized by helical trajectories around magnetic field lines, with the Larmor radius given by:

$$r_L = \frac{mv_\perp}{qB},$$

where $v_\perp$ is the velocity component perpendicular to the magnetic field. The Larmor motion effectively reduces the cross-field transport of particles, thereby confining them within magnetic surfaces.

A Tokamak utilizes a combination of a toroidal magnetic field and a generated by a plasma current to create a helical magnetic structure. The safety factor characterizes the number of toroidal transits a field line makes per poloidal revolution:

$$q = \frac{rB_t}{RB_p}.$$

For stable confinement, the safety factor must remain within an optimal range to prevent magnetohydrodynamic (MHD) instabilities, such as kink modes and ballooning modes.

A Stellarator relies purely on externally generated magnetic fields to create a twisted helical field structure, removing the need for a plasma current. This eliminates instabilities related to current-driven disruptions, making Stellarators inherently more stable but more complex to design.

Inertial Confinement

In inertial confinement fusion (ICF), a high-energy driver (e.g., laser or ion beam) compresses and heats a small plasma pellet, relying on rapid compression to confine the plasma long enough for fusion to occur. The confinement time, τ must satisfy:

$$\tau \geq \frac{R}{c_s},$$

where R is the pellet radius and c_s is the plasma sound speed. Rayleigh–Taylor and Richtmyer-Meshkov instabilities are major challenges, as they disrupt the uniformity of compression.

Electrostatic Confinement

Electrostatic confinement uses electrostatic potentials to trap charged particles. Devices like the Penning trap and Polywell fusion reactors employ spherically symmetric potential wells to confine ions electrostatically. However, electron losses due to Coulomb collisions and instabilities limit the efficiency of these systems.

2.5.2 Stability Considerations

The success of plasma confinement hinges on its stability under perturbations. Stability analysis is typically classified into fluid stability, kinetic stability, and wave-particle interactions.

Magnetohydrodynamic (MHD) Stability

Magnetohydrodynamics (MHD) treats plasma as a conducting fluid influenced by magnetic fields, governed by a set of fundamental equations that describe how mass, momentum, and magnetic fields evolve in space and time. These include the continuity equation for mass conservation, the momentum equation which incorporates the Lorentz force, and the induction equation that dictates the behaviour of the magnetic field within the plasma. Within this framework, various instabilities can arise, significantly impacting plasma confinement and stability in fusion devices and astrophysical settings [4, 5].

One key example is the kink instability, which emerges when the magnetic safety factor, a measure of how tightly magnetic field lines twist falls below a certain threshold. This leads to bending or distortion of the magnetic axis, especially problematic in cylindrical, or toroidal systems. The ballooning instability is another critical mode, typically observed in toroidal confinement devices like tokamaks, where steep pressure gradients can destabilize the magnetic field configuration and cause localized bulging of plasma. The tearing mode instability, driven by finite resistivity, allows for the reconnection of magnetic field lines and the formation of magnetic islands, potentially culminating in plasma disruptions. Meanwhile, the Rayleigh–Taylor instability frequently appears in inertial confinement fusion, where a heavier plasma layer accelerates a lighter one, resulting in turbulent mixing and loss of confinement symmetry. Understanding and mitigating these MHD instabilities is essential for achieving stable plasma conditions in both laboratory fusion experiments and astrophysical phenomena.

Kinetic Stability Considerations

While MHD treats plasma as a fluid, kinetic theory accounts for velocity distribution effects. The Vlasov equation governs kinetic stability:

$$\frac{\partial f}{\partial t} + \mathbf{v} \cdot \nabla f + \frac{q}{m}(\mathbf{E} + \mathbf{v} \times \mathbf{B}) \cdot \frac{\partial f}{\partial \mathbf{v}} = 0.$$

A key result from kinetic theory is Landau damping, where electrostatic waves are damped due to energy exchange with particles:

$$\omega_r - k v_{ph} = 0.$$

If $\frac{\partial f}{\partial v}$ is negative at v_{ph}, energy is transferred from the wave to the particles, leading to wave damping without particle collisions.

Transport and Turbulence Effects

In realistic plasma systems, turbulent fluctuations can cause anomalous transport, leading to particle and energy losses. The Bohm diffusion model provides an estimate for cross-field transport:

$$D_B \sim \frac{k_B T_e}{16eB}.$$

Mitigating turbulence through sheared flow stabilization, or zonal flows is crucial in maintaining confinement efficiency [6].

2.6 Charged Particle Trajectories in Electric and Magnetic Fields

The motion of a charged particle in electromagnetic fields is governed by the Lorentz force law:

$$m\frac{d\mathbf{v}}{dt} = q(\mathbf{E} + \mathbf{v} \times \mathbf{B})$$

where m is the particle mass, q is the charge, $\mathbf{E}$ is the electric field, and $\mathbf{B}$ is the magnetic field. This equation encapsulates the fundamental interactions governing charged particle dynamics in a plasma medium.

Motion in a Uniform Magnetic Field

For a charged particle moving in a uniform magnetic field $\mathbf{B} = B\widehat{z}$, the velocity components satisfy the coupled differential equations:

$$m\frac{dv_x}{dt} = qv_yB, \quad m\frac{dv_y}{dt} = -qv_xB.$$

Solving these equations leads to a circular motion with a characteristic gyrofrequency, or cyclotron frequency,

$$\omega_c = \frac{qB}{m},$$

and a Larmor radius,

$$r_L = \frac{v_\perp}{\omega_c},$$

where $v_\perp$ is the velocity component perpendicular to $\mathbf{B}$. The motion is helical when an additional parallel velocity component $v_{||}$ exists.

Motion in Crossed Electric and Magnetic Fields

If both electric and magnetic fields are present, a guiding center drift arises. For $\mathbf{E}\perp\mathbf{B}$, the particle undergoes an additional uniform drift velocity:

$$\mathbf{v}_E = \frac{\mathbf{E} \times \mathbf{B}}{B^2}.$$

This drift, known as the E × B, is independent of charge and mass, implying that electrons and ions drift collectively. Other significant drifts include the grad-B drift, curvature drift, and polarization drift, each arising from spatial variations in magnetic field strength, curved field lines, and time-dependent electric fields, respectively [7].

Equation of Motion for Charged Particles

A more general approach to charged particle motion considers the Lagrangian and Hamiltonian formulations. The Lagrangian for a charged particle in an electromagnetic field is:

$$L = \frac{1}{2}mv^2 + q(\mathbf{A} \cdot \mathbf{v} - \phi),$$

where $\mathbf{A}$ is the vector potential and ϕ is the scalar potential. The corresponding Hamiltonian is:

$$H = \frac{1}{2m}(\mathbf{p} - q\mathbf{A})^2 + q\phi.$$

The canonical equations of motion,

$$\frac{d\mathbf{r}}{dt} = \frac{\partial H}{\partial \mathbf{p}}, \frac{d\mathbf{p}}{dt} = -\frac{\partial H}{\partial \mathbf{r}},$$

recover the Lorentz force law and provide a basis for analytical and computational studies of particle dynamics, including stochasticity and chaos in non-uniform fields [7].

Boltzmann Equation and Distribution Functions

Plasma behaviour at a statistical level is described by the Boltzmann equation for the phase-space distribution function $f(\mathbf{r}, \mathbf{v}, t)$:

$$\frac{\partial f}{\partial t} + \mathbf{v} \cdot \nabla f + \frac{q}{m}(\mathbf{E} + \mathbf{v} \times \mathbf{B}) \cdot \nabla_{\mathbf{v}} f = \left(\frac{df}{dt}\right)_{\text{coll}}.$$

The left-hand side describes free-streaming motion under external fields, while the right-hand side accounts for collisions. In equilibrium, the Maxwell–Boltzmann distribution,

$$f_0(\mathbf{v}) = n\left(\frac{m}{2\pi k_B T}\right)^{3/2} e^{-mv^2/2k_B T},$$

describes the statistical state of a thermally relaxed plasma. Deviations from equilibrium, captured by perturbative expansions, lead to transport phenomena such as electrical conductivity and viscosity.

Fluid Description of Plasma (Two-Fluid Model)

The two-fluid model treats plasmas as consisting of separate electron and ion fluids, governed by the continuity, momentum, and energy equations for each species.

The continuity equation for species s (electrons, or ions) expresses charge and mass conservation:

$$\frac{\partial n_s}{\partial t} + \nabla \cdot (n_s \mathbf{v}_s) = 0.$$

This equation ensures that changes in density arise solely from divergence in particle flux. The momentum equation follows from Newton's second law:

$$m_s n_s \left(\frac{\partial \mathbf{v}_s}{\partial t} + \mathbf{v}_s \cdot \nabla \mathbf{v}_s \right) = -\nabla p_s + q_s n_s (\mathbf{E} + \mathbf{v}_s \times \mathbf{B}) - \nabla \cdot \Pi_s + \mathbf{R}_s,$$

where P_S is the pressure, Π_s is the stress tensor, and $\mathbf{R}_s$ represents collisional effects. The energy equation,

$$\frac{d}{dt}\left(\frac{3}{2} n_s k_B T_s \right) + p_s \nabla \cdot \mathbf{v}_s = Q_s,$$

describes the evolution of plasma temperature in response to heating, expansion, and collisional dissipation.

Magnetohydrodynamic (MHD) Equations

MHD provides a single-fluid approximation, valid when electron and ion inertia differences are negligible. The governing equations are derived by summing the two-fluid equations and using Ohm's law.

In ideal MHD, the plasma behaves as a perfect conductor, satisfying:

$$\frac{\partial \mathbf{B}}{\partial t} = \nabla \times (\mathbf{v} \times \mathbf{B}).$$

For resistive MHD, the inclusion of finite conductivity modifies Ohm's law:

$$\mathbf{E} + \mathbf{v} \times \mathbf{B} = \eta \mathbf{J},$$

where η is the resistivity. Equilibrium in a plasma requires that forces balance according to:

$$\rho \frac{d\mathbf{v}}{dt} = -\nabla p + \mathbf{J} \times \mathbf{B}.$$

Linear stability analysis considers $\mathbf{x} \to \mathbf{x} + \widetilde{\mathbf{x}} e^{i(\mathbf{k}\cdot\mathbf{x} - \omega t)}$ to determine growth rates. The MHD *stability criterion*, given by the energy principle,

$$\delta W = \frac{1}{2} \int (\delta \mathbf{J} \cdot \delta \mathbf{B} - \delta p \nabla \cdot \delta \mathbf{v}) dV,$$

predicts instabilities such as kink modes and ballooning modes, essential in magnetic confinement systems.

2.7 Classification of Plasma

Plasmas exhibit a wide range of physical properties depending on their ionization levels, temperature distributions, external interactions, and relativistic effects. The classification of plasma can be systematically divided into various categories based on key physical attributes. Each classification provides insight into the underlying dynamics, stability, and applications of plasma across different domains of physics.

2.7.1 Thermal Versus Non-thermal Plasma

The distinction between thermal and non-thermal plasmas is based on the energy distribution of charged particles. This classification is crucial in understanding energy transport, equilibrium properties, and collisional processes.

Thermal Plasma

A thermal plasma is characterized by local thermodynamic equilibrium (LTE), where the electrons, ions, and neutral particles possess approximately the same temperature ($T_e \approx T_i \approx T_n$). The velocity distribution of particles is typically described by a Maxwell–Boltzmann distribution:

$$f(v) = n \left(\frac{m}{2\pi k_B T} \right)^{3/2} e^{-mv^2/2k_B T}.$$

Thermal plasmas are characterized by high particle densities, leading to frequent collisions and rapid energy exchange among electrons, ions, and neutrals. This high collisionality ensures that the plasma remains in or near thermal equilibrium. Common examples of thermal plasmas include stellar interiors and fusion reactors, where extreme temperatures lead to almost complete ionization, as well as electric arcs and combustion flames, where collisions help sustain equilibrium conditions. Due to their strongly collisional nature, thermal plasmas are well-suited to macroscopic modeling approaches such as fluid dynamics and magnetohydrodynamics (MHD), which effectively capture their collective behaviour [8].

Non-thermal Plasma

Non-thermal plasmas are characterized by a significant disparity in temperature between electrons and heavier species ($T_e >> T_i, T_n$). The electron distribution in such plasmas often deviates from the Maxwellian form and may be better represented by a kappa distribution, or a Druyvesteyn function. The energy equation governing electron temperature evolution is:

$$\frac{3}{2}n_e k_B \frac{dT_e}{dt} = -\nabla \cdot \mathbf{q} + P - L,$$

where $\mathbf{q}$ is the heat flux, P is the power input, and L is the energy loss due to radiation, or recombination.

Non-thermal plasmas are typically found in environments where electrons are highly energetic while the ions and neutral particles remain relatively cool, resulting in a lack of thermal equilibrium between species. Examples include natural phenomena like auroras and lightning, where energetic electrons drive ionization without substantially heating the bulk plasma, as well as engineered systems such as glow discharges and plasma thrusters, where external electric fields selectively accelerate electrons. Due to their weakly collisional nature, non-thermal plasmas cannot be accurately described using fluid models alone. Instead, kinetic theory is often employed to capture the detailed velocity distributions and non-equilibrium dynamics that govern their behaviour [9].

2.7.2 Fully Ionized Versus Partially Ionized Plasma

Ionization fraction (α) is a fundamental parameter in classifying plasmas:

$$\alpha = \frac{n_i}{n_n + n_i},$$

where n_i and n_n are the number densities of ions and neutral particles, respectively.

Fully Ionized Plasma

A plasma is considered fully ionized when nearly all atoms have been stripped of their electrons, corresponding to an ionization fraction close to one ($\alpha \approx 1$). This state typically arises in environments where intense thermal or electromagnetic conditions inhibit the recombination of ions and electrons. Prominent examples include the Sun and stellar cores, where extremely high temperatures sustain continuous ionization, as well as in Tokamak fusion reactors, where hydrogen isotopes are heated to the point of complete ionization. In fully ionized plasmas, electromagnetic forces dominate the behaviour of the medium, while interactions involving neutral particles become negligible. This simplifies the plasma's transport equations by minimizing frictional effects and enabling more straightforward modeling of energy and momentum transfer [10].

Partially Ionized Plasma

In partially ionized plasmas, a substantial portion of the gas remains in a neutral state ($0 < \alpha < 1$), meaning that not all atoms are ionized. The presence of these neutral species introduces important physical effects that influence plasma behaviour. For instance, charge-exchange reactions can occur, where ions transfer their charge to neutral atoms, significantly affecting Transport properties and momentum balance. Additionally, electron-neutral collisions contribute to energy loss mechanisms and reduce the plasma's overall electrical conductivity. Common examples of partially ionized plasmas include the Earth's ionosphere, where a dynamic balance between solar ionization and recombination processes sustains partial ionization, and gas discharges such as those found in neon signs and fluorescent lamps, where applied electric fields maintain the plasma state. To accurately describe these systems, both kinetic and fluid models are employed, with modifications that incorporate the effects of ion-neutral interactions to capture the full range of plasma dynamics [11].

2.7.3 Magnetized Versus Unmagnetized Plasma

The interaction between plasma and magnetic fields determines fundamental transport properties, wave dynamics, and stability characteristics.

Magnetized Plasma

A plasma is considered magnetized when the gyrofrequency ($\omega_c = qB/m$) dominates over collisional and convective transport timescales. This condition is quantified by the Hall parameter:

$$\beta_H = \frac{\omega_c}{\nu_c} >> 1,$$

where ν_c is the collisional frequency. Magnetized plasmas exhibit anisotropic transport, where perpendicular and parallel motion relative to $\mathbf{B}$ follow different diffusion coefficients:

$$D_{\perp} << D_{||}.$$

Examples of magnetized plasmas include astrophysical environments like pulsars and the interstellar medium, where strong magnetic fields govern large-scale dynamics and particle behaviour. Similarly, in magnetic confinement fusion devices, such as Tokamaks and Stellarators, magnetic fields are used to control and confine the plasma, minimizing energy losses and maintaining stability for sustained fusion reactions [12, 13].

Unmagnetized Plasma

In unmagnetized plasmas, particle motion is predominantly governed by collisional interactions rather than magnetic forces ($\beta_H << 1$). These plasmas exhibit isotropic diffusion:

$$D_\perp \approx D_{||}.$$

Examples of unmagnetized plasmas include inertial confinement fusion (ICF), where plasma dynamics are primarily driven by thermal and radiation pressures instead of magnetic forces, and laser-induced plasmas, which are generated by short-duration laser pulses and typically exist in transient, unmagnetized states. In such cases, the absence of significant magnetic fields simplifies the governing equations, allowing the use of purely hydrodynamic models to describe plasma behaviour.

2.7.4 Natural Versus Laboratory Plasmas

Plasmas can be classified based on their origin and physical setting.

Natural Plasmas

Natural plasmas are found in various astrophysical and geophysical settings, such as solar and stellar interiors, where nuclear fusion maintains high-energy plasma states; interstellar and intergalactic regions, where magnetic reconnection and shock waves influence large-scale cosmic structures; and planetary magnetospheres, where interactions with the solar wind give rise to phenomena like auroras. Understanding these complex environments often demands the use of general relativistic frameworks, or advanced magnetohydrodynamic (MHD) models to accurately capture their behaviour.

Laboratory Plasmas

Artificially generated plasmas encompass a range of applications, including fusion plasmas produced in devices like Tokamaks and Stellarators, low-temperature plasmas utilized in semiconductor fabrication, and high-energy laser-induced plasmas used in inertial confinement fusion research. These laboratory-created plasmas offer a controlled environment, enabling detailed experimental validation of theoretical plasma models and facilitating advancements in both fundamental research and technological applications.

2.7.5 *Relativistic Versus Non-relativistic Plasma*

The classification of plasma based on relativistic effects is determined by the electron thermal velocity:

$$\gamma = \frac{1}{\sqrt{1 - v^2/c^2}}$$

Non-relativistic Plasma

For $v << c$, relativistic corrections are negligible. The kinetic energy follows the classical relation:

$$E = \frac{1}{2}mv^2$$

Non-relativistic plasmas are characterized by electron velocities that remain well below the speed of light. Examples include fusion plasmas operating at moderate temperatures, around 10^7 K, and various industrial plasmas used in applications such as material processing, where thermal energies are insufficient to induce relativistic effects.

Relativistic Plasma

When $v \sim c$, the relativistic energy expression:

$$E = \gamma mc^2$$

must be used. Relativistic effects play a significant role in plasmas found in extreme astrophysical environments. For instance, in blazar jets, particles are accelerated to near-light speeds within intense magnetic fields, while in pulsar magnetospheres, electrons can attain very high Lorentz factors ($\gamma >> 1$). In such regimes, classical plasma descriptions are no longer sufficient, and quantum electrodynamics (QED) corrections become essential. These relativistic plasmas exhibit unique phenomena, including pair production and intense synchrotron radiation, which are critical for understanding high-energy cosmic processes.

References

1. Downer, M. C., Zgadzaj, R., Debus, A., Schramm, U., & Kaluza, M. C. (2018). Diagnostics for plasma-based electron accelerators. *Reviews of Modern Physics, 90*(3), Article 035002.
2. Arjmand, S., Alesini, D., Anania, M. P., Biagioni, A., Chiadroni, E., Cianchi, A., et al. (2020). Characterization of plasma sources for plasma-based accelerators. *Journal of Instrumentation, 15*(09), C09055.

3. Dalimier, E., Oks, E., & Renner, O. (2014). Review of Langmuir-wave-caused dips and charge-exchange-caused dips in spectral lines from plasmas and their applications. *Atoms, 2*(2), 178–194.
4. Sakakibara, S., Watanabe, K. Y., Ohdachi, S., Narushima, Y., Toi, K., Tanaka, K., et al. (2010). Study of MHD stability in LHD. *Fusion Science and Technology, 58*(1), 176–185.
5. Markovskii, S. A., & Somov, B. V. (1996). Magnetohydrodynamic discontinuities in space plasmas: Interrelation between stability and structure. *Space Science Reviews, 78*(3), 443–506.
6. Gordeev, A. V., Kingsep, A. S., & Rudakov, L. I. (1994). Electron magnetohydrodynamics. *Physics Reports, 243*(5), 215–315.
7. Bittencourt, J. A. (2013). *Fundamentals of plasma physics*. Springer Science & Business Media.
8. Fasoli, A., Brunner, S., Cooper, W. A., Graves, J. P., Ricci, P., Sauter, O., & Villard, L. (2016). Computational challenges in magnetic-confinement fusion physics. *Nature Physics, 12*(5), 411–423.
9. Bruggeman, P. J., et al. (2017). Foundations of atmospheric pressure non-equilibrium plasmas. *Plasma Sources Science and Technology, 26*(12), 123002.
10. Barakat, A. R., & Schunk, R. W. (1982). Transport equations for multicomponent anisotropic space plasmas: A review. *Plasma Physics, 24*(4), 389.
11. Khomenko, E. (2016). On the effects of ion-neutral interactions in solar plasmas. *Plasma Physics and Controlled Fusion, 59*(1), Article 014038.
12. Ongena, J., Koch, R., Wolf, R., & Zohm, H. (2016). Magnetic-confinement fusion. *Nature Physics, 12*(5), 398–410.
13. Huang, C., & Li, L. (2018). Magnetic confinement fusion: A brief review. *Frontiers in Energy, 12*, 305–313.

Chapter 3
Electrodynamics in Plasmas

Electrodynamics serves as one of the cornerstones of classical and modern physics, governing the interaction of charged particles and electromagnetic fields. The study of electrodynamics in uncharged media and plasmas is particularly significant due to the diverse range of physical systems it describes, spanning astrophysical environments, controlled fusion devices, semiconductor plasmas, and exotic quantum plasmas. Unlike in classical vacuum electrodynamics, where Maxwell's equations take their simplest form, the presence of a medium, be it a dielectric, a conducting plasma, or an anisotropic quantum medium introduces modifications that necessitate a deeper investigation into field interactions, wave propagation, and energy–momentum transfer.

3.1 Maxwell's Equations in Vacuum and Media

Maxwell's equations form the cornerstone of classical electrodynamics, governing the behaviour of electric and magnetic fields in free space, in the presence of matter, and within complex plasmas. Their fundamental importance lies in their ability to describe electromagneticwave propagation, charge conservation, energy transfer, and interactions between charged particles and fields. These equations unify electrostatics and magnetostatics into a single framework of electrodynamics, incorporating both time-dependent and time-independent effects.

In plasma physics context, Maxwell's equations are not merely static laws, they serve as dynamical equations that dictate how electromagnetic waves propagate in a medium composed of free and bound charges, interact with collective plasma oscillations, and lead to phenomena such as wave dispersion, Landau damping, and instabilities. The standard form of Maxwell's equations provides a foundation upon which various modifications, including constitutive relations, gauge choices, and advanced electrodynamic models can be constructed.

P. Kumar, *Quantum Plasma*, https://doi.org/10.1007/978-981-95-0481-7_3

Maxwell's equations in their standard differential form are given by:

Gauss's Law for the Electric Field

$$\nabla \cdot \mathbf{E} = \frac{\rho}{\epsilon_0}.$$

This equation states that the divergence of the electric field **E** at a point is proportional to the local charge density $\boldsymbol{\rho}$, with $\boldsymbol{\varepsilon_0}$ being the permittivity of free space. Physically, this law expresses the fact that electric field lines originate from positive charges and terminate at negative charges, implying that charge acts as a source (or sink) of the electric field. In a plasma, where free electrons and ions exist, the charge density $\boldsymbol{\rho}$ is not a fixed quantity but varies due to charge motion, screening effects, and plasma oscillations.

Gauss's Law for Magnetism

$$\nabla \cdot \mathbf{B} = 0.$$

This equation asserts that there are no magnetic monopoles, magnetic field lines always form closed loops, meaning **B** has no divergence. This is a fundamental feature of magnetism, distinguishing it from electrostatics. In plasmas, where moving charges create magnetic fields, this equation ensures that the total magnetic flux through any closed surface remains zero, preserving the solenoidal nature of **B**.

Faraday's Law of Induction

$$\nabla \times \mathbf{E} = -\frac{\partial \mathbf{B}}{\partial t}.$$

This equation describes how a time-dependent magnetic field induces a circulating electric field. In plasma physics, this principle is crucial in explaining magnetohydrodynamic (MHD) waves, Alfvén waves, and reconnection processes. The presence of a rapidly changing **B** field in a plasma can generate induced currents, which in turn modify the plasma dynamics, leading to phenomena such as turbulent cascades in astrophysical plasmas.

Ampère's Law with Maxwell's Correction

$$\nabla \times \mathbf{B} = \mu_0 \mathbf{J} + \mu_0 \epsilon_0 \frac{\partial \mathbf{E}}{\partial t}.$$

This equation states that a circulating magnetic field **B** can be generated by an electric current **J**, or by a time-dependent electric field **E** (Maxwell's displacement current). The displacement current term, $\boldsymbol{\mu_0 \varepsilon_0\ \partial \mathbf{E}/\partial t}$, is fundamental for the self-consistent propagation of electromagnetic waves in vacuum and plasmas. In plasmas, where charged particles are free to move, this equation provides insight into current-driven instabilities, including Weibel instability and Kink instabilities in magnetized plasmas.

By taking the curl of Faraday's and Ampère's Laws and using vector identities, one can derive the wave equations for **E** and **B**:

$$\nabla^2 \mathbf{E} - \mu_0 \epsilon_0 \frac{\partial^2 \mathbf{E}}{\partial t^2} = \mu_0 \frac{\partial \mathbf{J}}{\partial t}.$$

$$\nabla^2 \mathbf{B} - \mu_0 \epsilon_0 \frac{\partial^2 \mathbf{B}}{\partial t^2} = -\mu_0 \nabla \times \mathbf{J}$$

In vacuum, where $\mathbf{J} = 0$, these reduce to homogeneous wave equations:

$$\nabla^2 \mathbf{E} - \frac{1}{c^2} \frac{\partial^2 \mathbf{E}}{\partial t^2} = 0, \nabla^2 \mathbf{B} - \frac{1}{c^2} \frac{\partial^2 \mathbf{B}}{\partial t^2} = 0.$$

which describe the propagation of electromagnetic waves at the speed **c**. In a plasma, however, these equations acquire source terms, leading to modified dispersion relations, cutoff frequencies, and resonance effects.

Maxwell's equations, in their standard form, describe the fundamental laws governing electric and magnetic fields. However, they are not by themselves sufficient to determine the field distributions in a given medium. The interplay between the electromagnetic field and the medium in which it propagates must be explicitly characterized through additional relations, known as constitutive relations. These relations establish how the medium responds to external electric and magnetic fields, incorporating microscopic polarization and magnetization effects into the macroscopic field equations.

In vacuum, Maxwell's equations are complete in their standard form, with the permittivity of free space $\boldsymbol{\varepsilon}_0$ and the permeability of free space $\boldsymbol{\mu}_0$ defining the medium properties. However, in a general medium, the response of bound and free charges must be accounted for, leading to modifications in field equations through polarization **P** and magnetization **M**. The mathematical framework of response functions generalizes these constitutive relations, particularly in frequency-dependent, dispersive, and non-local media, which are highly relevant in plasma physics and condensed matter electrodynamics.

3.1.1 Wave Propagation in Linear and Nonlinear Media

The propagation of electromagnetic waves in various media is a fundamental aspect of electrodynamics, with direct implications for plasma physics, optics, and wave-matter interactions in condensed matter systems. In a vacuum, wave propagation is governed purely by Maxwell's equations, leading to dispersion-free plane waves traveling at the speed of light **c**. However, in material media, wave dynamics are influenced by the medium's response, characterized by its permittivity **ε** and permeability **μ**, which introduce dispersion, absorption, and nonlinear interactions.

A rigorous understanding of wave propagation requires distinguishing between linearand nonlinear media. In linear media, the response of the medium to an applied field is proportional to the field strength, leading to well-defined normal modes with a predictable dispersion relation. In contrast, in nonlinear media, the material response depends nonlinearly on the electric and magnetic fields, resulting in phenomena such as harmonic generation, self-focusing, solitonic behaviour, and modulational instability. Plasma environments naturally exhibit both linear and nonlinear wave behaviours, making their study essential for applications ranging from controlled fusion to space and astrophysical plasma dynamics.

Electromagnetic Wave Propagation in Linear Media

In a homogeneous, isotropic, and linear medium, Maxwell's equations lead to the standard wave equation for the electric field **E**:

$$\nabla^2 \mathbf{E} - \mu\epsilon \frac{\partial^2 \mathbf{E}}{\partial t^2} = 0.$$

By assuming a plane-wave solution of the form:

$$\mathbf{E}(\mathbf{r}, t) = \mathbf{E}_0 e^{i(\mathbf{k}\cdot\mathbf{r} - \omega t)},$$

where **k** is the wave vector and **ω** is the angular frequency, substitution into the wave equation yields the dispersion relation:

$$k^2 = \mu\epsilon\omega^2.$$

The phase velocity of the wave is given by:

$$v_p = \frac{\omega}{k} = \frac{1}{\sqrt{\mu\epsilon}}.$$

For vacuum, where $\boldsymbol{\varepsilon} = \boldsymbol{\varepsilon}_o$ and $\boldsymbol{\mu} = \boldsymbol{\mu}_o$, we recover $v_p = c$, the speed of light. In a medium with frequency-dependent $\boldsymbol{\varepsilon}(\boldsymbol{\omega})$ and $\boldsymbol{\mu}(\boldsymbol{\omega})$, the phase velocity becomes dispersive, leading to frequency-dependent wave speeds and potential energy dissipation.

Wave Dispersion in Conducting and Plasma Media

In conducting media, the presence of free charge carriers introduces an additional frequency-dependent response. The complex permittivity, incorporating conductivity **σ**, modifies the dispersion relation as:

$$\epsilon_{\text{eff}}(\omega) = \epsilon(\omega) + \frac{i\sigma}{\omega}.$$

For a plasma medium, characterized by a free electron density n_e, the equation of motion for electrons under an oscillating electric field $\mathbf{E}$ leads to the well-known plasma permittivity:

$$\epsilon(\omega) = \epsilon_0\left(1 - \frac{\omega_p^2}{\omega^2}\right),$$

where ω_p is the plasma frequency:

$$\omega_p = \sqrt{\frac{n_e e^2}{\epsilon_0 m_e}}.$$

Substituting this into the dispersion relation for electromagnetic waves in a plasma gives:

$$k^2c^2 = \omega^2 - \omega_p^2.$$

This equation reveals a wave cutoff: when $\omega < \omega_p$, $\mathbf{k}$ becomes imaginary, leading to evanescent wave behaviour where waves are exponentially damped rather than propagating. This fundamental result governs radio wave reflection in the ionosphere and confinement mechanisms in fusion plasmas.

For a magnetized plasma, the wave equation becomes anisotropic, leading to distinct wave modes such as ordinary (O) waves, extraordinary (X) waves, and whistler waves, each obeying modified dispersion relations influenced by the external magnetic field $\mathbf{B}_0$. These effects introduce cyclotron resonances, leading to enhanced wave absorption at specific frequencies.

Nonlinear Wave Phenomena and Wave Mixing

In nonlinear media, the constitutive relations contain higher-order terms:

$$\mathbf{P} = \epsilon_0(\chi^{(1)}\mathbf{E} + \chi^{(2)}\mathbf{E}^2 + \chi^{(3)}\mathbf{E}^3 + \ldots).$$

Here, $\chi^{(1)}$ represents the usual linear susceptibility, while $\chi^{(2)}$ and $\chi^{(3)}$ describe nonlinear effects such as second-harmonic generation (SHG) and third-harmonic generation (THG), respectively.

In plasma physics, parametric wave interactions emerge due to nonlinearities, leading to three-wave mixing and modulational instability, where an initial wave of frequency $\boldsymbol{\omega}$ generates sidebands at $\boldsymbol{\omega} \pm \boldsymbol{\Delta\omega}$. Such effects contribute to plasma heating, wave turbulence, and the formation of coherent structures like solitons.

Solitons and Self-focusing in Nonlinear Plasmas

When the wave amplitude is sufficiently large, nonlinear effects lead to solitonic behaviour, where wave dispersion is exactly balanced by nonlinear self-modulation. The standard equation governing such dynamics is the nonlinear Schrödinger equation (NLSE):

$$i\frac{\partial E}{\partial t} + \frac{1}{2}\frac{\partial^2 E}{\partial x^2} + \gamma |E|^2 E = 0.$$

Here, γ is the nonlinear coefficient that dictates self-focusing or defocusing behaviour.

In plasmas, solitons are crucial for describing localized wave packets that propagate without dispersion. These are observed in both laboratory and astrophysical settings, such as Langmuir solitons in plasma turbulence and Alfvén solitons in magnetized plasmas [1].

Shock Waves and Wave Breaking in Plasma Environments

When nonlinear effects become extreme, wave steepening leads to shock wave formation, characterized by a sudden discontinuity in field amplitude. The governing equation in collisionless plasmas is the Korteweg–de Vries (KdV) equation:

$$\frac{\partial E}{\partial t} + E\frac{\partial E}{\partial x} + \frac{\partial^3 E}{\partial x^3} = 0.$$

This equation describes electrostatic shocks and ion-acoustic solitons, where dispersive corrections balance nonlinearity to prevent singularities. In extreme cases, wave breaking occurs, leading to energy dissipation and particle acceleration, as seen in laser-plasma interactions and space plasma shocks [2].

3.1.2 Energy–Momentum Conservation and Poynting's Theorem in Electrodynamics

The conservation of energy and momentum is a cornerstone of electrodynamics, ensuring that electromagnetic fields adhere to fundamental physical principles. In the context of plasmas, where charged particles interact dynamically with electromagnetic waves, energy–momentum conservation becomes essential for analyzing wave propagation, energy transfer, and field evolution. Poynting's theorem provides a rigorous mathematical framework for quantifying energy transport in electromagnetic fields and plays a crucial role in plasma dynamics, electrodynamically driven instabilities, and wave-particle interactions. The total energy in an electromagnetic system consists of the energy stored in the fields and the energy exchange with matter. Starting with Maxwell's equations in differential form:

$$\nabla \cdot \mathbf{E} = \frac{\rho}{\epsilon_0}, \nabla \cdot \mathbf{B} = 0,$$

$$\nabla \times \mathbf{E} = -\frac{\partial \mathbf{B}}{\partial t}, \quad \nabla \times \mathbf{B} = \mu_0 \mathbf{J} + \mu_0 \epsilon_0 \frac{\partial \mathbf{E}}{\partial t},$$

Taking the dot product of **E** with Faraday's Law and **B** with Ampère's Law, followed by vector identity manipulations, yields the energy balance equation:

$$\frac{\partial u}{\partial t} + \nabla \cdot \mathbf{S} = -\mathbf{J} \cdot \mathbf{E}.$$

Here,

$\left[u = \frac{1}{2}\left(\epsilon_0 |\mathbf{E}|^2 + \frac{1}{\mu_0} |\mathbf{B}|^2\right)\right]$ is the electromagnetic energy density,

$\mathbf{S} = \frac{1}{\mu_0} \mathbf{E} \times \mathbf{B}$ is the Poynting vector, representing the rate of energy flux,

$\mathbf{J} \cdot \mathbf{E}$ is the rate of work done on charges, describing energy dissipation or gain.

This equation states that the change in electromagnetic energy within a volume is equal to the net electromagnetic energy flow (divergence of **S**) minus the work done on free charges.

Poynting's theorem provides a clear physical interpretation of energy transfer in electromagnetic systems. The term involving the divergence of the Poynting vector ($\nabla \cdot S$) represents the flow of electromagnetic energy into or out of a region. In vacuum, it describes the propagation of electromagnetic waves, while in plasmas, it indicates the transport of energy through various wave modes. The term J·E signifies the work done on charged particles, which is a key mechanism in plasma heating processes such as ohmic dissipation and wave-particle interactions. The field energy density u denotes the energy stored in electric and magnetic fields, and its variation reflects the dynamic exchange of energy between fields and matter. In lossless media, where no energy is transferred to particles ($J \cdot E = 0$), the energy remains entirely within the fields and propagates as an undamped wave. However, in dissipative media like resistive plasmas, electromagnetic energy is continuously converted into heat or contributes to particle acceleration.

Beyond energy conservation, electromagnetic fields also carry momentum, which must be conserved in interactions with charged particles. The momentum density stored in the field is given by the electromagnetic momentum density vector:

$$\mathbf{g} = \frac{1}{\mu_0 c^2} \mathbf{E} \times \mathbf{B}.$$

Applying the stress-energy tensor formalism, the electromagnetic momentum flux tensor is given by:

$$T_{ij} = \epsilon_0 \left(E_i E_j - \frac{1}{2} \delta_{ij} |\mathbf{E}|^2 \right) + \frac{1}{\mu_0} \left(B_i B_j - \frac{1}{2} \delta_{ij} |\mathbf{B}|^2 \right).$$

The conservation equation for momentum takes the form:

$$\frac{\partial \mathbf{g}}{\partial t} + \nabla \cdot \mathbf{T} = -\rho \mathbf{E} - \mathbf{J} \times \mathbf{B}.$$

This equation highlights two key physical aspects: the **electromagnetic force density** (ρE + J × B), which corresponds to the Lorentz force acting on charged particles, and the **divergence of momentum flux**, which represents the transport and redistribution of electromagnetic momentum within the field.

Physical Significance in Plasma Systems

Wave Momentum and Radiation Pressure: Electromagnetic waves carry momentum, which can exert pressure on plasmas. This is significant in laser-plasma interactions, where high-intensity lasers induce radiation pressure-driven acceleration, a key concept in inertial confinement fusion (ICF).

Ponderomotive Forces in Plasmas: The nonlinear interaction of high-frequency waves with plasma leads to a ponderomotive force:

$$\mathbf{F}_{\text{pond}} = -\frac{e^2}{4m\omega^2} \nabla |\mathbf{E}|^2.$$

This force pushes plasma electrons away from high-intensity regions, crucial for plasma instabilities and wave guiding effects in plasma-based accelerators.

Magnetic Pressure and Plasma confinement: In magnetized plasmas, the magnetic pressure term:

$$P_B = \frac{B^2}{2\mu_0}$$

governs plasma confinement in magnetic fusion devices. The balance of magnetic pressure with plasma thermal pressure,

$$\nabla P + \mathbf{J} \times \mathbf{B} = 0,$$

determines equilibrium configurations such as Tokamaks and Stellarators.

Momentum Transfer and Wave-Particle Interactions

In collisionless plasmas, wave momentum can be transferred to charged particles through Landau damping and cyclotron resonance. The interaction condition,

$$\omega - \mathbf{k} \cdot \mathbf{v} = 0,$$

enables resonant energy and momentum exchange between waves and particles. This mechanism underlies the generation of energetic particle populations in astrophysical plasmas and space weather phenomena.

3.2 Gauge Choices in Electrodynamics

Maxwell's equations, expressed in terms of the vector potential **A** and the scalar potential ϕ, exhibit a fundamental redundancy due to the gauge invariance of electrodynamics. Specifically, the potentials are not uniquely determined by the fields, and different choices of gauge functions lead to equivalent physical descriptions. This redundancy is formalized through the gauge transformation:

$$\mathbf{A} \rightarrow \mathbf{A}' = \mathbf{A} + \nabla\chi, \quad \phi \rightarrow \phi' = \phi - \frac{\partial \chi}{\partial t},$$

where $\chi(x, t)$ is an arbitrary scalar function. To extract physically meaningful solutions from Maxwell's equations and simplify computations, different gauge conditions are imposed. The choice of gauge significantly influences the mathematical structure of electrodynamic problems and plays a crucial role in various formulations of plasma physics, wave propagation, and quantum electrodynamics.

3.2.1 *Lorenz Gauge*

The Lorenz gauge is defined by the condition:

$$\partial^{\mu} A_{\mu} = 0, \quad \text{or explicitly,} \quad \nabla \cdot \mathbf{A} + \frac{1}{c^2}\frac{\partial \phi}{\partial t} = 0.$$

This ensures the manifest Lorentz covariance of Maxwell's equations in terms of the four-potential, leading to the wave equations:

$$\Box\phi = \frac{\rho}{\epsilon_0}, \quad \Box\mathbf{A} = \mu_0\mathbf{J}.$$

The retarded potential formalism emphasizes the causal nature of electromagnetic interactions, as the wave-like solutions for the scalar potential ϕ and vector potential A propagate at the finite speed c. These retarded potentials arise naturally from the Green's function approach and are crucial for accurately describing radiation phenomena in plasmas, including synchrotron radiation.

In the framework of quantum field theory (QFT), particularly quantum electrodynamics (QED), the Lorenz gauge plays a vital role. It allows the photon's Feynman propagator to maintain a simple, Lorentz-covariant form, facilitating efficient perturbative calculations. This is especially beneficial in plasma quantum field theory, where photon interactions with plasma waves are analyzed at a fundamental, quantum level.

Regarding plasma wave dynamics, the resulting wave equations describe electromagnetic perturbations in a manner consistent with relativistic kinetic theory and the

Vlasov-Maxwell framework. The Lorenz gauge proves especially valuable in high-energy plasma contexts, such as laser-plasma interactions and astrophysical plasmas, where relativistic effects are significant and cannot be neglected.

3.2.2 Coulomb Gauge

The Coulomb gauge is given by:

$$\nabla \cdot \mathbf{A} = 0.$$

Applying this condition to Maxwell's equations leads to a Poisson's equation for the scalar potential:

$$\nabla^2 \phi = -\frac{\rho}{\epsilon_0},$$

indicating that ϕ is determined instantaneously by the charge distribution, whereas the vector potential satisfies:

$$\left(\frac{1}{c^2}\frac{\partial^2}{\partial t^2} - \nabla^2\right)\mathbf{A} = \mu_0 \mathbf{J} - \frac{1}{c^2}\nabla\frac{\partial \phi}{\partial t}.$$

In the Coulomb gauge, the Poisson equation for the scalar potential ϕ eliminates retardation effects, making this approach particularly well-suited for electrostatics and quasi-static plasma regimes. It is especially effective in describing phenomena like Debye shielding and plasma oscillations, as well as magnetostatic configurations found in magnetic confinement fusion systems and Tokamak equilibria [3]. This gauge also enables a clear separation between longitudinal and transverse field components, which is highly advantageous for studying plasma waves and instabilities. In Fourier space, the longitudinal component of the vector potential A vanishes, simplifying the analysis of wave modes and dispersion relations. Further, because ϕ is determined instantaneously, the Coulomb gauge is useful for investigating energy storage and electrostatic interactions in plasmas. It provides a natural framework for analyzing effects such as capacitance in plasma double layers and confinement mechanisms in electrostatic devices [4].

3.2.3 Axial Gauge

The axial gauge is defined by setting a specific component of the vector potential to zero, often:

$$A_3 = 0, \quad \text{or more generally } A_n = 0.$$

This condition is particularly convenient in problems involving waveguides, synchrotron radiation, and relativistic plasma waves, where electromagnetic waves propagate along a well-defined direction.

The axial gauge offers significant advantages in simplifying plane wave solutions and analyzing dispersion relations, particularly in structured plasma environments like plasma photonic crystals and metamaterials [5]. By aligning the gauge condition with a specific direction, certain components of the vector potential A are naturally eliminated, streamlining the mathematical treatment of wave propagation and dispersion characteristics.

In the context of quantum field theory, the axial gauge is instrumental in removing unphysical degrees of freedom from the photon propagator, often expressed as

$$D_{\mu\nu}(k) = \frac{g_{\mu\nu} - \frac{k_\mu n_\nu + k_\nu n_\mu}{k \cdot n}}{k^2 + i\epsilon}$$

where n^μ defines the chosen reference direction. This form is particularly advantageous in light-front field theory and in modeling high-energy plasma interactions, such as those occurring in astrophysical jets.

Additionally, the axial gauge is well-suited for exploring wave polarization and mode decomposition in magnetized plasmas. It provides a clear framework for analyzing the polarization characteristics of waves like whistler and Alfvén modes, and facilitates the study of plasma wave mode conversions, including the behaviour of extraordinary (X-mode) waves in strongly magnetized environments.

3.2.4 Gauge Redundancy

In classical electrodynamics, the physical observables, the electric field **E** and the magnetic field **B** are derived from the four-potential $A^\mu = (\phi, \mathbf{A})$, but this representation is not unique. The fundamental principle of gauge invariance states that an infinite family of four-potentials can describe the same electromagnetic fields, giving rise to gauge redundancy. While this freedom allows flexibility in choosing convenient mathematical formulations, it also necessitates additional constraints to remove unphysical degrees of freedom. Maxwell's equations can be expressed in terms of the electromagnetic four-potential as follows:

$$F_{\mu\nu} = \partial_\mu A_\nu - \partial_\nu A_\mu.$$

Since $F_{\mu\nu}$ is constructed from derivatives of A^μ, a transformation of the form:

$$A^\mu \to A'^\mu = A^\mu + \partial^\mu \Lambda,$$

where $\Lambda(x)$ is an arbitrary scalar function, leaves $F_{\mu\nu}$ and hence **E** and **B** unchanged. This expresses the fundamental gauge redundancy: infinitely many potential configurations can generate the same physical fields.

The non-uniqueness of electromagnetic potentials arises from the freedom to add an arbitrary scalar function Λ(x) to the four-potential without altering the physical fields. This gauge freedom introduces redundant degrees of freedom that do not correspond to observable quantities. To obtain well-defined and solvable equations for the potentials, both analytically and numerically appropriate gauge-fixing conditions must be imposed.

The causal structure of electromagnetic interactions is also influenced by the choice of gauge. In relativistic formulations, adopting the Lorenz gauge ensures that the resulting wave equations for the scalar and vector potentials preserve Lorentz invariance and uphold causality, particularly in problems involving radiation.

In the realm of quantum electrodynamics (QED), gauge redundancy significantly impacts the structure of the photon propagator and the renormalization behaviour of the theory. In plasma physics, gauge fixing plays a crucial role in determining how electromagnetic waves couple with plasma modes, directly influencing the formulation of dispersion relations and the evaluation of plasma susceptibilities.

Mathematical Constraints on Potentials

To remove redundant degrees of freedom from the four-potential, gauge conditions must be imposed. These conditions act as constraints that select a unique representative from the infinite set of potentials corresponding to the same electromagnetic fields.

Lorenz Gauge

The Lorenz gauge condition:

$$\partial^\mu A_\mu = 0 \quad \Rightarrow \quad \nabla \cdot \mathbf{A} + \frac{1}{c^2}\frac{\partial \phi}{\partial t} = 0,$$

ensures that each component of the potential satisfies the inhomogeneous wave equation:

$$\Box A^\mu = \mu_0 J^\mu.$$

This removes redundant degrees of freedom while maintaining explicit Lorentz covariance, making it essential for relativistic plasma descriptions. The Lorenz gauge allows solutions to be expressed in terms of retarded potentials, ensuring that electromagnetic interactions propagate causally in relativistic plasma environments.

Coulomb Gauge

The Coulomb gauge:

$$\nabla \cdot \mathbf{A} = 0,$$

eliminates longitudinal components of **A**, leading to a decomposition of the field into instantaneous electrostatic and transverse dynamical components:

$$\nabla^2\phi = -\frac{\rho}{\epsilon_0}, \quad \left(\frac{1}{c^2}\frac{\partial^2}{\partial t^2} - \nabla^2\right)\mathbf{A} = \mu_0\mathbf{J} - \frac{1}{c^2}\nabla\frac{\partial\phi}{\partial t}.$$

This gauge is useful in non-relativistic plasma physics, particularly in electrostatic plasma approximations, such as in Debye shielding and Langmuir wave analysis. It provides a clear separation between longitudinal and transverse wave modes, facilitating analysis of plasma wave dispersion relations.

Axial Gauge

In the axial gauge, a specific component of the vector potential is eliminated:

$$A_3 = 0.$$

This simplifies wave equations in problems where electromagnetic fields propagate in a preferred direction, such as waveguides, synchrotron radiation, and magnetized plasmas. The axial gauge also finds applications in canonical quantization of electrodynamics, particularly in the study of light-front field theory and high-energy plasma interactions.

Gauge constraints play a critical role in shaping the physical interpretation and mathematical formulation of electromagnetic fields. In the Coulomb gauge, the electrostatic potential ϕ satisfies an instantaneous Poisson's equation, which is particularly suited for analyzing quasi-static plasmas and magnetohydrodynamic (MHD) regimes. In contrast, the Lorenz gauge leads to wave equations for both ϕ and A, highlighting the time-dependent, dynamic behaviour of all field components.

These constraints also influence energy conservation and field decomposition. Restrictions on the four-potential A^μ determine how electromagnetic energy is partitioned between the fields and plasma matter. The gauge choice affects the derivation and interpretation of Poynting's theorem, especially in how energy flux is defined and analyzed in plasma interactions.

In the context of plasma-wave quantization, particularly in quantum plasma systems, different gauge choices alter how electromagnetic waves couple to plasma excitations. This directly impacts the form of the photon propagator, influencing the quantization of electromagnetic interactions in high-energy plasma theories.

While gauge redundancy in electrodynamics offers flexibility in choosing mathematical formulations, it necessitates the imposition of specific gauge conditions to eliminate non-physical degrees of freedom. Gauges such as Lorenz, Coulomb, and Axial not only define how electromagnetic potentials evolve, but also shape the characteristics of wave propagation and interaction in plasma physics and quantum electrodynamics. In relativistic scenarios, the Lorenz gauge preserves the causal structure of the theory, while the Coulomb gauge is invaluable for isolating longitudinal and transverse components in electrostatic and magnetized plasmas. Ultimately, these constraints not only deepen our conceptual understanding of electromagnetic theory, but also serve as essential tools for accurate modeling in plasma physics, radiation phenomena, and high-energy astrophysical environments [6].

3.3 Limitations of Maxwell's Equations in Media

3.3.1 Non-locality and Dispersion Relations

In classical electrodynamics, Maxwell's equations describe the behaviour of electromagnetic fields as functions of space and time, leading to local field equations where the fields at a given point depend only on the sources in their immediate vicinity. However, when dealing with wave propagation in dispersive and nonlocal media, such as plasmas, metamaterials, the response of the medium is often non-instantaneous and spatially extended. This gives rise to non-locality, wherein the electromagnetic response at a given point depends on fields at distant locations and prior times.

Non-locality fundamentally alters the dispersion relations of electromagnetic waves, affecting phase and group velocities, wave-particle interactions, and causality constraints. In plasma physics, dispersion relations characterize the propagation of electromagnetic and electrostatic waves, revealing key insights into instabilities, energy transport, and collective oscillations.

Non-locality in electrodynamics arises when the constitutive relations are expressed as convolution integrals instead of local differential equations. The generalized form of the electric displacement field and magnetization incorporating non-locality is:

$$D_i(\mathbf{x}, t) = \int d^3x' \int dt',$$
$$\epsilon_{ij}(\mathbf{x} - \mathbf{x}', t - t')E_j(\mathbf{x}', t'),$$

$$B_i(\mathbf{x}, t) = \int d^3x' \int dt',$$
$$\mu_{ij}(\mathbf{x} - \mathbf{x}', t - t')H_j(\mathbf{x}', t'),$$

Here $\epsilon_{ij}(\mathbf{x} - \mathbf{x}', t - t')$ and $\mu_{ij}(\mathbf{x} - \mathbf{x}', t - t')$ are the nonlocal permittivity and permeability tensors, which encode the medium's spatial and temporal memory effects.

Spatial non-locality arises in systems where the polarization at a given point depends on fields at other locations, as seen in spatially dispersive plasmas and metamaterials. Temporal non-locality appears when the medium exhibits a delayed response, leading to frequency-dependent dispersion relations and causal retardation effects.

These integral equations suggest that the standard local dielectric function $\epsilon(\omega)$ in frequency space must be replaced with a wavevector-dependent dielectric tensor $\epsilon_{ij}(\mathbf{k}, \omega)$, introducing wavevector-dependent dispersion relations.

In nonlocal media, the phase velocity $v_p = \omega/k$ may exceed **c**, but causality is preserved by ensuring that the group velocity $v_g = d\omega/dk$ remains subluminal.

Certain wave modes become unstable in plasmas due to nonlocal interactions, leading to beam-plasma instabilities and enhanced energy transfer mechanisms.

In quantum plasmas, nonlocality arises naturally from finite de Broglie wavelength corrections, leading to Bohm potential modifications in the dispersion relations. Nonlocality plays a crucial role in wave propagation across various plasma systems, from classical magnetized plasmas to quantum and modified electrodynamics settings. The dispersion relations derived in these cases encapsulate fundamental properties of wave modes, governing energy transport, stability, and wave-particle interactions. In high-energy plasmas, modified electrodynamics provides additional corrections, leading to observable deviations in extreme astrophysical and laboratory conditions. The rigorous mathematical framework developed here serves as a foundation for analyzing wave propagation, resonance effects, and instabilities in nonlocal plasmas.

3.3.2 Breakdown of Classical Maxwellian Theory at High Frequencies

The classical wave equation for the electric field in free space is derived from Maxwell's equations:

$$\left(\nabla^2 - \frac{1}{c^2}\frac{\partial^2}{\partial t^2}\right)\mathbf{E} = 0.$$

For an electromagnetic wave propagating in vacuum with wavevector k, the plane wave solutions take the form:

$$\mathbf{E}(\mathbf{x}, t) = \mathbf{E}_0 e^{i(\mathbf{k}\cdot\mathbf{x} - \omega t)}.$$

The classical dispersion relation follows as:

$$\omega = c|\mathbf{k}|.$$

However, when the frequency approaches quantum regimes (e.g., near plasma frequencies or in high-energy astrophysical scenarios), deviations arise.

In dense plasmas or structured media, the dielectric response becomes frequency-dependent, altering the dispersion relation as:

$$\epsilon(\omega) = 1 - \frac{\omega_p^2}{\omega^2},$$

leading to deviations from the simple wave equation and necessitating a quantum mechanical treatment.

At ultra-high frequencies, the electron response ceases to be continuous, leading to discrete Landau level transitions, or quantum recoil effects. The classical description of charge motion fails, necessitating the use of quantum kinetic equations.

At high frequencies, particularly in the ultraviolet (UV), X-ray, and gamma-ray regimes, the interaction of electromagnetic waves with matter cannot be adequately described using classical physics. Thus, high-frequency interactions necessitate quantum electrodynamics (QED) corrections.

3.3.3 Classical Versus Quantum Corrections

Classical electrodynamics, governed by Maxwell's equations, successfully describes electromagnetic interactions across a vast range of scales, from macroscopic plasma dynamics to wave propagation in dielectric media. However, at sufficiently high field strengths, small length scales, or high frequencies, classical theory fails to capture the discrete nature of interactions, necessitating quantum corrections. These deviations manifest in scenarios such as:

Quantum Plasmas, where electron degeneracy, wave-particle duality, and Fermi–Dirac statistics alter collective behaviour.

Strong-field electrodynamics, where vacuum polarization and nonlinearity arise due to quantum fluctuations in the photon field.

High-energy particle interactions, requiring quantum electrodynamics (QED) rather than Maxwellian field theory.

In classical electrodynamics, the fields **E** and **B** satisfy Maxwell's equations in vacuum:

$$\nabla \cdot \mathbf{E} = \frac{\rho}{\epsilon_0}, \nabla \times \mathbf{B} - \frac{1}{c^2}\frac{\partial \mathbf{E}}{\partial t} = \mu_0 \mathbf{J}.$$

These equations assume that charge and current densities evolve continuously, and electromagnetic waves propagate according to the classical dispersion relation:

$$\omega^2 = c^2 k^2.$$

However, classical electrodynamics fails under extreme conditions where:

Wave-Particle Duality Becomes Significant: The assumption that charge carriers can be described as classical point particles breaks down when their de Broglie wavelength becomes comparable to system dimensions:

$$\lambda_{\mathrm{dB}} = \frac{h}{p}.$$

Energy Quantization Is Non-negligible: Classical fields assume continuous energy exchange, while quantum electrodynamics imposes energy quanta, given by:

$$E = \hbar\omega.$$

Vacuum Fluctuations Modify Free-Space Propagation: In classical theory, free-space propagation follows the simple wave equation, but in quantum electrodynamics, vacuum polarization and virtual particle interactions introduce nonlinear effects.

Plasma Interactions Exhibit Quantum Modifications: In dense astrophysical and laboratory plasmas, Fermi–Dirac statistics alter collision and wave dynamics, modifying classical dispersion relations.

Quantum Corrections in Electrodynamics

a. *Energy Quantization and Photon-Matter Interactions*

One of the fundamental corrections when transitioning from Maxwellian theory to quantum electrodynamics is the quantization of the electromagnetic field. Classical electrodynamics treats the field as a continuous wave, whereas QED describes it as a superposition of photon states. The quantization procedure starts with promoting the classical vector potential A^{μ} to an operator:

$$\hat{A}^{\mu}(x) = \sum_{k} \left(\epsilon_{k}^{\mu} a_{k} e^{-ikx} + \epsilon_{k}^{\mu *} a_{k}^{\dagger} e^{ikx} \right).$$

This leads to the concept of photons as field quanta, replacing the classical notion of smooth electromagnetic waves. The total field energy is then expressed as:

$$H = \sum_{k} \hbar\omega_{k} \left(a_{k}^{\dagger} a_{k} + \frac{1}{2} \right),$$

introducing fundamental discreteness in energy transfer, essential in photoelectric effects and Compton scattering phenomena that classical electrodynamics cannot describe.

b. *Vacuum Polarization and Nonlinear Corrections*

In high-field regimes, quantum vacuum fluctuations generate corrections to Maxwell's equations, leading to nonlinear electrodynamics. The Euler-Heisenberg Lagrangian introduces an effective action correcting Maxwell's equations:

$$\mathcal{L} = -\frac{1}{4} F_{\mu\nu} F^{\mu\nu} + \frac{\alpha^2}{90 m_e^4} \left[(F_{\mu\nu} F^{\mu\nu})^2 + 7 (F_{\mu\nu} \tilde{F}^{\mu\nu})^2 \right].$$

This additional term modifies the wave equation, allowing vacuum birefringence and photon-photon scattering effects absent in classical theory.

3.3.4 Linear Response Theory (LRT)

Linear response theory (LRT) serves as the foundation for understanding how a system reacts to external perturbations. In plasma physics and electrodynamics, it provides an analytical framework to derive susceptibilities, dielectric functions, dispersion relations, and transport coefficients. The central assumption of LRT is that the system's response is proportional to the applied perturbation, enabling Fourier-space techniques such as the dielectric formalism and Green's function methods. Mathematically, this is expressed through the susceptibility tensor $\chi_{ij}(\mathbf{k}, \omega)$, linking the external fields to the system's response via:

$$P_i(\mathbf{k}, \omega) = \sum_j \chi_{ij}(\mathbf{k}, \omega) E_j(\mathbf{k}, \omega).$$

However, under certain conditions, such as strong field interactions, turbulence, kinetic instabilities, and non-equilibrium dynamics this linear approximation fails, necessitating nonlinear formulations. The breakdown of LRT is crucial in various plasma regimes, including:

1. Strongly coupled plasmas, where interactions between particles become highly nonlinear.
2. Turbulent plasmas, where small perturbations induce large-scale fluctuations.
3. Quantum plasmas, where correlations and entanglement modify the response.
4. Ultra-intense field electrodynamics, where vacuum nonlinearities emerge.

1. *Generalized Formulation of Linear Response Theory*

LRT is formulated using causal Green's functions, where the system's response to a small external perturbation $F_{\text{ext}}(t)$ is given by:

$$R(t) = \int_{-\infty}^{\infty} \chi(t - t') F_{\text{ext}}(t') dt'.$$

In frequency space, this becomes:

$$R(\omega) = \chi(\omega) F_{\text{ext}}(\omega),$$

where $\chi(\omega)$ is the susceptibility function, encoding the medium's response properties. For an electromagnetic plasma, the dielectric response function is:

$$\epsilon(\mathbf{k}, \omega) = 1 + \frac{4\pi i}{\omega} \sigma(\mathbf{k}, \omega),$$

where σ is the conductivity tensor. This formulation assumes:

- Superposition principle holds, i.e., responses to different inputs sum linearly.

- Causal structure of the response, implying Kramers–Kronig relations must be satisfied:

$$\mathrm{Re}[\chi(\omega)] = \frac{1}{\pi}\mathcal{P}\int_{-\infty}^{\infty} \frac{\mathrm{Im}[\chi(\omega')]}{\omega' - \omega} d\omega'.$$

- Small perturbation limit, ensuring that second and higher-order terms in the expansion remain negligible.

However, these conditions break down under nonlinear field interactions, requiring modifications in the theoretical framework [7].

2. *Strongly Coupled Plasmas and Nonlinear Susceptibility*

In a strongly coupled plasma, the interparticle Coulomb potential energy dominates over kinetic energy, leading to nonlinear field interactions. The conventional dielectric function is no longer valid, and higher-order susceptibilities $\chi^{(2)}, \chi^{(3)}$ emerge:

$$P_i = \sum_j \chi_{ij}^{(1)} E_j + \sum_{jk} \chi_{ijk}^{(2)} E_j E_k + \sum_{jkl} \chi_{ijkl}^{(3)} E_j E_k E_l + \dots$$

These terms lead to frequency mixing, wave coupling, and energy transfer beyond LRT. Such nonlinearities appear in [8]:

Warm dense matter, where electron screening modifies plasma response.
Laser-plasma interactions, where field intensities approach Relativistic levels.
Astrophysical plasmas, where turbulence generates non-Maxwellian distributions.

3. *High-Field Electrodynamics and Relativistic Nonlinearities*

In ultra-intense fields, such as those found in laser-plasma interactions, the classical assumption that charge motion remains linear in the applied field breaks down. The equation of motion for a charge in a strong field,

$$m\frac{d\mathbf{v}}{dt} = -e(\mathbf{E} + \mathbf{v} \times \mathbf{B}),$$

becomes highly nonlinear due to relativistic corrections:

$$\gamma m\frac{d\mathbf{v}}{dt} = -e(\mathbf{E} + \mathbf{v} \times \mathbf{B}),$$

where $\gamma = (1 - v^2/c^2)^{-1/2}$ introduces velocity-dependent modifications. This leads to self-induced transparency, ponderomotive forces, and high-harmonic generation, which linear theory fails to describe.

4. *Quantum Plasma Nonlinearity and Nonlocal Response*

In quantum plasmas, LRT assumes that susceptibility remains local in space and time. However, quantum diffraction and Bohm potential corrections introduce nonlocality:

$$\chi(\mathbf{k}, \omega) = \int d^3x'dt' \chi(\mathbf{x} - \mathbf{x}', t - t')E(\mathbf{x}', t').$$

This manifests in the quantum-modified dispersion relation:

$$\omega^2 = \omega_p^2 + \frac{\hbar^2 k^4}{4m^2},$$

which deviates from classical plasma oscillations. Such nonlocal and nonlinear interactions invalidate linear susceptibility approaches in quantum-scale plasmas [9].

The breakdown of Linear Response Theory (LRT) in high-energy and turbulent plasma environments necessitates the adoption of advanced theoretical frameworks. These include nonlinear dielectric formalisms that account for higher-order susceptibilities, kinetic approaches involving direct solutions of the Boltzmann, Vlasov, or Wigner equations, nonlinear wave equations such as the Zakharov system used in modeling Langmuir turbulence, and quantum field theoretic methods where propagators incorporate interaction corrections. These sophisticated approaches are crucial for accurately describing a wide range of phenomena, including laser-plasma interactions under extreme field conditions, turbulence and particle acceleration in astrophysical plasmas, the behaviour of nonlinear photonic structures in metamaterials and plasmonics, and quantum electrodynamical (QED) effects in strong fields. Therefore, while LRT continues to serve as a foundational tool in weakly interacting systems, its limitations in more complex regimes underscore the need for these more comprehensive and nuanced theoretical models.

References

1. Brodin, G., & Marklund, M. (2007). Spin solitons in magnetized pair plasmas. *Physics of Plasmas, 14*(11).
2. Klimo, O., & Tikhonchuk, V. T. (2013). Laser–plasma interaction studies in the context of shock ignition: The regime dominated by parametric instabilities. *Plasma Physics and Controlled Fusion, 55*(9), Article 095002.
3. Stewart, A. M. (2003). Vector potential of the Coulomb gauge. *European Journal of Physics, 24*(5), 519.
4. Brill, O. L., & Goodman, B. (1967). Causality in the Coulomb gauge. *American Journal of Physics, 35*(9), 832–837.
5. Burland, J. B., & Symes, M. J. P. R. (1982). A simple axial displacement gauge for use in the triaxial apparatus. *Geotechnique, 32*(1), 62–65.
6. Wang, T., Monteiro, R. D., & Pang, J. S. (1996). An interior point potential reduction method for constrained equations. *Mathematical Programming, 74*(2), 159–195.

7. Charbonneau, M., Van Vliet, K. M., & Vasilopoulos, P. (1982). Linear response theory revisited III: One-body response formulas and generalized Boltzmann equations. *Journal of Mathematical Physics, 23*(2), 318–336.
8. Yoon, P. H. (2005). Nonlinear electromagnetic susceptibilities of unmagnetized plasmas. *Physics of Plasmas, 12*(11).
9. Fedele, R., Tanjia, F., De Nicola, S., Jovanović, D., & Shukla, P. K. (2012). Quantum ring solitons and nonlocal effects in plasma wake field excitations. *Physics of Plasmas, 19*(10).

Chapter 4
Introduction to Quantum Plasma

The emergence of quantum plasma physics as a distinct domain of study represents a natural evolution in the theoretical and applied understanding of plasma systems when quantum mechanical effects become non-negligible in describing the collective behaviour of charged particles. Classical plasma physics, based on Maxwell's equations coupled with fluid, or kinetic descriptions of particles, suffices under conditions where the thermal de Broglie wavelength $\lambda_{dB} = h/\sqrt{2\pi m k_B T}$ is much smaller than the average inter-particle distance $d \sim n^{-1/3}$, with n being the particle number density. However, as density increases, or temperature decreases such that $\lambda_{dB} \sim d$, quantum effects, most notably wavefunction overlap, degeneracy pressure, and quantum coherence, begin to dominate, thereby necessitating a quantum mechanical reformulation of plasma theory.

Historically, the necessity for a quantum plasma theory was first hinted at in the context of astrophysical systems. As early as the 1920s, Fowler, followed by Chandrasekhar, recognized the importance of quantum degeneracy pressure in explaining the equilibrium and stability of white dwarfs, extremely dense stellar remnants composed primarily of electron-degenerate matter. The degeneracy pressure, arising from the Pauli's exclusion principle, prevents further gravitational collapse by introducing a quantum mechanical pressure independent of temperature, fundamentally altering the standard ideal gas law.

In condensed matter systems, the increasing miniaturization of electronic components and the emergence of semiconductor plasmas in solid-state devices further motivated the refinement of plasma theory to include quantum effects. The development of quantum wells, quantum wires, and quantum dots, systems where the motion of electrons is confined to zero, one, or two spatial dimensions, respectively highlighted scenarios where classical treatments fail due to spatial quantization, tunneling, and quantum coherence. In such media, the quantum confinement effect modifies energy levels and density of states, and the Coulomb interaction must be treated with exchange-correlation corrections derived from many-body quantum theory. From the 1990s onward, with the advent of ultrafast lasers and the ability

P. Kumar, *Quantum Plasma*, https://doi.org/10.1007/978-981-95-0481-7_4

to probe warm dense matter (WDM) a regime characterized by high density and moderate temperature where both quantum degeneracy and strong coupling coexist quantum plasma gained prominence as a distinct and necessary framework [1].

On the theoretical front, significant developments emerged from the formal generalization of classical kinetic theory to its quantum counterpart. The Wigner functionformalism, introduced by Eugene Wigner in 1932, provides a phase-space quasi-probability distribution that incorporates quantum interference and diffraction effects. Another notable development was the Quantum Hydrodynamic (QHD) model, which reinterprets the Schrödinger equation in the hydrodynamic language via the Madelung transformation [2].

The development of quantum plasma physics is not merely a theoretical artifact, but is tightly coupled with empirical and technological motivations. High-intensity laser–plasma interactions, quantum Hall devices, and topological insulators are modern platforms where quantum plasma effects manifest experimentally. Moreover, recent work in quantum electrodynamic (QED) plasmas, particularly in the presence of ultra-intense laser fields exceeding 10^{22} $\mathrm{W/cm^2}$, has necessitated the consideration of not only quantum mechanics, but also quantum field theoretical corrections, such as vacuum polarization and pair production.

The motivation for developing quantum plasma theory is multifaceted, driven by pressing needs across various domains of science and technology. In astrophysics, it is essential for accurately describing degenerate stellar matter found in white dwarfs and neutron stars. From a technological standpoint, the miniaturization of devices in nanostructures and microelectronics demands models that incorporate quantum effects at microscopic scales. On the experimental front, advances in high energy density physics push the boundaries of traditional theories, necessitating quantum descriptions to interpret results from ultra-intense laser and particle beam experiments. At a deeper level, quantum plasma theory is crucial for fundamental physics, as it enables the understanding of non-classical interactions, quantum tunneling, and collective excitations that cannot be explained by classical frameworks.

The physical transition from classical to quantum plasma can therefore be understood as a regime shift governed by fundamental length and energy scales most critically, the inter-particle spacing 'd', the de Broglie wavelength λ_{dB}, and the Fermi energy E_F. Quantum plasma physics then forms the cornerstone for understanding and modeling systems where these scales intersect, enabling the construction of predictive theories grounded in quantum statistical mechanics, non-equilibrium quantum dynamics, and many-body quantum field theory.

This historical development underscores the necessity for a rigorous, foundational, and versatile treatment of plasma that accommodates quantum principles and serves both as a theoretical framework and a practical toolkit for addressing current and future challenges in diverse scientific domains.

4.1 Defining Features

The fundamental departure of quantum plasmas from classical plasma behaviour is rooted in the emergence and dominance of certain intrinsic quantum mechanical length and energy scales, which become relevant in dense, low-temperature, or strongly confined systems. These characteristic scales most prominently the thermal de Broglie wavelength and the Fermi energy establish the domain of quantum statistical and wave effects in plasma dynamics. Their magnitude determines when classical descriptions break down and quantum corrections become indispensable for accurately modeling plasma thermodynamics, transport, and collective excitations.

4.1.1 *Thermal De Broglie Wavelength*

The thermal de Broglie wavelength λ_{dB} serves as a critical quantum scale that quantifies the spatial extent of a particle's wavefunction due to thermal motion. For a plasma particle of mass m at temperature T, this length scale is defined as:

$$\lambda_{\mathrm{dB}} = \frac{h}{\sqrt{2\pi m k_B T}}$$

Here, h denotes Planck's constant and k_B is Boltzmann's constant. Physically, λ_{dB} delineates the regime in which quantum diffraction, originating from Heisenberg's uncertainty principle becomes significant. When the mean inter particle distance $\overline{d} \sim n^{-1/3}$, becomes comparable to, or smaller than λ_{dB}, wave functions begin to overlap, leading to quantum interference effects that alter phase space occupation and invalidate classical Boltzmann statistics. The condition:

$$\lambda_{\mathrm{dB}}^3 \gtrsim 1$$

thus marks the onset of quantum degeneracy. In this regime, the phase space density exceeds unity, and the indistinguishability of fermions necessitates Fermi–Dirac statistics. This inequality is not merely a heuristic, but a manifestation of the fact that the thermal kinetic energy is not sufficient to localize particles within a spatial volume smaller than their quantum uncertainty volume. Consequently, phenomena such as Pauli blocking, zero-point pressure, and quantum tunneling become dominant.

The inclusion of λ_{dB} in governing equations leads to non-local terms such as the quantum Bohm potential in quantum hydrodynamics:

$$V_{\mathrm{Bohm}} = -\frac{\hbar^2}{2m}\frac{\nabla^2\sqrt{n}}{\sqrt{n}}$$

This term encodes the quantum pressure arising from spatial variation of the particle density and embodies the effect of quantum diffraction. It modifies wave dispersion relations and supports quantum solitons, modulational instabilities, and non-classical collective modes, particularly in systems with strong inhomogeneities [3].

4.1.2 Fermi Energy and Fermi Temperature

The second critical quantum scale is the Fermi energy E_F, which arises due to the Pauli's exclusion principle governing fermions such as electrons. In a zero-temperature limit, all available single-particle states up to E_F are filled. For a three-dimensional homogeneous electron gas of number density n, the Fermi energy is given by:

$$E_F = \frac{\hbar^2}{2m}\left(3\pi^2 n\right)^{2/3}.$$

The Fermi temperature T_F is defined via the relation $T_F = E_F/k_B$. These quantities set the energy and temperature scale below which the fermionic system becomes degenerate. In a degenerate plasma where $T \ll T_F$, thermal excitations are confined to a narrow energy shell of width $\sim k_B T$ around E_F, and the specific heat becomes linear in temperature rather than constant (as in classical systems), following:

$$C_V \propto \frac{T}{T_F}$$

In quantum plasmas, this statistical degeneracy introduces a non-thermal pressure, known as degeneracy pressure, given in the non-relativistic limit by:

$$P = \frac{\left(3\pi^2\right)^{2/3}}{5}\frac{\hbar^2}{m}n^{5/3}$$

This pressure arises purely from quantum statistics, independent of temperature, and is critical for maintaining hydrostatic balance in dense astrophysical objects like white dwarfs and neutron stars, as well as influencing the structure and dynamics of dense laboratory and condensed matter plasmas. In the relativistic regime, this pressure scales with $n^{4/3}$, modifying the equation of state and contributing to the formulation of mass-radius relations.

The ratio $x = T/T_F$ serves as a key diagnostic parameter for classifying quantum plasmas. Systems with $x \ll 1$ are strongly degenerate, those with $x \sim 1$ are partially degenerate, and those with $x \gg 1$ revert to classical behaviour [4].

4.1.3 *Quantum Coupling*

To fully characterize quantum plasmas, one must also consider the dimensionless coupling parameter Γ, which measures the ratio of potential to kinetic energy:

$$\Gamma = \frac{e^2}{4\pi\varepsilon_0 a k_B T}$$

where $a = (3/4\pi n)^{1/3}$ is the Wigner–Seitz radius. While in classical systems $\Gamma \ll 1$ signals weak coupling, in quantum systems the temperature in the denominator must be interpreted in terms of the Fermi temperature when degeneracy is high. Thus, the quantum coupling parameter becomes:

$$\Gamma_Q = \frac{e^2}{4\pi\varepsilon_0 a k_B T_F} \propto n^{-1/3}$$

This expression reveals that quantum plasmas can be weakly, or strongly coupled depending on density and Fermi energy, and that quantum corrections to screening, correlation, and collective modes must be incorporated via quantum kinetic theory or quantum molecular dynamics for accurate modeling.

4.1.4 *Quantum Degeneracy*

The emergence of quantum plasma behaviour in dense and cold environments is determined primarily by the relationship between the interparticle spacing and the quantum length and energy scales of the system. The fundamental criterion for quantum degeneracy is that the mean interparticle spacing $\overline{d} \sim n^{-1/3}$ becomes smaller than, or comparable to the thermal de Broglie wavelength λ_{dB}. As derived previously, the de Broglie wavelength is given by:

$$\lambda_{\mathrm{dB}} = \frac{h}{\sqrt{2\pi m k_B T}}.$$

The degeneracy condition $n\lambda_{\mathrm{dB}}^3 \gtrsim 1$ implies that, as the temperature decreases, or the density increases, quantum effects such as wavefunction overlap and indistinguishability become dominant. More quantitatively, the dimensionless degeneracy parameter:

$$x = \frac{T}{T_F} \quad \text{where} \quad \mathrm{T_F} = \frac{\hbar^2}{2mk_B}\left(3\pi^2 n\right)^{2/3}$$

determines the plasma's statistical nature. For $x \ll 1$, the system is fully degenerate, and the Fermi–Dirac distribution governs the occupation of quantum states, rather than the classical Maxwell–Boltzmann distribution. In this regime, thermal energy is insufficient to excite particles far above the Fermi level, and collective excitations are shaped predominantly by quantum statistical pressures and exchange interactions.

4.1.5 *Zero-Temperature Limit*

In the idealized limit of zero temperature, the plasma's electron component forms a degenerate Fermi gas with all states filled up to the Fermi energy E_F. The equilibrium pressure in such a system does not vanish, unlike classical gases, but instead arises from the quantum kinetic energy associated with the Pauli's exclusion principle, commonly referred to as *degeneracy pressure*. For a three-dimensional non-relativistic electron gas, this pressure is derived as:

$$P = \frac{(3\pi^2)^{2/3}}{5}\frac{\hbar^2}{m_e}n_e^{5/3}$$

This relation, which scales as $n^{5/3}$, differs from classical ideal gas pressure that scales linearly with temperature and density. It has a direct impact on the compressibility, acoustic modes, and stability of quantum plasmas. In white dwarfs and other compact astrophysical objects, this degeneracy pressure counters gravitational collapse.

If relativistic corrections become significant at extreme densities, the pressure transitions to the relativistic form:

$$\mathrm{P}_{\mathrm{rel}} = \frac{(3\pi^2)^{1/3}}{4}\frac{\hbar c}{n_e^{4/3}}$$

highlighting the breakdown of non-relativistic approximations at ultra-high densities [5].

4.2 Physical Characteristics of Quantum Plasma

4.2.1 *Wave–Particle Duality and Quantum Coherence*

A defining feature of quantum plasma, particularly under conditions of low temperature and high density, is the inseparability of its wave-like and particle-like aspects, manifesting in the *wave–particle duality* and *quantum coherence* of its constituents. In contrast to classical plasma, where particles follow distinguishable trajectories and collective behaviour arises from statistical averaging and long-range Coulomb interactions, the dynamics of quantum plasma are inherently governed by the coherent superposition of matter-wave amplitudes, quantum interference, and non-local correlations. At the heart of quantum plasma physics lies the principle of wave–particle duality, which states that microscopic entities such as electrons and ions exhibit both particle-like and wave-like behaviour depending on the nature of their interaction and observation. In the context of quantum plasma, this duality is operationalized through the wavefunction $\psi(\mathbf{r}, t)$, which encodes thc probability amplitude of a particle's presence at a location $\mathbf{r}$ and time t. The squared modulus $|\psi(\mathbf{r}, t)|^2$ represents the probability density, while the phase and gradient structure of ψ determines its local momentum and current density.

The key quantum scale characterizing this duality is the thermal de Broglie wavelength, λ_{dB}, already defined as:

$$\lambda_{\mathrm{dB}} = \frac{h}{\sqrt{2\pi m k_B T}}.$$

When this length becomes comparable to, or exceeds the mean interparticle spacing $\overline{d} \sim n^{-1/3}$, individual matter waves overlap, and interference phenomena emerge. In this regime, particle trajectories cease to have objective meaning, the plasma behaves as a quantum ensemble where collective effects arise from phase-coherent superpositions.

Such overlapping wavefunctions alter the kinetic description of the system. Rather than using the classical Boltzmann distribution $f(\mathbf{r}, \mathbf{v}, t)$, one must employ the Wigner distribution function, $f_W\left[(\mathbf{r}, \mathbf{p}, t)\right]$, defined via the Wigner transform:

$$f_W(\mathbf{r}, \mathbf{p}, t) = \frac{1}{(2\pi\hbar)^3} \int e^{i\mathbf{p}\cdot\mathbf{y}/\hbar} \psi^*\left(\mathbf{r} - \frac{\mathbf{y}}{2}, t\right) \psi\left(\mathbf{r} + \frac{\mathbf{y}}{2}, t\right) d^3 y$$

This quasi-probability distribution encodes both local and non-local quantum correlations and exhibits negative regions that signal genuine quantum interference. The evolution of f_W obeys the Wigner–Moyal equation, which generalizes the Vlasov equation and reveals how wave–particle duality modifies phase-space dynamics.

Quantum Coherence

Quantum coherence refers to the maintenance of definite phase relationships between components of a quantum system's wavefunction. In many-body quantum plasmas, coherence manifests not only within single-particle amplitudes, but also among collective excitations plasma waves, density oscillations, and spin modes leading to macroscopic quantum phenomena.

At the mathematical level, coherence is reflected in the off-diagonal elements of the density matrix $\rho\left(\mathbf{r}, \mathbf{r}', t\right)$. For a pure, fully coherent state, this matrix factorizes as:

$$\rho\left(\mathbf{r}, \mathbf{r}', t\right) = \psi^*\left(\mathbf{r}', t\right) \psi(\mathbf{r}, t)$$

and the degree of coherence is quantified by the normalized first-order coherence function:

$$g^{(1)}(\mathbf{r}, \mathbf{r}') = \frac{\langle \psi^*(\mathbf{r}')\psi(\mathbf{r})\rangle}{\sqrt{\langle |\psi(\mathbf{r}')|^2\rangle\langle |\psi(\mathbf{r})|^2\rangle}}$$

In a fully coherent quantum plasma, $|g^{(1)}| \approx 1$ over long spatial separations, indicating robust phase correlations. This coherence underlies the emergence of collective quantum phenomena such as Bose–Einstein condensation (in bosonic systems), super-radiance, and quantum phase transitions. Even in fermionic systems, such as degenerate electron plasmas, long-range coherence enables the emergence of quantum collective modes, including plasmons and zero sound, governed by phase-correlated particle–hole excitations near the Fermi surface.

The dual wave–particle character and intrinsic coherence of quantum plasma constituents represent a fundamental departure from classical paradigms, imposing new structures on the collective dynamics of charged systems. The mathematical formalism required to describe such behavior Wigner functions, Bohm potentials, density matrices, and coherence functions captures the rich tapestry of quantum nonlocality, interference, and statistical indistinguishability. Physically, wave–particle duality and coherence unify the microscopic (single-particle) and macroscopic (collective) pictures, allowing phenomena such as quantum tunneling, zero-point pressure, and entanglement-enhanced transport to be incorporated into a unified theory of plasma dynamics.

The incorporation of these quantum features into plasma models is not merely an academic exercise, but is essential for accurately describing cutting-edge systems such as warm dense matter, ultracold plasmas, metallic nanostructures, quantum devices, and astrophysical compact objects. Their study continues to illuminate the boundary between classical statistical physics and fully quantum field-theoretical treatments, providing a rigorous and profound framework for understanding collective quantum matter.

4.2.2 Landau Damping and Phase Space Description

The transition from classical to quantum plasma regimes necessitates a fundamental re-evaluation of many core physical concepts. Notably, the classical notions of Landau damping and phase space dynamics, which are cornerstones in the kinetic theory of classical plasma, undergo significant modifications or even breakdowns in the quantum regime. These departures arise due to the emergence of quantum mechanical principles such as wave–particle duality, non-commutative phase space structures, and quantum coherence.

In classical plasma theory, Landau damping is a collisionless damping mechanism whereby collective electrostatic oscillations typically Langmuir waves lose energy through resonant interaction with particles moving at velocities near the phase velocity of the wave. The seminal result arises from the linearized Vlasov equation:

$$\frac{\partial f_1}{\partial t} + v\frac{\partial f_1}{\partial x} + \frac{e}{m}E_1\frac{\partial f_0}{\partial v} = 0$$

where $f_0(v)$ is the equilibrium distribution, f_1 is the perturbation, and E_1 is the perturbed electric field. Solving the self-consistent system using Laplace transforms and contour integration yields a complex dispersion relation for the dielectric function $\epsilon(k, \omega)$, whose imaginary part leads to an exponential decay of the wave amplitude, signifying damping without collisions. The key requirement for Landau damping is the presence of particles with velocities near the wave's phase velocity $v_\phi = \omega/k$ and a negative slope of the distribution function at that velocity, i.e., $\partial f_0/\partial v|_{v=v_\phi} < 0$..

However, in quantum plasmas, this mechanism undergoes profound alteration due to the granularity of phase space, Pauli's exclusion, and wave-function coherence. The classical resonance condition is blurred by the Heisenberg uncertainty principle, and the phase space picture must be replaced with a quasi-probability distribution, leading to a qualitatively different damping behaviour [6].

The appropriate kinetic framework for quantum plasma is the Wigner–Moyal formalism, which defines the Wigner function $f_W(x, p, t)$ as a quantum analogue of the classical distribution function. Its evolution is governed by the Wigner equation:

$$\frac{\partial f_W}{\partial t} + \frac{p}{m}\frac{\partial f_W}{\partial x} - \Theta\big[f_W, V(x)\big] = 0.$$

where $\Theta\big[f_W, V\big]$ is the Moyal bracket, expressed as:

$$\Theta\big[f_W, V(x)\big] = \frac{1}{i\hbar}\left(V\left(x + \frac{i\hbar}{2}\frac{\partial}{\partial p}\right) - V\left(x - \frac{i\hbar}{2}\frac{\partial}{\partial p}\right)\right)f_W$$

This operator replaces the classical force term and introduces higher-order derivatives that account for non-local quantum effects. The structure of the Moyal bracket implies that the Wigner equation is non-local in both space and momentum, leading to dispersion and interference phenomena absent in classical systems.

The corresponding dielectric function in the quantum regime, for a linear perturbation of the form $\sim e^{i(kx-\omega t)}$, becomes:

$$\epsilon(k, \omega) = 1 + \frac{e^2}{\varepsilon_0 k^2}\int \frac{f_0(p + \hbar k/2) - f_0(p - \hbar k/2)}{\hbar k(\omega - kp/m)}\frac{dp}{2\pi\hbar}$$

This expression reveals the quantum generalization of the Landau integral, incorporating the finite momentum resolution due to the de Broglie wavelength. In particular, the resonance condition becomes smeared over a quantum momentum width $\Delta p \sim \hbar k$, reflecting the inherent uncertainty in defining exact resonant velocities.

Consequently, the sharp resonant energy transfer of classical Landau damping is replaced by a broader, phase-diffusive mechanism in the quantum case.

Moreover, in degenerate Fermi plasmas, the Pauli's exclusion principle prevents occupation of already filled states near the Fermi level, thereby suppressing resonant energy transfer and potentially eliminating Landau damping altogether for perturbations with phase velocities below the Fermi velocity v_F. This suppression becomes especially pronounced at zero temperature, where the distribution becomes a Heaviside step function.

Phase Space Description

The concept of phase space $\Gamma \equiv \{x, p\}$, fundamental to classical statistical mechanics and kinetic theory, becomes fundamentally altered in quantum plasmas. In classical mechanics, phase space is continuous and points in it correspond to well-defined particle states. This underpins the Liouville equation and the Boltzmann, or Vlasov equations used to describe collective dynamics.

Quantum mechanics, however, imposes a fundamental constraint on phase space resolution: no physical state can be localized beyond the minimal cell $\Delta x \Delta p \geq \hbar/2$. This uncertainty renders the classical phase space picture invalid at quantum scales. The Wigner function, though defined over a classical phase space, can take on negative values and does not represent a true probability distribution, reflecting the incompatibility between simultaneous position and momentum observables.

Classical phase mixing responsible for Landau damping assumes fine-scale filamentation of the distribution function in phase space. However, in quantum mechanics, this process is curtailed by the uncertainty principle, which prohibits arbitrarily fine structures. Instead, damping or decoherence arises from interference patterns in the Wigner function, and the evolution is dominated by the spectral properties of the Liouville super-operator (Wigner–Liouville dynamics).

4.3 Dimensionless Parameters in Quantum Plasma

A rigorous understanding of quantum plasmas requires not only the formulation of governing equations, but also the identification of key dimensionless parameters that delineate distinct physical regimes. These parameters determine whether a system exhibits classical or quantum behaviour, weak or strong coupling, collective or individual dynamics, and ultimately whether standard approximations (e.g., Vlasov, fluid, Thomas–Fermi) are valid. Among the most fundamental of these are the quantum degeneracy parameter χ, the coupling parameter Γ, and the related temperature–density phase diagrams which map the domain of classical plasmas, quantum plasmas, and strongly coupled systems. These parameters are not merely diagnostic, but serve as guides to the dominant physical processes and effective theoretical descriptions.

4.3.1 Quantum Degeneracy Parameter

The parameter $x \equiv n\lambda_{\mathrm{dB}}^3$ encapsulates the degree of quantum degeneracy of a plasma system. It measures the number of particles per cubic thermal de Broglie wavelength, effectively quantifying the degree to which the wavefunctions of neighboring particles overlap. Here, the thermal de Broglie wavelength is defined as:

$$\lambda_{\mathrm{dB}} = \frac{h}{\sqrt{2\pi m k_B T}}.$$

Thus, the degeneracy parameter becomes:

$$\chi = n\left(\frac{h}{\sqrt{2\pi m k_B T}}\right)^3 = \frac{nh^3}{(2\pi m k_B T)}$$

Physically, $\chi \gg 1$ signifies that the interparticle spacing is smaller than the typical quantum uncertainty in position due to thermal motion, leading to significant wavefunction overlap. In this regime, quantum statistical effects such as Pauli's exclusion (for fermions), or Bose–Einstein condensation (for bosons) become crucial, and classical Maxwell–Boltzmann statistics breakdown. Conversely, $\chi \ll 1$ corresponds to the classical dilute regime, where quantum effects are negligible.

From a thermodynamic perspective, χ controls the equation of state in the degenerate regime, pressure and energy density are dominated by quantum statistics (Fermi, or Bose pressures), whereas in the classical regime, thermal kinetic contributions dominate. In the hydrodynamic formulation, this transition corresponds to the replacement of classical ideal gas pressure with expressions derived from Fermi–Dirac or Bose–Einstein integrals. Importantly, χ also sets the scale for validity of quantum hydrodynamic (QHD) models, as it determines the relative importance of non-local and exchange–correlation terms [7].

4.3.2 Coupling Parameter

The coupling parameter Γ quantifies the relative strength of interparticle potential energy to thermal kinetic energy. For a one-component plasma of particles of charge "q" and average interparticle spacing $a = (3/4\pi n)^{1/3}$, it is given by:

$$\Gamma = \frac{q^2}{4\pi\varepsilon_0 a k_B T} = \frac{q^2 n^{1/3}}{(4\pi\varepsilon_0)k_B T}\left(\frac{3}{4\pi}\right)^{1/3}$$

The physical interpretation of Γ is straightforward:

- $\Gamma \ll 1$: weakly coupled regime. Kinetic energy dominates, and the plasma behaves like an ideal gas with small Coulomb corrections. Most astrophysical and fusion plasmas fall into this category.
- $\Gamma \gtrsim 1$: strongly coupled regime. Potential energy becomes comparable to or exceeds kinetic energy. The plasma exhibits short-range order, possibly forming liquid-like or crystalline states (Wigner crystallization).
- $\Gamma \gg 1$: ultra-strong coupling. This regime is relevant in high-energy–density physics, compact stars (e.g., white dwarfs), dusty plasmas, and ultracold systems.

It must be noted that Γ is classically defined, but can be extended into the quantum domain by appropriately replacing the thermal energy with the Fermi energy E_F in degenerate systems:

$$\Gamma = \frac{q^2}{4\pi \varepsilon_0 a E_F}$$

This quantum coupling parameter enables one to examine whether Coulomb interactions dominate even in a degenerate Fermi gas, an essential consideration in dense matter physics [8].

4.3.3 Temperature–Density Phase Diagrams

A comprehensive visualization of the combined behaviour of χ and Γ is provided by the temperature–density (T–n) phase diagram, which maps the fundamental plasma regimes (Fig. 4.1). The structure of the diagram arises from the functional dependencies:

$$\chi \propto nT^{-3/2}$$
$$\Gamma \propto n^{1/3}T^{-1}$$

These scaling laws define distinct boundaries in the (T,n)-plane:

The degeneracy line $\chi = 1$ defines the quantum–classical transition. For densities below this line, the system behaves classically.

The coupling line $\Gamma = 1$ separates weak and strong coupling. Above this line, collective Coulomb effects cannot be treated perturbatively.

The thermal to quantum pressure transition occurs where $k_{\mathrm{dB}} \sim E_F$.

This diagram reveals that dense systems (high n) at low temperatures (low T) are both degenerate and strongly coupled, exemplified by solid-density laser-compressed targets and white dwarf interiors. At the opposite end, low-density, high-temperature systems (e.g., solar corona) reside in the classical weakly coupled domain.

The phase diagram also contains domains of liquid-like and solid-like ordering (Wigner crystal phases), regions of quantum turbulence, and even possible excitonic condensates, or superconducting behaviour in exotic plasma compositions.

The introduction and analysis of these dimensionless parameters provide valuable insights into the behaviour of quantum plasmas.

Scaling and model selection are directly influenced by these parameters, as they determine which physical effects, such as the quantum Bohm potential, Fermi pressure, or classical electromagnetic forces dominate the system. This, in turn, guides the development of appropriate reduced models.

From a computational perspective, the choice of theoretical framework, whether classical molecular dynamics (MD), quantum hydrodynamics (QHD), or quantum kinetic theory depends critically on the values of parameters like χ and Γ.

In terms of experimental relevance, these parameters enable a consistent classification and interpretation of a wide range of phenomena observed in laser-compressed plasmas, ultracold plasma experiments, and astrophysical environments.

Most significantly, they provide a unifying framework that connects diverse plasma systems from semiconductor plasmas in laboratories and inertial confinement fusion experiments to compact astrophysical objects by revealing fundamental similarities across systems that may initially appear unrelated.

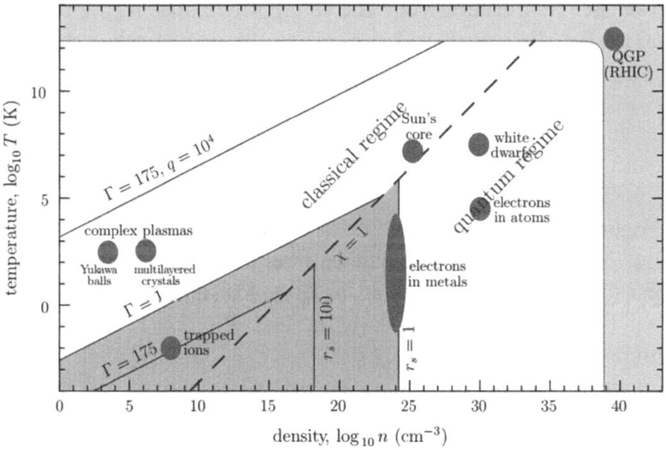

Fig. 4.1 Density–temperature phase diagram of plasma in both natural and laboratory environments. The dashed line $\chi = 1$ delineates the boundary between classical and quantum regimes. For singly charged particles, the shaded region enclosed by the curves $\Gamma = 1$ and $r_s = 1$, represents the domain where strong correlations and structural formations occur (*Source* Bonitz, M., Horing, N., & Ludwig, P. (Eds.). (2010). Introduction to complex plasmas (Vol. 59). Springer Science & Business Media)

4.4 Regimes in Plasma Physics

A thorough classification of physical regimes in plasma physics necessitates the delineation between classical, semiclassical, and fully quantum domains. Each of these regimes embodies a distinct physical picture, mathematical formalism, and hierarchy of relevant effects. Within the context of quantum plasma theory, this classification is not merely a matter of nomenclature, but a critical scaffold upon which the appropriate model, fluid, kinetic, or field-theoretic is selected. These regimes are defined by the relative importance of quantum effects, such as wavefunction overlap, Heisenberg uncertainty, quantum statistics, and exchange interactions compared to thermal and electromagnetic classical effects.

4.4.1 Classical Regime

The classical regime of plasma physics is defined by the negligible influence of quantum effects. This occurs when the interparticle spacing is much larger than the thermal de Broglie wavelength, i.e., when the quantum degeneracy parameter $\chi = n\lambda_{\text{dB}}^3 \ll 1$. Mathematically, this implies:

$$\lambda_{\text{dB}} = \frac{h}{\sqrt{2\pi m k_B T}} \ll n^{-1/3} \Rightarrow \chi \ll 1$$

In this regime:

Statistics: The Maxwell–Boltzmann distribution accurately describes the particle velocity distribution, and Fermi–Dirac, or Bose–Einstein corrections are negligible.

Pressure: The pressure is dominantly thermal, $P \approx nk_BT$, without contributions from quantum degeneracy pressure.

Wavefunctions: The wavefunctions are spatially localized and non-overlapping; hence, quantum interference and exchange effects are absent.

Electrodynamics: The electromagnetic interactions are described by classical Maxwell theory, with Coulomb interactions treated as pointwise and instantaneous.

Models: Classical fluid equations (e.g., magnetohydrodynamics), or kinetic models (e.g., Vlasov–Poisson–Maxwell) are appropriate.

This regime encompasses most high-temperature, low-density plasmas such as solar wind, fusion plasmas, and ionospheric environments. Quantum corrections are of higher order and contribute insignificantly to thermodynamic, or transport properties [9].

4.4.2 Semiclassical Regime

The semiclassical regime lies at the transitional interface between the classical and fully quantum domains. It is characterized by the onset of significant quantum effects such as quantum diffraction and tunneling without the full enforcement of quantum statistics, or exchange symmetry. This regime typically arises when $\chi \sim \mathcal{O}(1)$, i.e., when the interparticle spacing is comparable to the thermal de Broglie wavelength:

$$\lambda_{\mathrm{dB}} \sim n^{-1/3} \Rightarrow \chi \sim 1$$

Key characteristics of this regime include:

Wavefunction Overlap: Particles begin to experience delocalization, leading to measurable quantum interference effects and tunneling. However, antisymmetrization (fermions), or symmetrization (bosons) of the many-body wavefunction may not be necessary for all observables.

Quantum Corrections: The inclusion of gradient terms, such as the quantum Bohm potential V_B, becomes essential in the momentum equations:

$$V_B = -\frac{\hbar^2}{2m}\frac{\nabla^2\sqrt{n}}{\sqrt{n}}$$

This term emerges from the Madelung transformation of the Schrödinger equation and captures quantum diffraction particularly important in the context of wave-packet spreading and boundary layer dynamics.

Semiclassical Kinetics: Wigner–Moyal-type quantum kinetic theories or quantum corrected Vlasov equations are appropriate. The Wigner function $f_W(\mathbf{r}, \mathbf{p})$, while retaining a phase-space formulation, exhibits non-classical behaviour via negative probability regions and quantum corrections through the Moyal expansion.

Statistical Transition: Thermodynamic quantities may require partial inclusion of Fermi–Dirac, or Bose–Einstein corrections, particularly when $T \sim T_F$ or $\mu \sim k_B T$.

Applicability: This regime is particularly relevant in laser-compressed plasma, metallic and semiconductor plasmas at moderate temperatures, and quantum wells, or wires in condensed matter systems. It also captures essential physics in under dense, low-temperature quantum plasma environments, where quantum diffraction dominates over degeneracy pressure.

4.4.3 Fully Quantum Regime

The fully quantum regime manifests when both wavefunction overlap and quantum statistics govern the system's behaviour. This is the domain where $\chi \gg 1$ and, frequently, where the system approaches or exceeds the Fermi temperature T_F:

$$T_F = \frac{E_F}{k_B} = \frac{\hbar^2}{2mk_B}\left(3\pi^2 n\right)^{2/3} \Rightarrow \frac{T}{T_F} \ll 1$$

The following hallmarks define this regime:

Quantum statistics: Fermions obey Fermi–Dirac statistics and are governed by the Pauli's exclusion principle, leading to the emergence of degeneracy pressure even at $T \to 0$. For bosons, Bose–Einstein condensation may occur if conditions permit.

Collective Quantum Effects: Exchange–correlation potentials, many-body quantum fluctuations, and non-local effects dominate. These require advanced models such as quantum kinetic theory, density functional theory (DFT), Hartree–Fock theory, and even quantum field-theoretic approaches when particles are relativistic or undergo particle creation/annihilation.

Pressure Law: For a degenerate Fermi gas at zero temperature:

$$P = \frac{\left(3\pi^2\right)^{2/3}\hbar^2}{5m} n^{5/3}.$$

which significantly differs from the classical ideal gas law, reflecting the purely quantum mechanical origin of this pressure.

Quantum Hydrodynamics: The full inclusion of quantum pressure (degeneracy), quantum Bohm potential, and exchange–correlation stress tensor is required. Models must be supplemented with thermodynamically consistent closures, and spin effects (e.g., magnetization, spin–orbit interaction) may need to be included via Pauli, or Dirac formalism.

Electrodynamics: Quantum electrodynamics (QED), or generalized formulations such as Podolsky electrodynamics become relevant for field quantization, vacuum polarization, and radiation reaction effects.

This regime governs the behaviour of white dwarf interiors (electron degeneracy), neutron stars (nuclear degeneracy), solid-state plasmas in metals at cryogenic temperatures, and ultracold plasmas near the critical quantum degeneracy point. Moreover, extreme laser-matter interactions may transiently produce plasma states within this domain.

4.5 Quantum Effects in Plasma Dynamics

4.5.1 Bohm Potential

In the regime of quantum plasmas, where the microscopic behaviour of constituent particles is dominated by quantum mechanical effects such as wave-particle duality, dispersion, and phase coherence, the standard hydrodynamic descriptions must be generalized to incorporate non-classical phenomena. Among the most critical modifications is the inclusion of the Bohm potential, a term originating from the quantum mechanical phase evolution of matter waves, which encapsulates the inherently nonlocal and delocalized nature of quantum particles. When combined with the analysis of quantum tunneling, this potential becomes central to understanding transport, localization, and barrier penetration in dense plasmas [10, 11].

The concept of the Bohm potential emerges naturally from the Madelung transformation of the Schrödinger equation. For a single-particle system, the time-dependent Schrödinger equation reads:

$$i\hbar\frac{\partial\psi(\mathbf{r},t)}{\partial t}=\left(-\frac{\hbar^2}{2m}\nabla^2+V(\mathbf{r},t)\right)\psi(\mathbf{r},t)$$

Decomposing the wavefunction into amplitude and phase:

$$\psi(\mathbf{r},t)=\sqrt{n(\mathbf{r},t)}\exp\left(\frac{iS(\mathbf{r},t)}{\hbar}\right)$$

where $n(\mathbf{r},t)=|\psi|^2$ is interpreted as a probability density, and S($\mathbf{r}$,t) is the action-like phase function. Substituting this into the Schrödinger equation and separating real and imaginary parts leads to two coupled equations: the quantum Hamilton–Jacobi equation and the continuity equation.

The quantum Hamilton–Jacobi equation is:

$$\frac{\partial S}{\partial t}+\frac{(\nabla S)^2}{2m}+V(\mathbf{r},t)+V_B(\mathbf{r},t)=0$$

where

$$V_B(\mathbf{r},t)=\frac{\hbar^2}{2m}\frac{\nabla^2\sqrt{n}}{\sqrt{n}}$$

is the Bohm potential. It is essential to emphasize that this term has no classical analogue. Physically, it arises due to the spatial variation of the quantum amplitude $\sqrt{n}$, encoding the nonlocal dispersive nature of quantum particles. The Bohm potential introduces an effective quantum pressure arising purely from quantum diffraction, which becomes significant in regions of rapidly varying density, such as at interfaces, boundaries, or regions of strong confinement.

In quantum plasma hydrodynamics, this potential is incorporated into the generalized momentum equation:

$$m\left(\frac{\partial\mathbf{v}}{\partial t}+\mathbf{v}\cdot\nabla\mathbf{v}\right)=-P-ne\nabla\phi+n\nabla V_B$$

Here, P is the pressure (which may include degeneracy contributions), ϕ is the electrostatic potential, and $\mathbf{v}$ is the fluid velocity. The term $n\nabla V_B$ accounts for quantum dispersion and coherence, critical for accurately describing wave propagation and stability in dense quantum plasmas [12].

In classical plasmas, dispersion arises primarily due to finite temperature effects and collective oscillations (e.g., Landau damping). However, in quantum plasmas, dispersion persists even at zero temperature due to the wave-like nature of electrons, encapsulated by the Bohm potential. This term inherently accounts for quantum tunneling, interference, and diffraction effects on the hydrodynamic scale.

The form of the Bohm potential implies a negative curvature response of the density field: when $n(\mathbf{r})$ exhibits sharp spatial variation (such as at boundaries or in double layers), the second derivative $\nabla^2\sqrt{n}$ contributes significantly, indicating that electrons 'spill over' classical barriers, a manifestation of quantum leakage. The corresponding force:

$$\mathbf{F}_B = -\nabla V_B = \frac{\hbar^2}{2m}\nabla\left(\frac{\nabla^2\sqrt{n}}{\sqrt{n}}\right)$$

acts as a quantum dispersive correction that resists compression and supports spreading of the density distribution, similar to the effect of a Fermi pressure, but derived from the phase coherence of the wavefunction rather than Pauli exclusion. This becomes particularly important in low-temperature, high-density plasmas such as metallic hydrogen, white dwarf interiors, and ultra-cold laboratory plasmas.

4.5.2 Quantum Tunneling

Quantum tunneling, the ability of particles to penetrate classically forbidden regionsis an intrinsic aspect of quantum mechanics and is profoundly modified in collective plasma systems. In a quantum plasma, the process of tunneling is not merely governed by single-particle quantum mechanics, but is influenced by many-body interactions, degeneracy effects, and the Bohm potential.

To illustrate, consider a particle encountering a one-dimensional potential barrier V(x). In the WKB approximation, the tunneling probability is given by:

$$T \sim \exp\left(-\frac{2}{\hbar}\int_{x_1}^{x_2}\sqrt{2m(V(x)-E)}dx\right)$$

where x_1 and x_2 are the classical turning points. In a quantum plasma, the effective potential includes not only the electrostatic component, but also the Bohm potential and possibly a degeneracy pressure contribution. Hence, the total barrier may be modified as:

$$v_{eff}(x) = V(x) + V_B(x) + V_{\text{deg}}(x)$$

where V_{deg} arises from Fermi–Dirac statistics. This modifies the barrier shape, enhancing or suppressing tunneling depending on the spatial gradients and density profile. Importantly, in dense plasmas, collective screening effects and quantum diffraction modify the width and height of potential barriers, influencing electron transport, resistivity, and wave damping.

In particular, field emission, thermionic emission, and quantum well resonance in plasmas confined in semiconductor, or nanostructured geometries must account for the influence of the Bohm potential. Moreover, in strongly coupled plasmas, tunneling between adjacent potential wells (as in Wigner crystals, or layered plasma systems) is heavily influenced by quantum delocalization governed by this potential [13].

4.5.3 Wigner Function Formalism

In the comprehensive description of quantum plasmas, where collective effects coalesce with inherently quantum mechanical properties such as coherence, interference, and tunneling, the traditional kinetic theory based on classical phase space concepts becomes inadequate. The Wigner function formalism provides a rigorous quantum mechanical generalization of the classical distribution function and serves as the quantum analog of the Vlasov equation, the cornerstone of collisionless kinetic plasma theory. Unlike the classical approach, the Wigner formalism allows for the encoding of quantum interference and nonlocality in a phase space picture, thereby providing a framework in which both statistical and wave-like aspects of quantum plasma dynamics can be treated on equal footing [14].

The Wigner function was introduced by Eugene Wigner in 1932 as a quasi-probability distribution function in phase space. For a single-particle system described by the density matrix $\hat{\rho}(t)$, the Wigner function $f_W(\mathbf{r}, \mathbf{p}, t)$ is defined as:

$$f_W(\mathbf{r}, \mathbf{p}, t) = \frac{1}{(2\pi\hbar)^3} \int d^3\mathbf{s} e^{-i\mathbf{p}\cdot s/\hbar} \left\langle \mathbf{r} + \frac{\mathbf{s}}{2} | \hat{\rho}(t) | \mathbf{r} - \frac{\mathbf{s}}{2} \right\rangle.$$

This definition effectively maps the density matrix onto a phase space representation. The Wigner function maintains real-valuedness, but may acquire negative values in certain regions of phase space, reflecting the underlying quantum nature and precluding its interpretation as a true probability density. Nonetheless, it permits the computation of quantum expectation values analogously to classical statistical mechanics:

$$\langle A \rangle = \int d^3\mathbf{r} d^3\mathbf{p} A_W(\mathbf{r}, \mathbf{p}) f_W(\mathbf{r}, \mathbf{p}, t)$$

where $A_W(\mathbf{r}, \mathbf{p})$ is the Weyl transform of the operator $\hat{A}$, ensuring consistency between operator-based quantum mechanics and phase-space integration.

The dynamical evolution of the Wigner function is governed by the Wigner–Moyal equation, which plays a role analogous to the Vlasov equation in classical plasma physics. For a quantum particle of mass m in a potential V(**r**), the Wigner–Moyal equation is derived from the von Neumann equation $i\hbar\partial_t\hat{\rho} = \left[\hat{H}, \hat{\rho}\right]$, and reads:

$$\frac{\partial f_W}{\partial t} + \frac{\mathbf{p}}{m} \cdot \nabla_{\mathbf{r}} f_W - \Theta\left[f_W, V\right] = 0$$

Here, $\Theta\left[f_W, V\right]$ is the Moyal bracket, a quantum correction term to the classical force, defined as:

$$\Theta\left[f_W, V\right] = \frac{1}{i\hbar}\left(V\left(\mathbf{r} + \frac{i\hbar}{2}\nabla_{\mathbf{p}}\right) - V\left(\mathbf{r} - \frac{i\hbar}{2}\nabla_{\mathbf{p}}\right)\right) f_W$$

which can be expanded as:

$$\Theta\left[f_W, V\right] = \sum_{n=1}^{\infty} \frac{(-1)^n \hbar^{2n}}{2^{2n}(2n+1)!} \nabla_{\mathbf{r}}^{2n+1} V \cdot \nabla_{\mathbf{p}}^{2n+1} f_W$$

In the classical limit $\hbar \to 0$, this expression reduces to:0

$$\frac{\partial f}{\partial t} + \frac{\mathbf{p}}{m} \cdot \nabla_{\mathbf{r}} f - \nabla_{\mathbf{r}} V \cdot \nabla_{\mathbf{p}} f = 0$$

which is precisely the collisionless Vlasov equation. Thus, the Wigner–Moyal equation is the Natural quantum generalization, incorporating nonlocal and higher-order spatial derivative terms that encode quantum diffraction, interference, and tunneling effects.

The Wigner function describes a quasi-probabilistic flow in phase space where quantum effects manifest through deviations from classical trajectories. The higher-order derivatives in the Moyal expansion account for the nonlocal response of the quantum system, implying that the plasma's distribution function at a point in phase space is influenced by values at neighboring points, an explicit violation of classical locality due to the uncertainty principle.

Physically, this nonlocality becomes essential in regions of sharp gradients, such as in double layers, Debye sheaths, and quantum wells. The Wigner function inherently incorporates quantum coherence and thus permits the study of interference patterns, resonant tunneling, and wave-packet spreading without invoking separate wavefunction treatments.

Moreover, the evolution of the Wigner function reveals the quantum suppression of classical instabilities. For instance, in the case of Landau damping, quantum corrections may alter the resonance condition $\omega - \mathbf{k} \cdot \mathbf{v} = 0$, modifying damping rates and even stabilizing certain modes under extreme density and temperature conditions.

For charged particles in self-consistent electromagnetic fields, the Wigner–Moyal formalism extends to include the Lorentz force and field equations. The Wigner function $f_W(\mathbf{r}, \mathbf{p}, t)$ for electrons in an electromagnetic field described by scalar and vector potentials (ϕ,$\mathbf{A}$) evolves according to:

$$\frac{\partial f_W}{\partial t} + \frac{1}{m}\left(\mathbf{p} - \frac{e}{c}\mathbf{A}\right) \cdot \nabla_{\mathbf{r}} f_W + e\left[\mathbf{E} + \frac{1}{c}\left(\frac{\mathbf{p} - \frac{e}{c}\mathbf{A}}{m} \times \mathbf{B}\right)\right] \cdot \nabla_{\mathbf{p}} f_W + \mathcal{Q}[f_W] = 0$$

where $\mathcal{Q}[f_W]$ contains the quantum correction terms arising from the Moyal bracket. This equation must be solved self-consistently with Maxwell's equations, or their generalized versions (e.g., Podolsky electrodynamics) to describe wave-particle interactions, electromagnetic instabilities, and radiative phenomena in quantum plasmas [15].

4.5.4 Spin Effects and Spin-Polarized Plasmas

The inclusion of spin degrees of freedom in the theoretical framework of quantum plasma physics marks a profound departure from the traditional spinless or scalar plasma models. While classical plasma physics treats particles as point charges with no intrinsic angular momentum, the electron spin, a fundamental quantum property associated with intrinsic magnetic moment introduces entirely new categories of interactions, couplings, and collective phenomena in a plasma medium. These effects are not merely additive, they restructure the dynamical equations, enable new modes of propagation, and influence transport, stability, and magnetization in high-density or low-temperature quantum plasmas. The study of spin effects and spin-polarized plasmas thereby forms a critical domain within quantum plasma theory, especially in environments where relativistic degeneracy, intense magnetic fields, or spin-dependent quantum statistics play pivotal roles.

From a fundamental perspective, the electron spin is a manifestation of relativistic quantum mechanics, emerging naturally from the Dirac equation. It leads to the intrinsic magnetic moment of the electron:

$$\boldsymbol{\mu}_s = -g\mu_B\frac{\mathbf{s}}{\hbar}, \quad \mu_B = \frac{e\hbar}{2m_e}, \quad g \approx 2.0023,$$

where μ_B is the Bohr magneton and $\mathbf{s}$ is the spin vector with Eigen values $\pm\hbar/2$. This magnetic moment interacts with external and self-consistent magnetic fields, introducing magnetization currents, spin forces, and spin–spin interactions that do not arise in scalar quantum plasma models.

The Pauli's exclusion principle, in conjunction with the spin degree of freedom, dictates the statistical behaviour of fermions, leading to spin-polarized, Fermi–Dirac distributions in thermal and non-equilibrium contexts. In fully Spin-polarized plasmas, a majority of particles occupy a specific spin state, leading to anisotropies in pressure, magnetization, and transport coefficients. These effects become particularly important in magnetized white dwarfs, spintronic plasmas, magnetized semiconductor quantum wells, and laser-magnetized nanoplasmas.

In spin-polarized plasmas, the population imbalance between spin-up and spin-down electrons leads to net magnetization $\mathbf{M} = \mu_B(n_\uparrow - n_\downarrow)\hat{\mathbf{z}}$. The degree of spin polarization $\xi = (n_\uparrow - n_\downarrow)/(n_\uparrow + n_\downarrow)$ serves as a critical parameter controlling magnetic susceptibility, pressure anisotropy, and transport coefficients.

This polarization can be induced externally via applied magnetic fields (Zeeman splitting), spontaneously through exchange interactions (in dense degenerate plasmas), or dynamically via laser–plasma coupling. In strongly magnetized quantum plasmas, where the cyclotron energy $\hbar\omega_c$ exceeds $k_B T$, spin alignment becomes energetically favorable, and the plasma enters a quantum Hall-type regime, characterized by quantized Landau levels and non-trivial topological properties.

The magnetization current $\mathbf{j}_M = \nabla \times \mathbf{M}$ introduces additional terms in Maxwell's equations, modifying the propagation of electromagnetic waves and enabling spin-induced birefringence and nonlinear magneto-optical effects.

Spin effects and spin-Polarized plasmas are highly relevant across a wide range of physical systems. In astrophysical compact objects such as white dwarfs and neutron star crusts, extreme densities and strong magnetic fields induce significant spin polarization, impacting the equation of state, cooling behaviour, and magnetohydrodynamic stability. In magnetized semiconductor plasmas, spin-polarized electron gases confined in quantum wells and heterostructures exhibit tunable spin transport properties, offering promising applications in plasma-based spintronics and quantum information technologies. In laser-magnetized nanoplasmas, ultra-intense laser–plasma interactions can generate short-lived magnetic fields that dynamically polarize the plasma, resulting in transient spin currents and enabling novel magnetic confinement schemes. Additionally, spinor Bose–Einstein condensates, despite being composed of bosons, demonstrate analogous spin hydrodynamic behaviour, allowing for cross-disciplinary application of theoretical models and experimental insights between quantum plasmas and condensed matter systems.

The inclusion of spin effects in quantum plasma theory is not a superficial extension but a qualitative enrichment of the physical framework. It introduces new degrees of freedom, novel collective modes, magnetization dynamics, and spin–field couplings that dramatically alter the system's behaviour. Mathematically, it demands a transition from scalar to spinorial, or tensorial formalisms, while physically, it brings plasma theory into alignment with the broader realm of quantum many-body physics, magnetized quantum fluids, and non-Abelian plasma dynamics.

The rigorous development of spin quantum hydrodynamics and kinetic models lays a foundation for exploring nonlinear spin wave interactions, topologically protected plasma states, and magneto-spin turbulence, among other phenomena. In sum, the spin degree of freedom is a central organizing principle for the next generation of quantum plasma theories, linking microscopic quantum coherence to macroscopic collective dynamics in a magnetized and charged medium [16, 17].

4.5.5 Role of Exchange–Correlation Potential

In the theoretical framework of quantum plasma physics, particularly when dealing with systems of high particle density and low temperature where quantum degeneracy is significant, the exchange–correlation potential emerges as a fundamental component governing inter-particle interactions. Its inclusion ensures that the model correctly captures the many-body quantum effects that cannot be described by classical Coulomb interactions alone. Exchange and correlation effects stem from the indistinguishability of fermions and the antisymmetric structure of the many-body wavefunction, which are central to quantum statistics. These effects become pronounced when the de Broglie wavelength of particles becomes comparable to their mean separation typical of degenerate plasmas found in astrophysical compact objects, semiconductor plasmas, or ultracold laboratory plasmas.

The exchange–correlation potential serves to effectively capture the non-local quantum mechanical interactions arising due to both the Pauli's exclusion principle (exchange) and dynamical Coulomb interactions beyond mean-field theory (correlation). In a system of fermions, such as an electron gas, the exchange interaction arises due to the antisymmetrization of the total many-body wavefunction. This leads to an effective interaction even in the absence of explicit Coulomb forces and manifests itself in the so-called Fock term of quantum field-theoretic formulations. The correlation contribution, on the other hand, refers to the residual interactions between particles beyond what is captured by the average-field approximation arising due to instantaneous Coulomb repulsion and resulting in short-range ordering.

In the context of quantum plasma, these effects are not perturbative corrections but integral components of the system's equilibrium and dynamical properties. The magnitude of exchange–correlation effects is governed by the Wigner–Seitz radius:

$$r_s = \left(\frac{3}{4\pi n}\right)^{1/3} \frac{1}{a_0}, \quad \text{where } a_0 = \frac{4\pi\varepsilon_0\hbar^2}{m_e e^2},$$

which measures the ratio of inter-particle spacing to the Bohr radius. For $r_s \ll 1$, the system is weakly correlated; for $r_s \gtrsim 1$, exchange–correlation effects dominate and must be included explicitly.

4.6 Multi-species and Component-Specific Quantum Plasmas

In the domain of quantum plasma physics, various physical realizations arise depending upon the nature of constituent particles, their mass and charge asymmetries, and the background medium. Among these, electron–ion, electron–positron, electron–hole, and dusty quantum plasmas represent four physically distinct, but conceptually interconnected subclasses of quantum plasmas. Each possesses unique quantum statistical properties, interaction mechanisms, and collective behaviours. In high-density, low-temperature regimes, or under intense confinement or magnetization, these plasmas exhibit quantum degeneracy, nonlocal interactions, and wave–particle duality, necessitating rigorous quantum treatment through quantum hydrodynamics, Wigner formalism, or density functional theory.

Electron–Ion Quantum Plasmas

Electron–ion quantum plasmas are the most canonical form of quantum plasmas, typically encountered in metallic nanostructures, white dwarfs, intense laser–plasma interactions, and semiconductor nanodevices. The system consists of degenerate electrons and non-degenerate or partially degenerate ions, with mass asymmetry $m_i \gg m_e$, and charge neutrality condition $n_e \approx Zn_i$, where Z is the ion charge state [18].

Electron–Positron Quantum Plasmas

Electron–positron (e–p) plasmas are composed of equal-mass, equal-density, and oppositely charged fermionic species. Such systems are charge symmetric and mass symmetric, a condition which leads to the vanishing of the electrostatic field in equilibrium and unique dispersion characteristics. These plasmas arise naturally in astrophysical environments (e.g., pulsar magnetospheres, early universe), and in laboratory under ultra-intense gamma-ray lasers, or pair production via Breit–Wheeler process.

The pair plasmas are free of ion-related instabilities (e.g., ion-acoustic), and possess purely quantum electrostatic oscillations, relativistic transparency, and fast thermalization. Quantum entanglement, non-Abelian plasma oscillations, and quantum decoherence processes become relevant in modeling such systems under extreme field conditions [19].

Electron–Hole Quantum Plasmas

Electron–hole plasmas are central to semiconductor physics, especially in low-dimensional quantum structures such as quantum wells, nanowires, and quantum dots. They consist of conduction electrons and valence-band holes, where holes behave as positively charged quasi-particles with effective mass m_h. These systems are governed by band-structure modified dispersion, and subject to non-parabolicity, anisotropy, and quantum confinement effects.

The exchange–correlation interaction is influenced by the material-dependent dielectric screening and strongly affects the exciton formation and polariton coupling. Quantum statistical corrections are significant at cryogenic temperatures, or high excitation densities, where the plasma transitions from a non-degenerate gas to a degenerate electron–hole liquid. The dielectric function and resulting screened Coulomb potential determine the dynamics of collective excitations such as plasmons, excitons, and Bogoliubov modes [20].

Dusty Quantum Plasmas

Dusty Quantum Plasmas (DQP) are a more recent and exotic class where nanometer to micrometer-sized particles (dust grains) are embedded within a quantum plasma environment, and the background plasma itself exhibits quantum degeneracy. These systems are relevant to quantum nanotechnology, semiconductor fabrication, and strongly coupled quantum complex systems.

Unlike classical dusty plasmas, quantum dusty plasmas involve quantum-degenerate electrons (and occasionally ions), while the dust particles are modeled either classically, or as quantum mesoscopic entities. The pronounced mass and charge disparities, along with size inhomogeneity, give rise to several novel phenomena, including quantum-modified interactions between ions, electrons, and dust grains, quantum-corrected Debye shielding around dust particles, and the emergence of dust-acoustic and dust-Langmuir wave modes that incorporate quantum mechanical effects.

Each subclass of quantum plasma whether electron–ion, electron–positron, electron–hole, or dusty quantum presents a unique configuration of quantum-statistical, collective, and interactional effects. Their theoretical modeling necessitates a careful synthesis of quantum hydrodynamics, kinetic theory, and field-theoretic approaches, each tailored to the underlying mass, charge, and degeneracy characteristics. From highly symmetric electron–positron systems to asymmetrical dusty quantum media, these plasmas provide a diverse platform for exploring quantum coherence, screening, wave dispersion, and nonlinear dynamics in systems where quantum mechanics and plasma physics converge profoundly [21].

4.7 Theoretical Models of Quantum Plasma

4.7.1 Fluid Models in Quantum Plasmas

Quantum plasmadynamics, though fundamentally governed by the many-body quantum field equations, can be elegantly approximated under certain conditions by fluid models, which transform the microscopic quantum information into macroscopic, continuum-level evolution equations. These fluid models are not merely analogues of classical fluid descriptions, but encode essential quantum effects such as the uncertainty principle, Fermi degeneracy pressure, quantum tunneling, and coherent matter-wave interference. The primary framework in this regard is the quantum hydrodynamic (QHD) model, extended where appropriate to include electromagnetic field effects through Quantum Magnetohydrodynamics (Q-MHD).

Quantum Hydrodynamic (QHD) Model

The QHD model is a macroscopic, Eulerian description of quantum plasmas, wherein quantum–mechanical features of charged particles (typically electrons) are captured through collective field variables: particle density $n(\mathbf{r}, t)$, velocity field u(r,t), and self-consistent electrostatic, or electromagnetic fields. The model is derived from a microscopic quantum description, notably through the Madelung representationof the Schrödinger equation, and enriched with quantum corrections such as the quantum pressureand the Bohm potential, which encapsulate effects due to quantum coherence and non-locality [22].

Let the system be governed by the single-particle time-dependent Schrödinger equation:

$$i\hbar \frac{\partial \psi(\mathbf{r}, t)}{\partial t} = \left(-\frac{\hbar^2}{2m}\nabla^2 + V(\mathbf{r}, t)\right)\psi(\mathbf{r}, t),$$

where $\psi(\mathbf{r}, t)$ is the complex-valued wavefunction and V includes self-consistent interactions such as electrostatic potential ϕ. The Madelung transformation re-expresses the wavefunction in polar form: $\psi(\mathbf{r}, t) = \sqrt{n(\mathbf{r}, t)}e^{iS(\mathbf{r},t)/\hbar}$, where $n = |\psi|^2$ is the probability (number) density and $\mathbf{u} = \frac{1}{m}\nabla S$ is interpreted as a velocity field. Inserting this form into the Schrödinger equation and separating real and imaginary parts yields two coupled hydrodynamic equations:

i. Continuity equation:

$$\frac{\partial_n}{\partial_t} + \nabla \cdot (n\mathbf{u}) = 0,$$

which ensures local particle number conservation.

ii. Momentum (Euler-like) equation:

$$m\left(\frac{\partial \mathbf{u}}{\partial t} + (\mathbf{u} \cdot \nabla)\mathbf{u}\right) = -\nabla V - \nabla Q,$$

where Q, the *quantum Bohm potential*, is defined as:

$$Q = -\frac{\hbar^2}{2m}\frac{\nabla^2\sqrt{n}}{\sqrt{n}}.$$

This term represents quantum pressure due to wavefunction curvature, encapsulating non-local quantum coherence and dispersion. When V is replaced by the electrostatic potential ϕ, sourced via Poisson's equation $\nabla^2\phi = -4\pi e(n - n_0)$, the QHD system becomes self-consistently closed.

Additional terms such as thermal pressure $P(n)$, which may take the form of Fermi pressure in degenerate regimes $P_F = \frac{(3\pi^2)^{2/3}\hbar^2}{5m}n^{5/3}$, are often included phenomenologically, leading to:

$$m\left(\frac{\partial \mathbf{u}}{\partial t} + (\mathbf{u} \cdot \nabla)\mathbf{u}\right) = -\nabla P(n)/n - e\nabla\phi + \nabla Q.$$

This combination of quantum and classical pressure terms permits an effective description of both coherent quantum dynamics and statistical effects.

The quantum pressure and Bohm potential emerge as second-order derivative terms that capture the delocalization and tunneling effects fundamental to quantum mechanics. Physically, the Bohm potential encapsulates the impact of quantum interference and diffraction, generating a force, via its gradient that resists sharp density variations, effectively expressing Heisenberg's uncertainty principle in a hydrodynamic framework. Meanwhile, quantum pressure arises from Fermi–Dirac statistics, representing the zero-point kinetic energy of fermions enforced by the Pauli exclusion principle. This pressure becomes especially significant in dense, low-temperature plasmas such as those found in white dwarf interiors, or laser-compressed matter. When incorporated into the Quantum Hydrodynamic (QHD) model, these terms enable the prediction of distinctly quantum phenomena including quantum solitons, nonlinear wave breaking at sub-Fermi length scales, and quantum-modified collective modes, none of which are captured by classical fluid descriptions.

Quantum Magnetohydrodynamics (Q-MHD)

In the presence of magnetic fields, the QHD model extends naturally into the Quantum Magnetohydrodynamic (Q-MHD) regime. The momentum equation now includes the Lorentz force term and magnetic pressure contributions [23]:

$$m\left(\frac{\partial \mathbf{u}}{\partial t} + (\mathbf{u} \cdot \nabla)\mathbf{u}\right) = -\frac{1}{n}\nabla P(n) - e(\nabla\phi + \mathbf{u} \times \mathbf{B}) + \nabla Q.$$

The dynamics of the magnetic field is governed by the induction equation:

$$\frac{\partial \mathbf{B}}{\partial t} = \nabla \times (\mathbf{u} \times \mathbf{B}) - \nabla \times (\eta \nabla \times \mathbf{B}),$$

where η is the magnetic diffusivity. The magnetic field couples nontrivially with the quantum Bohm potential, especially in highly anisotropic, or quantized environments, such as Landau-quantized systems. In relativistic, or strongly magnetized contexts, the quantum corrections modify Alfvén wave dispersion, introduce non-classical magnetic reconnection scales, and affect stability criteria of magnetized quantum fluids, including Weibel and mirror-mode instabilities.

Although Quantum hydrodynamics (QHD) and Quantum Magnetohydrodynamics (Q-MHD) offer tractable frameworks for investigating macroscopic quantum plasma phenomena, their applicability is limited by several important factors. These models rely on the mean-field approximation, assuming negligible correlation effects beyond the average field, which becomes inadequate in strongly coupled, or turbulence-dominated plasmas that demand quantum kinetic, or many-body field-theoretic approaches. The underlying Madelung transformation presupposes a single-particle wavefunction, thereby excluding essential quantum features such as multi-particle entanglement, exchange interactions, and quantum decoherence. Additionally, quantum corrections are often treated perturbatively, restricting the models' ability to describe highly nonlinear quantum effects like entanglement-induced coherence, or quantum phase transitions. Their accuracy is further constrained by thermal and degeneracy assumptions, making them most reliable in fully degenerate, or weakly non-degenerate regimes in contrast, high-temperature, relativistic, or mixed-statistics plasmas may require extended or hybrid models. Moreover, standard QHD formulations neglect spin dynamics and relativistic corrections, which are vital for understanding magnetized relativistic plasmas, such as those present in pulsar magnetospheres or during ultra-intense laser–plasma interactions. These limitations define the practical scope of QHD and Q-MHD and underscore the necessity of more comprehensive models in extreme or strongly quantum plasma environments.

Despite these caveats, QHD and Q-MHD models offer a rich and computationally accessible framework for capturing the collective quantum behaviour in a wide variety of plasma contexts. Their strength lies in bridging the microscopic and macroscopic scales through physically intuitive, hydrodynamic variables while retaining key quantum mechanical signatures.

In the broader landscape of quantum plasma theory, fluid models derived from first principles such as the QHD and Q-MHD formalisms serve as powerful intermediate tools between microscopic quantum field descriptions and macroscopic experimental observables. They provide essential insight into wave propagation, nonlinearity, coherence, and stability in quantum plasma environments ranging from nanostructured materials to astrophysical compact objects. When appropriately applied within their domain of validity, these models contribute to a foundational understanding of quantum fluid behaviour and serve as a theoretical scaffold for more sophisticated kinetic and field-theoretic explorations [24].

4.7.2 *Kinetic Models in Quantum Plasma Theory*

In the regime where collective behaviour and microscopic quantum effects coexist yet fluid approximations become insufficient, kinetic models provide a rigorous and indispensable framework for capturing the evolution of the distribution function in phase space. In contrast to hydrodynamic models that operate with macroscopic moments, quantum kinetic theory retains full resolution of momentum and position distributions, thus allowing for a finer and more complete description of both linear and nonlinear quantum plasma dynamics [25].

The Wigner Formalism: A Quantum Analogue of Classical Phase Space

The foundation of quantum kinetic theory is the Wigner distribution function $f_W(\mathbf{r}, \mathbf{p}, t)$, introduced by Eugene Wigner, which is the quantum mechanical analogue of the classical distribution function in phase space. For a single-particle density matrix $\hat{\rho}(\mathbf{r}, \mathbf{r}', t)$, the Wigner function is defined by the Weyl transform:

$$(\mathbf{p}, t) = \frac{1}{(2\pi\hbar)^3} \int d^3 s e^{-i\mathbf{p}\cdot\mathbf{s}/\hbar} \rho\Big(\mathbf{r} + \frac{\mathbf{s}}{2}, \ \mathbf{r} - \frac{\mathbf{s}}{2}, t\Big).$$

Though real-valued, f_W is quasi-probabilistic, as it can assume negative values due to quantum interference, a feature absent in classical systems. It captures the full quantum state in phase space and is thus the centerpiece of kinetic quantum plasma modeling [26].

Wigner–Poisson and Wigner–Maxwell Systems

The dynamics of f_W is governed by the Wigner equation, a quantum kinetic equation derived from the von Neumann equation for the density matrix. In the absence of magnetic fields, the evolution equation is:

$$\frac{\partial f_w}{\partial t} + \frac{\mathbf{p}}{m} \cdot \Delta_{\mathbf{r}} f_w - \frac{i}{(2\pi\hbar)^3} \int d^3 s d^3 p \prime e^{i(\mathbf{p}-\mathbf{p}')\cdot\mathbf{s}/\hbar}$$
$$\left[V\left(\mathbf{r}+\frac{\mathbf{s}}{2}\right) - V\left(\mathbf{r}-\frac{\mathbf{s}}{2}\right)\right] f_w\left(\mathbf{r}, \mathbf{p}', t\right) = 0.$$

In this form, the Wigner equation generalizes the classical Vlasov equation to include non-local quantum corrections. For a self-consistent electrostatic potential $\phi(\mathbf{r}, t)$, sourced by Poisson's equation $\nabla^2\phi = -4\pi e(n - n_0)$, the Wigner–Poisson system is closed. The electron number density is obtained by integrating the Wigner function over momentum:

$$n(\mathbf{r}, t) = \int f_w(\mathbf{r}, \mathbf{p}, t) d^3 p.$$

In the relativistic or magnetized regimes, the Wigner–Poisson system generalizes into the Wigner–Maxwell system, where the self-consistent fields are governed by the full set of Maxwell's equations. The magnetic field enters the Wigner equation via a gauge-invariant modification of the phase space and introduction of the vector potential A, leading to more intricate quantum–electromagnetic coupling.

The non-local pseudo-differential operator in the Wigner equation embodies quantum tunneling, diffraction, and coherence. Wave–particle resonances, quantum Landau damping, and higher-order dispersive features in the collective behaviour of dense plasmas can be accurately captured using this formalism, which is particularly relevant for ultra-fast laser–solid interactions, semiconductor plasmas, and strongly confined degenerate systems [27].

Hartree and Hartree–Fock Approximations

The Wigner formalism, as derived, assumes a single-particle picture. However, for a plasma consisting of interacting electrons, a many-body quantum kinetic formulation is necessary. The most tractable approximation is the Hartree approximation, in which the total wavefunction is assumed to be a direct product of single-particle wavefunctions:

$$\Psi(\mathbf{r}_1, \ldots, \mathbf{r}_N, t) = \prod_{i=1}^{N} \psi_i(\mathbf{r}_i, t),$$

and the interactions are replaced by a self-consistent mean-field potential. In this approximation, exchange effects due to the antisymmetry of the fermionic wavefunction are neglected. To include these effects, the Hartree–Fock approximation considers an antisymmetrized Slater determinant for the total wavefunction, leading to a modified self-consistent field that incorporates exchange interactions. In the kinetic description, this translates to the inclusion of a non-local exchange term in the potential energy, yielding a modified Wigner equation:

$$\left(\frac{\partial f_W}{\partial t}\right)_{\text{exchange}} \quad C_{\text{HF}}[f_w],$$

where C_{HF} represents a complex functional integral operator that accounts for the Pauli's exclusion principle beyond mere Fermi pressure.

The Hartree–Fock corrections are critical in quantum plasmas where inter-particle distances are comparable to the thermal de Broglie wavelength. They account for antisymmetry-induced phase-space exclusion and lead to observable corrections in screening lengths, dispersion relations, and collective mode structures in systems such as dense astrophysical plasmas and low-dimensional electron gases [28].

Quantum Vlasov Models with Exchange–Correlation

In analogy with the classical Vlasov equation, the quantum Vlasov model describes the evolution of a single-particle distribution function in the presence of a self-consistent mean field, but now with quantum corrections:

$$\frac{\partial f}{\partial t} + \frac{\mathbf{p}}{m} \cdot \nabla_{\mathbf{r}} f - \nabla_{\mathbf{r}} U_{\text{eff}} . \nabla_{\mathbf{p}} f = 0.$$

Here, $U_{\text{eff}} = \phi + V_{XC}$, where $V_{xc}(n)$ denotes the exchange–correlation potential, a concept adapted from quantum density functional theory (DFT). In dense systems, particularly where electrons are highly degenerate, or correlations are strong (e.g., warm dense matter), exchange–correlation effects modify the potential landscape experienced by particles. Typical functional forms are parametrized based on homogeneous electron gas calculations, or Kohn–Sham DFT approximations:

$$V_{\text{xc}}(n) = \frac{d}{dn}[n\epsilon_{\text{xc}}(n)],$$

where $\epsilon_{\text{xc}}(n)$ is the exchange–correlation energy per particle.

Exchange–correlation potentials contribute to effective screening, influence the dielectric function, and modify nonlinear wave structures such as quantum double layers, or shock waves. They are essential for modeling phenomena in degenerate laboratory plasmas and strongly correlated nanoplasmonic systems where local field effects are significant [29].

Comparative Structure and Application Regimes

Each of the kinetic models presented has specific domains of validity and operational strengths:

- *Wigner–Poisson systems* excel in describing fully coherent quantum plasmas where tunneling and interference dominate, such as laser–solid interactions, quantum beam dynamics, and electron transport in nanostructures.
- *Hartree–Fock approximations* are essential in cold, dense fermionic systems where exchange effects cannot be neglected, as in white dwarfs, metallic hydrogen, or quantum wells.
- *Quantum Vlasov with Exchange–Correlation* terms allows a tractable yet accurate modeling of Warm dense matter and non-equilibrium systems under extreme conditions, bridging kinetic and DFT approaches.

Kinetic models form the bedrock of a precise and microscopic description of quantum plasma behaviour. Unlike fluid models that average out details, kinetic approaches account for the full distributional character of particles in phase space, while faithfully retaining quantum coherence and correlations through Wigner functions, exchange integrals, and self-consistent field formulations.

4.7.3 Field-Theoretical Approaches to Quantum Plasma

As quantum plasma systems reach regimes of high density, low temperature, or relativistic energies, the particle-based description of matter becomes inadequate without a framework capable of systematically incorporating particle creation and annihilation, collective quantum fluctuations, and correlation effects beyond the mean field. In this context, quantum field theory (QFT) emerges as the most natural and rigorous language for analyzing quantum plasma. This section provides a comprehensive and physically grounded exposition of the key field-theoretical methods applicable to quantum plasma, specifically: (i) second quantization and plasma field theory, (ii) path integral and Green's function approaches in dense plasma, and (iii) the Linear response theory formulated via the polarization tensor. These frameworks do not merely extend existing quantum kinetic, or fluid theories but recast the plasma state as a dynamical medium of interacting quantum fields.

Second Quantization and Plasma Field Theory

In quantum plasma physics, the second quantization formalism is indispensable for constructing a many-body theory that respects the indistinguishability and statistics of particles, fermions (electrons, protons) and bosons (photons, plasmon modes, etc.). The field operators $\hat{\psi}^{\dagger}(\mathbf{r}, t)$ and $\hat{\psi}(\mathbf{r}, t)$ serve as creation and annihilation operators acting on a Fock space, with the fundamental (anti)commutation relations:

$$\left[\hat{\psi}(\mathbf{r}, t), \hat{\psi}^{\dagger}\left(\mathbf{r}', t\right)\right]_{\pm} = \delta\left(\mathbf{r}, \mathbf{r}'\right), \text{(fermions: minus, bosons: plus.)}$$

The Hamiltonian of the system is expressed in terms of these fields. For instance, a basic second-quantized Hamiltonian for an interacting electron gas reads:

$$\hat{H} = \int d^3r \hat{\psi}^\dagger(\mathbf{r}) \left(\frac{\hbar^2}{2m} \nabla^2 \right) \hat{\psi}(\mathbf{r}) + \frac{1}{2} \int d^3r d^3r' \hat{\psi}^\dagger(\mathbf{r}) \hat{\psi}^\dagger(\mathbf{r}') V(\mathbf{r} - \mathbf{r}') \hat{\psi}(\mathbf{r}') \hat{\psi}(\mathbf{r}).$$

This formalism enables systematic inclusion of pair creation, exchange interactions, and collective bosonic modes through bosonization, or canonical transformations. Crucially, the second-quantized description naturally interfaces with gauge field theories when electromagnetic fields are quantized, as in quantum electrodynamics (QED), thus enabling the treatment of plasma under conditions where photon-plasma interactions, vacuum polarization, and gauge invariance become relevant.

In quantum plasmas, particles are excitations of underlying fields. Second quantization allows plasmon, photon, and particle–hole modes to be treated on equal footing as dynamic quantum degrees of freedom. It forms the basis for computing thermodynamic quantities via partition functions in grand canonical ensembles and is indispensable in degenerate regimes, or when quantum fluctuations dominate [30].

Path Integral and Green's Function Methods in Dense Plasma

A powerful complement to operator-based field theory is the path integral formalism, wherein the quantum partition function and correlation functions are formulated as functional integrals over fields. For fermions, the generating functional is given by:

$$Z[\eta, \overline{\eta}] = \int D[\overline{\psi}, \psi] e^{\frac{i}{\hbar} S[\overline{\psi}, \psi] + \int d^4x (\overline{\eta}\psi + \overline{\psi}\eta)},$$

where $\eta, \overline{\eta}$ are Grassmann-valued external sources, and S is the action including kinetic and interaction terms. For quantum plasmas, this is often extended to include electromagnetic gauge fields A^μ through minimal coupling.

The Green's functions (or propagators) are obtained by functional differentiation of the generating functional. The central objects are:

i. Fermionic Green's function:

$$G(x, x') = -i \langle T \psi(x) \overline{\psi}(x') \rangle,$$

which describes the amplitude for particle propagation and encodes spectral and thermodynamic properties.

ii. Gauge field propagator:

$$D_{\mu\nu}(x, x') = -i \langle T A_\mu(x) A_\nu(x') \rangle,$$

governing the response of the field and mediating interactions.

In finite-temperature quantum plasmas, the Matsubara formalism reformulates time evolution over imaginary time $\tau = it$, leading to compact expressions for thermodynamic quantities via Matsubara Green's functions. The inclusion of interaction effects yields the Dyson equation:

$$G = G_0 + G_0 \sum G,$$

where Σ is the self-energy, encapsulating exchange–correlation and interaction effects. The plasmon pole, electron damping, and screening phenomena are encoded in the analytic structure of the Green's functions in frequency space.

The path integral and Green's function formalisms provide systematic access to non-perturbative, finite-temperature, and real-time quantum dynamics in dense plasmas. These methods underpin the calculation of response functions, dielectric tensors, and many-body dispersion relations, and are essential for analyzing quantum plasma oscillations, relativistic degeneracy, and collective excitations such as phonons and polaritons [31].

Linear Response and Polarization Tensor in Quantum Plasma

The linear response theory establishes the framework for determining how a quantum plasma responds to small perturbations, such as external electromagnetic fields. Central to this is the polarization tensor $\Pi^{\mu\nu}(x, x')$, which quantifies the current induced in response to an applied gauge field:

$$J^{\mu}(x) = \int d^4x' \Pi^{\mu\nu}(x, x') A_{\nu}(x').$$

In Fourier space, for a homogeneous and isotropic medium, this becomes:

$$J^{\mu}(k) = \Pi^{\mu\nu}(k) A_{\nu}(k),$$

where $k^{\mu} = (\omega, \mathbf{k})$ is the four-momentum of the external perturbation. The polarization tensor is calculated via the one-loop fermionic diagram in field theory:

$$\Pi^{\mu\nu}(k) = -ie^2 \int \frac{d^4p}{(2\pi)^4} \mathrm{Tr}\left[\gamma^{\mu} G(p) \gamma^{\nu} G(p + k)\right],$$

where *G*(*p*) is the fermion propagator. In a medium, Lorentz invariance is broken, and $\Pi^{\mu\nu}$ splits into longitudinal and transverse components with respect to k, leading to distinct modes:

- Longitudinal response yields the dielectric function $\varepsilon_L(\omega, \mathbf{k})$,
- Transverse response relates to magnetic permeability $\mu_T(\omega, \mathbf{k})$.

The dielectric function can be obtained via the scalar component:

$$\varepsilon(\omega, \mathbf{k}) = 1 - \frac{\Pi_{00}(\omega, \mathbf{k})}{\mathbf{k}^2}.$$

The polarization tensor is a complete linear descriptor of the plasma's quantum electromagnetic response. It governs wave propagation, damping (e.g., Landau and quantum damping), and instability thresholds. The poles of the dressed photon propagator $D_{\mu\nu}(k) = \left[k^2 g_{\mu\nu} - k_\mu k_\nu - \Pi^{\mu\nu}(k)\right]^{-1}$ yield collective modes plasmons, photon–plasmon hybrids, and in certain regimes, photonic band gaps, or effective photon masses.

Field-theoretical methods constitute the most fundamental and general framework for quantum plasma physics. They allow a unified treatment of quantum statistical effects, collective excitations, and relativistic corrections in both equilibrium and non-equilibrium settings. Second quantization lays the algebraic groundwork for quantum statistics and interactions, the path integral approach connects quantum mechanics with statistical field theory, while the Green's function and polarization tensor formalisms provide computational machinery for obtaining physical observables.

These techniques are essential in various advanced plasma regimes, including relativistic dense plasmas where effects like vacuum polarization and electron–positron pair creation become significant, degenerate Fermi plasmas found in astrophysical environments such as neutron stars, strongly coupled quantum plasmas characterized by dominant correlation effects, and quantum electrodynamical plasmas, which arise in scenarios involving ultra-intense laser–matter interactions. When seamlessly integrated with kinetic and hydrodynamic models, the field-theoretical approaches presented here serve as the capstone of a complete quantum plasma theory. Their rigorous mathematical structure ensures consistency with fundamental symmetries, while their physical interpretability allows for a deep understanding of collective quantum behaviour across scales.

4.8 Applications and Emerging Domains

Dense Astrophysical Objects (White Dwarfs, Neutron Stars)

The application of quantum plasma physics to dense astrophysical objects such as white dwarfs and neutron stars represents one of the most profound utilizations of the quantum kinetic and field-theoretical formalisms discussed throughout this chapter. The thermodynamic and dynamical behaviour of such systems is dominated by quantum degeneracy, relativistic effects, strong interactions, and magnetization phenomena. A rigorous understanding of the plasma in these stellar remnants requires a quantum field-theoretic treatment wherein the electron gas, ion lattice, and in some regimes, neutron and proton fluids are governed by Fermi–Dirac statistics, quantum exchange–correlation effects, and, in the case of neutron stars, the behaviour of quark-gluon plasma and superfluid nuclear matter.

White dwarfs are composed primarily of a degenerate electron gas counterbalanced by electrostatic and gravitational forces from a heavy ion lattice (typically carbon, or oxygen nuclei). The electron number density n_e ranges from $10^{29} - 10^{32}$ cm^{-3}, placing the system in a regime where the electrons form an ultra-relativistic, zero-temperature Fermi gas. The pressure of this gas is responsible for halting gravitational collapse in the absence of nuclear burning.

Neutron stars extend the domain of quantum plasma theory into relativistic and strongly interacting regimes. At densities exceeding nuclear saturation, the matter is composed predominantly of neutrons, with a small admixture of protons, electrons, and possibly muons to maintain charge neutrality and beta equilibrium. The electrons form an ultra-relativistic, degenerate quantum plasma, while the neutrons and protons form Fermi liquids exhibiting superfluid and superconducting phases.

Both white dwarfs and neutron stars possess extremely strong magnetic fields, often in excess of 10^9G $- 10^{15}$G. At such strengths, the quantization of Landau levels becomes significant, and the motion of charged particles is confined transversely, giving rise to quantum magnetoplasma phenomena (Fig. 4.2) [32].

Semiconductor and Nanostructure Plasmas

The exploration of semiconductor and nanostructure plasmas marks a fundamental shift in the application of quantum plasma theory toward mesoscopic and nanoscale systems, wherein quantum effects become indispensable even at room temperature due to the scale of confinement. These systems provide a rich platform for observing the interplay of quantum statistical mechanics, quantum electrodynamics, collective plasma phenomena, and electron correlation effects in a condensed matter framework. The study of such plasmas forms a crucial component of quantum plasmonics, nanoelectronics, quantum information processing, and next-generation photonic and optoelectronic devices. Within this domain, plasma behaviour is exhibited not only in traditional bulk semiconductor materials but also in quantum wells, quantum wires,

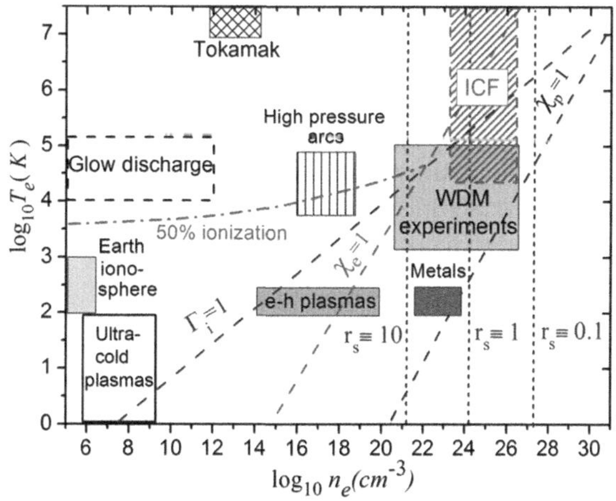

Fig. 4.2 The density–temperature diagram illustrating various plasma types along with their characteristic parameters. ICF and WDM represent inertial confinement fusion and warm dense matter, respectively. The terms 'metals' refer to the electron gas in metals [*Source* Bonitz, M., Moldabekov, Z. A., & Ramazanov, T. S. (2019). Quantum hydrodynamics for plasmas—Quo vadis?. Physics of Plasmas, 26(9)]

quantum dots, and graphene-like system s, all of which confine charge carriers in reduced dimensions and facilitate novel emergent phenomena due to spatial quantization.

In semiconductor materials, the free charge carriers (electrons and holes) can be treated as a degenerate Fermi gas at sufficiently high doping levels, or low temperatures, leading to behaviour characteristic of a quantum plasma. Unlike classical plasmas, where the collective response is governed primarily by thermal agitation, here the Fermi energy and quantum diffraction effects dominate.

In one-dimensional systems, such as quantum wires, or carbon nanotubes, the plasmon dispersion becomes even more anomalous, typically exhibiting a logarithmic correction due to strong Coulombic interactions in 1D, while in quantum dots, discrete excitations resembling quantum-confined plasmons arise, which are sensitive to the particle number and dot geometry. Such modes are probed using electron energy loss spectroscopy (EELS) and terahertz spectroscopy, and they offer potential for plasmonic sensing, quantum light sources, and single-electron transistors [33].

The small spatial extent and high carrier densities in semiconductor nanostructures amplify the role of exchange–correlation (XC) effects, which cannot be neglected as in ideal Fermi gas models. These effects are accounted for via quantum fluid models supplemented with local field corrections and XC potentials derived from quantum Monte Carlo simulations, or generalized gradient approximations (GGAs). The total pressure and chemical potential of the electron gas thereby include significant corrections, influencing both screening behaviour and plasma oscillations.

Quantum plasmas in semiconductor nanostructures underpin a wide range of modern technologies. In plasmonic field-effect transistors (FETs), collective electron oscillations are harnessed for ultra-fast switching and terahertz detection. In quantum cascade lasers (QCLs) and nanowire photodetectors, plasma resonances are engineered to enhance light–matter interactions at subwavelength scales [34].

Moreover, graphene, a truly two-dimensional material, exhibits relativistic-like dispersion and supports tightly confined Dirac plasmons, whose behaviour can be tuned via electrostatic gating and doping. The quantum plasma framework extended to graphene and other 2D materials provides theoretical insight into the design of metamaterials, topological photonic devices, and quantum sensors.

Importantly, with the increasing miniaturization of electronics, quantum decoherence, quantum entanglement of plasmonic modes, and quantum tunneling between adjacent Nanostructures are emerging as dominant features that can only be understood through a quantum plasmalens. Hybrid quantum plasmas, involving strong light–matter coupling, e.g., in exciton–plasmon polariton systems, represent the forefront of research in quantum optoelectronics [35].

Inertial Confinement Fusion and Warm Dense Matter

The intersection of inertial confinement fusion (ICF) and warm dense matter (WDM) represents a critical frontier in plasma physics where quantum mechanical, electrodynamic, and collective phenomena coalesce under extreme conditions of density and temperature. These domains are particularly salient for the study of quantum plasma, as they manifest in parameter regimes where classical approximations break down and quantum degeneracy, electron–ion correlation, and quantum kinetic effects become non-negligible. As such, a rigorous theoretical framework rooted in quantum plasma physics becomes indispensable for a consistent description of their structure, dynamics, and thermodynamic behaviour.

In the ICF scenario, a fuel pellet typically composed of deuterium and tritium is rapidly compressed by high-energy laser or ion beams. The resulting matter achieves densities on the order of 10–1000 times that of solid hydrogen and temperatures ranging from a few to hundreds of electronvolts (1 eV $\approx 1.16 \times 10^4$ K). This places the system in a transitional regime between condensed matter and plasma physics, thereby necessitating a quantum treatment [36].

Warm dense matter, often formed as a precursor, or by-product in ICF targets, is characterized by electron degeneracy and partial ionization. In this regime, neither the ideal gas laws nor fully ionized plasma models suffice. The critical parameters that delineate the WDM regime include the electron degeneracy parameter $\theta = T/T_F$ and the Coulomb coupling parameter $\Gamma = \frac{e^2}{4\pi\varepsilon_0 a k_B T}$, where T_F is the Fermi temperature and a is the Wigner–Seitz radius. Quantum plasma theory becomes essential for $\theta \lesssim 1$ and $\Gamma \gtrsim 1$, indicating significant degeneracy and strong coupling.

Applications and Emerging Domains

The progression of quantum plasma physics into complex and highly engineered materials systems, such as quantum beam–plasma interactions and plasmas in low-dimensional, or topological media, represents a profound extension of classical plasma theory. These domains are fundamentally governed by quantum statistical effects, relativistic electron dynamics, collective excitations, and intricate boundary conditions imposed by crystal symmetries, spin–orbit interactions, and non-trivial topologies. As such, their theoretical formulation demands the confluence of quantum field theory, relativistic fluid dynamics, and condensed matter physics, integrated within the canonical framework of quantum plasma dynamics [37].

Quantum beam–plasma systems emerge in high-intensity laser–plasma interactions, astrophysical jet phenomena, and compact beam–target experiments. These systems consist of relativistic degenerate electron beams propagating through a dense quantum plasma, where both Fermi pressure and quantum diffraction govern the collective response of the background medium. The beam electrons often obey a distribution function near the zero-temperature Fermi–Dirac limit due to the ultra-short pulse durations and high number densities. Applications of these quantum beam–plasma systems include plasma-based accelerators, inertial fusion fast ignition, and relativistic current drivers in compact high-energy facilities. In astrophysics, similar configurations may model pulsar magnetospheres, accretion shock fronts, and quasi-relativistic outflows in AGNs [38].

Graphene and related two-dimensional (2D) materials have emerged as natural laboratories for quantum plasmas confined in reduced-dimensional spaces. The Dirac-like dispersion of charge carriers in monolayer graphene, characterized by a linear energy–momentum relation $\epsilon(k) = \hbar v_F |k|$, engenders massless relativistic fermionic behaviour even at low energies. Exotic quantum materials such as Weyl semimetals, super lattices, and moiré heterostructures further enrich the landscape of quantum plasma theory. In Weyl systems, Berry curvature effects induce anomalous velocities and modify the continuity and momentum balance equations. This leads to chiral magnetic waves, quantum anomalies, and topological magnetoplasmons.

The study of quantum beam–plasma systems and plasmas in exotic quantum materials signifies a transformative extension of traditional plasma physics into the realm of relativistic quantum many-body systems, topologically non-trivial geometries, and low-dimensional matter. The mathematical formalism underpinning these systems from Dirac–Wigner transport equations to quantum fluid dynamics with spin and topology has introduced new paradigms for understanding collective excitations, stability criteria, and transport processes. Physically, these systems exemplify the inseparability of quantum mechanics, electrodynamics, and statistical mechanics in governing modern plasma-like behaviour across a multitude of high-energy, condensed matter, and engineered platforms. They not only serve as testbeds for novel quantum theories, but also have profound technological implications in quantum optics, plasmonics, spintronics, and next-generation computation and energy transport systems [39].

References

1. Tajima, T. (2018). *Plasma astrophysics*. CRC Press.
2. Ragheb, M. (2023). The nuclear, plasma, and radiation universe.
3. Riley, D. (2021). *Warm dense matter: Laboratory generation and diagnosis*. IOP Publishing.
4. Dornheim, T., Moldabekov, Z. A., Ramakrishna, K., Tolias, P., Baczewski, A. D., Kraus, D., Preston, T.R., Chapman, D.A., Böhme, M.P., Döppner, T., Graziani, F., Bonitz, M., Cangi, A., & Vorberger, J. (2023). Electronic density response of warm dense matter. *Physics of Plasmas, 30*(3).
5. Bergeson, S. D., & Spencer, R. L. (2003). Neutral-plasma oscillations at zero temperature. *Physical Review E, 67*(2), Article 026414.
6. Thorne, R. M., & Summers, D. (1991). Landau damping in space plasmas. *Physics of Fluids B: Plasma Physics, 3*(8), 2117–2123.
7. Haas, F., & Mahmood, S. (2022). Linear and nonlinear waves in quantum plasmas with arbitrary degeneracy of electrons. *Reviews of Modern Plasma Physics, 6*(1), 7.
8. Vishnyakov, V. I., & Dragan, G. S. (2007). Coupling parameter for low-temperature plasma with condensed phase. *Condensed Matter Physics*.
9. Brodin, G., Marklund, M., & Manfredi, G. (2008). Quantum plasma effects in the classical regime. *Physical Review Letters, 100*(17), Article 175001.
10. Harris, V. G., & Andalib, P. (2024). Modern quantum materials. *Frontiers in Materials, 11*, 1343005.
11. Musser, G. (2022). *Emergence in condensed matter and quantum gravity: A nontechnical review*. Springer Nature.
12. Misra, A. P., & Shukla, P. K. (2012). Stability and evolution of wave packets in strongly coupled degenerate plasmas. *Physical Review E—Statistical, Nonlinear, and Soft Matter Physics, 85*(2), 026409.
13. Shukla, P. K., & Eliasson, B. (2010). Recent developments in quantum plasma physics. *Plasma Physics and Controlled Fusion, 52*(12), Article 124040.

14. Tabatabaee, S. M. A., & Sadooghi, N. (2020). Wigner function formalism and the evolution of thermodynamic quantities in an expanding magnetized plasma. *Physical Review D, 101*(7), Article 076022.
15. Al-Naseri, H., Zamanian, J., & Brodin, G. (2021). Spacetime-dependent electric field effects in vacuum and plasma using the Wigner-formalism. *arXiv preprint* arXiv:2104.05370.
16. Cowley, S. C., Kulsrud, R. M., & Valeo, E. (1984). *Kinetic equation for spin-polarized plasmas* (No. PPPL-2119). Princeton Plasma Physics Lab. (PPPL), Princeton, NJ (United States).
17. Seipt, D., & Thomas, A. G. (2023). Kinetic theory for spin-polarized relativistic plasmas. *Physics of Plasmas, 30*(9).
18. Mahmood, S., & Mushtaq, A. (2008). Quantum ion acoustic solitary waves in electron–ion plasmas: A Sagdeev potential approach. *Physics Letters A, 372*(19), 3467–3470.
19. Tsintsadze, N. L., Tsintsadze, L. N., Hussain, A., & Murtaza, G. (2011). New longitudinal waves in electron-positron-ion quantum plasmas. *The European Physical Journal D, 64*, 447–452.
20. Cingolani, R., Lage, H., Tapfer, L., Kalt, H., Heitmann, D., & Ploog, K. (1991). Quantum confined one-dimensional electron-hole plasma in semiconductor quantum wires. *Physical Review Letters, 67*(7), 891.
21. Ali, S., & Shukla, P. K. (2006). Dust acoustic solitary waves in a quantum plasma. *Physics of Plasmas, 13*(2).
22. Moldabekov, Z. (2018). *Quantum fluid description of dense plasmas* (Doctoral dissertation).
23. Shukla, N., Brodin, G., Marklund, M., Shukla, P. K., & Stenflo, L. (2009). Nonlinear electromagnetic wave equations for superdense magnetized plasmas. *Physics of Plasmas, 16*(7).
24. Momeni, M. (2013). Linear study of Rayleigh-Taylor instability in a diffusive quantum plasma. *Physics of Plasmas, 20*(8).
25. Melrose, D. B. (2020). Quantum kinetic theory for unmagnetized and magnetized plasmas: A tutorial review of quantum plasma theory. *Reviews of Modern Plasma Physics, 4*(1), 8.
26. Bonitz, M. (2012). Kinetic theory for quantum plasmas. In *AIP conference proceedings* (Vol. 1421, No. 1, pp. 135–155). American Institute of Physics.
27. Akbari-Moghanjoughi, M. (2015). Hydrodynamic limit of Wigner-Poisson kinetic theory: Revisited. *Physics of Plasmas, 22*(2).
28. Langhoff, P. W., Karplus, M., & Hurst, R. P. (1966). Approximations to Hartree—Fock perturbation theory. *The Journal of Chemical Physics, 44*(2), 505–514.
29. Shahmansouri, M. (2015). The exchange-correlation effects on surface plasmon oscillations in semi-bounded quantum plasma. *Physics of Plasmas, 22*(9).
30. Mendonça, J. T., Martins, A. M., & Guerreiro, A. (2000). Field quantization in a plasma: Photon mass and charge. *Physical Review E, 62*(2), 2989.
31. Shevkunov, S. V. (2002). A dense hydrogen plasma modeled by the path integral-Monte Carlo method. *Journal of Experimental and Theoretical Physics, 94*, 943–965.
32. Piron, R., & Blenski, T. (2018). Average-atom model calculations of dense-plasma opacities: Review and potential applications to white-dwarf stars. *Contributions to Plasma Physics, 58*(1), 30–41.
33. Sharma, K., & Deka, U. (2021). Comprehensive review on various instabilities in semiconductor quantum plasma. *Brazilian Journal of Physics, 51*(6), 1944–1955.
34. Khanna, V. K. (2012). Plasma synthesis of quantum dots. *Nanoscience and Nanotechnology Letters, 4*(3), 275–290.
35. Eliasson, B., & Shukla, P. K. (2009). Nonlinear aspects of quantum plasma physics: Nanoplasmonics and nanostructures in dense plasmas. *Plasma and Fusion Research, 4*, 032–032.

36. Hu, S. X., Collins, L. A., Boehly, T. R., Kress, J. D., Goncharov, V. N., & Skupsky, S. (2014). First-principles thermal conductivity of warm-dense deuterium plasmas for inertial confinement fusion applications. *Physical Review E, 89*(4), Article 043105.
37. Clericò, V. (2020). Fabrication and characterization of quantum materials: Graphene heterostructures and topological insulators.
38. Bozhevolnyi, S. I., & Mortensen, N. A. (2017). Plasmonics for emerging quantum technologies. *Nanophotonics, 6*(5), 1185–1188.
39. Passian, A., & Imam, N. (2019). Nanosystems, edge computing, and the next generation computing systems. *Sensors, 19*(18), 4048.

Chapter 5
Theoretical Models

The study of quantum plasmas has undergone a fundamental evolution, transitioning from a specialized offshoot of many-body physics to a robust theoretical and applied discipline intersecting quantum mechanics, field theory, and plasma physics. While prior chapters in this book have offered a systematic exposition of the foundational principles underlying quantum theory, classical and quantum plasma physics, as well as their astrophysical, technological, and high-energy manifestations, the present chapter inaugurates a formal departure from general expositions to a focused development of theoretical modeling frameworks.

The necessity of such a formal construction arises from a core challenge unique to quantum plasma systems, the simultaneous relevance of quantum statistics, electromagnetic collective effects, and wave–particle duality. Classical models whether fluid, or kinetic fail in high-density, or low-temperature regimes where quantum coherence, Fermi pressure, tunneling, and entanglement effects dominate. Similarly, quantum mechanical models that omit long-range Coulomb interactions, or collective electromagnetic fluctuations remain incomplete in capturing the macroscopic dynamical evolution of such systems. This dual demand to respect quantum microphysics while retaining plasma macrostructure renders the theoretical modeling of quantum plasmas inherently multiscale and nontrivial.

Historically, efforts to construct a unified theoretical picture of quantum plasmas began with linear quantum kinetic theories, such as those extending the Vlasov–Poisson system via the Wigner–Moyal formalism. These approaches illuminated how quantum dispersion, coherence, and uncertainty could be embedded into phase-space dynamics. Simultaneously, quantum hydrodynamic theories provided insights into collective excitations Langmuir waves, solitons, and shock structures through fluid-like equations augmented by quantum potentials and spin corrections. However, both of these frameworks are effective theories in a specific sense they are derived under assumptions that limit their range of applicability, often excluding strong coupling, relativistic effects, and nonlinear quantum electrodynamical (QED) interactions.

P. Kumar, *Quantum Plasma*, https://doi.org/10.1007/978-981-95-0481-7_5

This realization necessitates a systematic categorization of quantum plasma theories according to the hierarchy of physical scales they resolve and the degree of microscopic detail they retain. From this perspective, one can broadly classify the modeling strategies into four primary regimes, each forming a pillar of this chapter: (1) quantum kinetic theories based on the Wigner–Vlasov, or quantum Boltzmann formalisms, (2) quantum hydrodynamic models emphasizing collective field behaviour, (3) density functional approaches prioritizing electronic structure and ground-state properties under confinement, or high compression, and (4) quantum field-theoretic frameworks leveraging second quantization, perturbative expansions, and effective actions to study relativistic, strong-field, and nonequilibrium regimes.

5.1 Kinetic Theory Approaches to Quantum Plasma

5.1.1 Quantum Generalization of Vlasov Equation

The classical kinetic description of plasma, encapsulated in the Vlasov equation, offers a robust framework for understanding collisionless collective dynamics in weakly coupled plasma environments. It is based on the evolution of a distribution function $f(\mathbf{r}, \mathbf{p}, t)$, defined over the classical phase space, under the influence of self-consistently determined electromagnetic fields. However, when the underlying constituents of the plasma enter a regime governed by quantum statistical mechanics and non-negligible wave-particle duality, the classical framework exhibits intrinsic limitations that necessitate a systematic generalization.

At its core, the Vlasov equation is derived under assumptions valid only in the limit where the de Broglie wavelength $\lambda_{\text{dB}} = h/p$ is much smaller than the characteristic spatial scale L of the system, and where quantum interference between trajectories may be ignored. In a high-density, low-temperature plasma, or equivalently when the thermal de Broglie wavelength becomes comparable to interparticle separation, quantum statistical effects such as exchange symmetry, tunneling, quantum recoil, and entanglement of dynamical modes begin to dominate. This marks a departure from the classical Liouvillian evolution and calls into question the adequacy of classical phase space trajectories and deterministic particle dynamics.

Moreover, classical distribution functions are positive-definite and commute with the notion of a definite trajectory in phase space. In contrast, quantum systems are characterized by uncertainty relations, i.e., $\Delta x \, \Delta p \geq \hbar/2$, which disallow the simultaneous sharp definition of position and momentum, thus invalidating the foundational premise of the Vlasov picture. Additionally, the neglect of spin, quantum correlations, and statistical antisymmetrization for fermionic species, or bosonic enhancement for photons and composite quasiparticles further divorces classical kinetic theory from quantum plasma phenomena.

Further, in the context of high-energy plasmas, or ultrafast processes governed by intense laser–plasma interactions, relativistic quantum corrections including the

Compton recoil, Zitterbewegung, and pair production thresholds necessitate kinetic equations that extend beyond the scope of classical Vlasov formalism. These effects become particularly relevant in plasmas permeated by high-frequency or high-momentum transfer processes, such as those involving nonlinear Compton scattering, bremsstrahlung, or quantum radiation reaction.

Mathematically, the classical Vlasov equation can be written as:

$$\frac{\partial f}{\partial t} + \frac{\mathbf{p}}{m} \cdot \nabla_{\mathbf{r}} f + \mathbf{F} \cdot \nabla_{\mathbf{p}} f = 0,$$

where $\mathbf{F}$ is the Lorentz force acting on a particle of momentum $\mathbf{p}$ at position $\mathbf{r}$. This equation embodies a conservation law for phase space density in the absence of collisions. However, such a conservation law presumes that phase space volume is invariant under Hamiltonian evolution, an assumption that quantum mechanics amends through non-commutative geometry and operator-based formulations.

From a physical standpoint, quantum corrections manifest as non-localities in both configuration and momentum space, encoded via phase correlations and non-Markovian memory kernels [1, 2]. These corrections are particularly vital for capturing quantum tunneling phenomena in barrier penetration, superposition-driven wave packet interference, and the delocalized nature of quantum states that do not possess classical analogues.

In light of these limitations, a more fundamental approach to kinetic theory becomes imperative one which does not rely on classical trajectories, but rather on the statistical properties of quantum operators, their correlations, and the dynamics of quantum distribution functions. Such an approach must be capable of encompassing both the coherent evolution governed by the Schrödinger or, von Neumann equations and the self-consistent feedback from collective excitations, radiation fields, and quantum many-body interactions. These generalizations culminate in the formalism of quantum kinetic theory, beginning with the Wigner distribution function and extending to Moyal brackets, Kadanoff–Baym equations, and Schwinger–Keldysh non-equilibrium Green's function methods [3, 4].

5.1.2 Wigner Distribution Function: Derivation from the von Neumann Equation

The transition from classical to quantum kinetic theory necessitates a redefinition of the distribution function in phase space, since the classical notion of a probability density $f(\mathbf{r}, \mathbf{p}, t)$ is incompatible with the foundational tenets of quantum mechanics, specifically the uncertainty principle. In this regard, the Wigner distribution function provides a pivotal formulation, representing a quasi-probability distribution in phase space that encapsulates the full quantum dynamics of a system while preserving contact with its classical correspondence.

The Wigner function is constructed from the quantum density operator $\hat{\rho}(t)$, which governs the statistical state of the system via the von Neumann equation,

$$i\hbar\frac{\partial\hat{\rho}(t)}{\partial t} = [\hat{H}, \hat{\rho}(t)],$$

where $\hat{H}$ is the many-body Hamiltonian operator. For simplicity and clarity, we begin with a single-particle system in one spatial dimension, and then later generalize to higher-dimensional and many-body quantum plasmas. The key idea is to project the operatorial dynamics of $\hat{\rho}(t)\}$ onto a quasi-classical phase space by utilizing a symmetrized Fourier transform of the position-space matrix elements of the density operator.

Let $\rho(x_1, x_2, t) = \langle x_1|\hat{\rho}(t)|x_2\rangle$ denote the density matrix in the position representation. The Wigner function $f_W(x, p, t)$ is then defined by the Wigner–Weyl transformation:

$$f_W(x, p, t) = \frac{1}{2\pi\hbar}\int_{-\infty}^{\infty} dy e^{-ipy/\hbar}\rho\left(x + \frac{y}{2}, x - \frac{y}{2}, t\right).$$

This function is real, but not necessarily positive definite, a reflection of the intrinsic quantum interference and non-commutativity of phase space observables [5]. Nevertheless, its marginals reproduce correct quantum mechanical probability densities in position and momentum space:

$$\int dpf_W(x, p, t) = \rho(x, x, t), \quad \int dxf_W(x, p, t) = \langle p|\hat{\rho}(t)|p\rangle.$$

To derive the time evolution equation for $f_W(x, p, t)$, we start by substituting the von Neumann equation into the Wigner transform. Assuming a Hamiltonian of the form

$$\hat{H} = \frac{\hat{p}^2}{2m} + V(\hat{x}),$$

the commutator $[\hat{H}, \hat{\rho}]$ becomes

$$[\hat{\rho}] = \left[\frac{\hat{p}^2}{2m}, \hat{\rho}\right] + [V(\hat{x}), \hat{\rho}].$$

Each of these terms can be projected into phase space via the Wigner transform. The kinetic term yields a straightforward contribution due to the quadratic dependence on momentum:

$$\mathcal{W}\left(\left[\frac{\hat{p}^2}{2m}, \hat{\rho}\right]\right) = -\frac{p}{m}\frac{\partial f_W}{\partial x}.$$

However, the potential term introduces quantum corrections, which are encapsulated in the Moyal expansion. Specifically, the potential term contributes:

$$\mathcal{W}([V(\hat{x}), \hat{\rho}]) = \frac{2}{\hbar}\sin\left(\frac{\hbar}{2}\left(\overleftarrow{\partial}_x \overrightarrow{\partial}_p - \overleftarrow{\partial}_p \overrightarrow{\partial}_x\right)\right)V(x)f_W(x, p, t).$$

This expression defines the Moyal bracket, which is a deformation of the classical Poisson bracket. Expanding the sine function in a power series yields:

$$\{V, f_W\}_{\text{Moyal}} = \{V, f_W\}_{\text{PB}} + \sum_{n=1}^{\infty} \frac{(-1)^n(\hbar/2)^{2n}}{(2n+1)!} \partial_x^{2n+1} V(x) \partial_p^{2n+1} f_W(x, p, t),$$

where $\{V, f_W\}_{\text{PB}} = \frac{\partial V}{\partial x}\frac{\partial f_W}{\partial p}$ is the classical Poisson bracket. Combining the kinetic and potential contributions, the Wigner equation governing the evolution of f_W becomes:

$$\frac{\partial f_W}{\partial t} + \frac{p}{m}\frac{\partial f_W}{\partial x} - \frac{\partial V}{\partial x}\frac{\partial f_W}{\partial p} = \sum_{n=1}^{\infty} \frac{(-1)^n(\hbar/2)^{2n}}{(2n+1)!} \partial_x^{2n+1} V(x) \partial_p^{2n+1} f_W(x, p, t).$$

The left-hand side reproduces the classical Liouville, or Vlasov equation in the limit $\hbar \to 0$, while the right-hand side embodies quantum corrections in an infinite gradient expansion. These terms account for the wave-like features of the plasma constituents, including tunneling, zero-point motion, and quantum interference [6].

From a physical perspective, the Wigner function approach offers several advantages for quantum plasma modeling. First, it facilitates a semiclassical interpretation of quantum transport by maintaining a phase-space structure akin to classical distributions. Second, it allows the incorporation of self-consistent electromagnetic interactions through Maxwell's equations, where the charge and current densities are calculated from quantum-corrected distribution functions:

$$\rho(x, t) = \int dp f_W(x, p, t), \quad j(x, t) = \int dp \frac{p}{m} f_W(x, p, t).$$

Third, it provides a natural foundation for incorporating quantum statistical effects such as Fermi-Dirac, or Bose-Einstein occupation via initial conditions, or by embedding Wigner functions within many-body frameworks, such as the BBGKY hierarchy, or the Kadanoff–Baym equations.

The Wigner formalism also admits extensions to relativistic plasmas, spinor fields, and Gauge interactions, though these require the incorporation of Dirac spinor structures, spin Wigner functions, and Gauge-covariant phase-space distributions. The Wigner distribution function, derived rigorously from the von Neumann equation, establishes a foundational link between quantum operator dynamics and classical phase space intuition. It serves as the bedrock for quantum kinetic theory, enabling the description of quantum plasma phenomena ranging from coherent collective excitations and quantum tunneling to non-equilibrium evolution under ultrafast or strong-field conditions. The richness of this formalism lies in its ability to incorporate both the structure of quantum mechanics and the intuitive power of classical kinetic theory, rendering it indispensable in the theoretical modeling of quantum plasma systems [7].

5.1.3 Linearized Wigner–Poisson and Wigner–Maxwell Systems

In quantum plasma physics, the kinetic description based on phase space distributions undergoes significant modification due to the presence of nonlocality and quantum coherence effects inherent to microscopic dynamics. The Wigner formalism, serving as a quantum analog of the classical distribution function approach, enables one to cast quantum statistical evolution in a phase-space representation reminiscent of the classical Vlasov picture.

The foundation of the Wigner kinetic theory rests on the Wigner function $f_W(\mathbf{r}, \mathbf{p}, t)$, which is a quasi-distribution function defined via the Wigner–Weyl transform of the single-particle density matrix. For a spinless, nonrelativistic particle of charge 'q' and mass mm, interacting with an electrostatic potential $\phi(\mathbf{r}, t)$, the evolution of the Wigner function in the collisionless limit is governed by the Wigner equation:

$$\frac{\partial f_W}{\partial t} + \frac{\mathbf{p}}{m} \cdot \nabla_{\mathbf{r}} f_W - q \int \frac{d^3\mathbf{s}}{(2\pi\hbar)^3} e^{\frac{i}{\hbar}\mathbf{s}\cdot\mathbf{p}} \left[\phi\left(\mathbf{r} + \frac{\mathbf{s}}{2}, t\right) - \phi\left(\mathbf{r} - \frac{\mathbf{s}}{2}, t\right)\right] f_W(\mathbf{r}, \mathbf{p}, t) = 0.$$

The nonlocality in the force term, represented as a convolution integral, encodes quantum interference and diffraction effects. In the classical limit $\hbar \to 0$, this reduces to the Vlasov equation for collisionless classical plasmas. The electrostatic potential is self-consistently determined by the Poisson's equation:

$$\nabla^2 \phi(\mathbf{r}, t) = -\frac{q}{\varepsilon_0} \left[\int f_W(\mathbf{r}, \mathbf{p}, t) d^3\mathbf{p} - n_0 \right],$$

where n_0 is the equilibrium background density and ε_0 is the vacuum permittivity. This constitutes the nonlinear Wigner–Poisson system.

To extract collective mode dynamics, the Wigner–Poisson system is linearized around a spatially homogeneous and stationary equilibrium:

$$f_W(\mathbf{r}, \mathbf{p}, t) = f_0(\mathbf{p}) + \delta f(\mathbf{r}, \mathbf{p}, t), \quad \phi(\mathbf{r}, t) = \delta\phi(\mathbf{r}, t),$$

where $f_0(\mathbf{p})$ is a Fermi–Dirac distribution in thermal equilibrium, and $\delta f \ll f_0, \delta\phi \ll 1$. Substituting into the Wigner equation and retaining terms linear in perturbations yields:

$$\frac{\partial \delta f}{\partial t} + \frac{\mathbf{p}}{m} \cdot \nabla_{\mathbf{r}} \delta f - q \int \frac{d^3\mathbf{s}}{(2\pi \hbar)^3} e^{\frac{i}{\hbar}\mathbf{s}\cdot\mathbf{p}} \Big[\delta\phi\Big(\mathbf{r} + \frac{\mathbf{s}}{2}, t\Big) - \delta\phi\Big(\mathbf{r} - \frac{\mathbf{s}}{2}, t\Big)\Big] f_0(\mathbf{p}) = 0.$$

Taking the spatial Fourier transform $(\mathbf{r}, t) \mapsto (\mathbf{k}, \omega)$, and using the identity

$$\int d^3\mathbf{s} e^{\frac{i}{\hbar}\mathbf{s}\cdot\mathbf{p}} \Big[\delta\phi\Big(\mathbf{r} + \frac{\mathbf{s}}{2}\Big) - \delta\phi\Big(\mathbf{r} - \frac{\mathbf{s}}{2}\Big)\Big] = 2i\hbar \sin\Big(\frac{\hbar}{2}\mathbf{k} \cdot \nabla_{\mathbf{p}}\Big) \delta\phi(\mathbf{k}) e^{i\mathbf{k}\cdot\mathbf{r}},$$

we obtain the linearized Wigner equation in Fourier space,

$$-i\Big(\omega - \frac{\mathbf{k} \cdot \mathbf{p}}{m}\Big) \delta f(\mathbf{k}, \mathbf{p}, \omega) = -iq\delta\phi(\mathbf{k}, \omega) \frac{2}{\hbar} \sin\Big(\frac{\hbar\mathbf{k} \cdot \nabla_{\mathbf{p}}}{2}\Big) f_0(\mathbf{p}).$$

The Poisson equation becomes:

$$k^2 \delta\phi(\mathbf{k}, \omega) = -\frac{q}{\varepsilon_0} \int \delta f(\mathbf{k}, \mathbf{p}, \omega) d^3\mathbf{p}.$$

This system of equations describes the linear electrostatic response of a quantum plasma and generalizes the dielectric formalism of classical plasmas.

To obtain the dielectric function, one defines the longitudinal susceptibility $\chi(\mathbf{k}, \omega)$ via the relation:

$$\varepsilon(\mathbf{k}, \omega) = 1 + \chi(\mathbf{k}, \omega),$$

which relates the induced charge density to the potential perturbation. Solving for δf and substituting into the Poisson's equation yields:

$$\chi(\mathbf{k}, \omega) = \frac{q^2}{\varepsilon_0 \hbar} \int \frac{1}{\omega - \frac{\mathbf{k}\cdot\mathbf{p}}{m} + i0^+} \Big[\frac{2}{\hbar} \sin\Big(\frac{\hbar\mathbf{k} \cdot \nabla_{\mathbf{p}}}{2}\Big) f_0(\mathbf{p})\Big] d^3\mathbf{p}.$$

In the limit of small $\hbar k$, this reduces to:

$$\omega \approx -\frac{q^2}{\varepsilon_0} \int \frac{\mathbf{k} \cdot \nabla_{\mathbf{p}} f_0(\mathbf{p})}{\omega - \frac{\mathbf{k}\cdot\mathbf{p}}{m} + i0^+} d^3\mathbf{p} + \mathcal{O}(\hbar^2),$$

recovering the Lindhard susceptibility in the appropriate degenerate limit. The dielectric function governs the dispersion relation for electrostatic modes via the condition:

$$\varepsilon(\mathbf{k}, \omega) = 0,$$

yielding a spectrum that includes quantum corrections such as Fermi pressure effects, quantum recoil, and wave–particle phase coherence [8]. Notably, the dispersion relation for longitudinal oscillations in a quantum plasma with zero temperature and homogeneous density n_0 is:

$$\omega^2 = \omega_p^2 + \frac{3}{5}v_F^2 k^2 + \frac{\hbar^2 k^4}{4m^2},$$

where $\omega_p^2 = q^2 n_0/(m\varepsilon_0)$ is the plasma frequency, and v_F is the Fermi velocity. The last term, absent in classical theory, represents quantum dispersion due to wave–particle uncertainty.

Extending to the electromagnetic case, the Wigner–Maxwell system incorporates the full set of Maxwell's equations with quantum modifications encoded in the Wigner equation for the Gauge-invariant Wigner function $f_W(\mathbf{r}, \mathbf{p}, t)$. The minimal-coupling prescription leads to:

$$\frac{\partial f_W}{\partial t} + \left[\frac{\mathbf{p} - q\mathbf{A}}{m}\right] \cdot \nabla_{\mathbf{r}} f_W + q\left[\mathbf{E} + \frac{1}{m}(\mathbf{p} - q\mathbf{A}) \times \mathbf{B}\right] \cdot \nabla_{\mathbf{p}} f_W + \mathcal{Q}\left[f_W\right] = 0,$$

where $\mathbf{A},\ \mathbf{E},\ \mathbf{B}$ are the vector potential, electric field, and magnetic field respectively, and $\mathcal{Q}\left[f_W\right]$ represents higher-order quantum corrections involving Gauge fields and derivatives. The source terms in Maxwell's equations are constructed from phase-space moments of f_W, yielding:

$$\nabla \cdot \mathbf{E} = \frac{q}{\varepsilon_0}\int f_W d^3\mathbf{p},\ \nabla \times \mathbf{B} = \mu_0\left(q\int \frac{\mathbf{p}}{m} f_W d^3\mathbf{p} + \varepsilon_0 \frac{\partial \mathbf{E}}{\partial t}\right)$$

which, when linearized, couple to the perturbative evolution of the Wigner function. In the long-wavelength and low-temperature regime, this system reproduces the dielectric tensor structure familiar from linear response theory, while accommodating quantum mechanical corrections to wave propagation and damping.

The linearized Wigner–Poisson and Wigner–Maxwell systems provide a unified framework that captures quantum dispersive effects, Pauli blocking, and tunneling-induced mode shifts in degenerate or coherent plasmas. Physically, these corrections manifest as nonlocal response, Landau-like damping suppression, and enhanced sensitivity to wavevector-scale phenomena [9, 10]. In particular, quantum kinetic modeling becomes indispensable in ultra-cold plasmas, nanostructured media, and plasmonic systems where the de Broglie wavelength becomes comparable to system dimensions.

From a modern perspective, the Wigner formalism bridges quantum hydrodynamic and field-theoretic approaches, providing a mesoscopic resolution of plasma dynamics. It accommodates coherent effects absent in stochastic descriptions and enables the derivation of quantum dispersion relations beyond the semiclassical expansion. Moreover, in contexts such as quantum radiation transport, nonequilibrium electroweak plasmas, and strongly coupled light–matter systems, the Wigner–Maxwell framework offers an essential, first-principles route to quantifying transport, damping, and instabilities under quantum constraints [11].

5.1.4 Pauli Equation-Based Kinetic Models for Spin-½ Plasmas

The dynamical behaviour of spin-½ charged particles embedded in a quantum plasma medium, such as electrons, mandates the incorporation of both translational and intrinsic spin degrees of freedom in the evolution equations. The Pauli equation, being a non-relativistic limit of the Dirac equation retaining spin coupling through minimal electromagnetic interaction, forms a foundational basis for constructing kinetic models of such systems.

The starting point is the single-particle Pauli equation for a spin-½ fermion of charge qq and mass m in an external electromagnetic field $(\mathbf{E}, \mathbf{B})$, given by:

$$i\hbar\frac{\partial\Psi}{\partial t} = \left[\frac{1}{2m}(-i\hbar\nabla - q\mathbf{A})^2 + q\phi - \mu_B\boldsymbol{\sigma}\cdot\mathbf{B}\right]\Psi.$$

Here, $\mu_B = \frac{q\hbar}{2m}$ is the Bohr magneton, $\boldsymbol{\sigma}$ are the Pauli matrices, ϕ is the scalar potential, and $\mathbf{A}$ is the vector potential. The spin interaction term $-\mu_B\boldsymbol{\sigma}\cdot\mathbf{B}$ encapsulates the Zeeman energy, establishing the central role of spin–magnetic field coupling in the plasma evolution. Notably, this equation already encodes spin precession dynamics, an essential component in spin-polarized plasma systems [12].

To proceed toward a kinetic description, the quantum wavefunction $\Psi(\mathbf{r}, t) \in \mathbf{C}^2$ is interpreted statistically over an ensemble, and its corresponding Wigner transform is computed to derive a phase-space quasi-probability distribution function incorporating spin.

The Wigner function formalism for spin-½ systems introduces a matrix-valued distribution function in the extended phase space $(\mathbf{r}, \mathbf{p}, t)$, namely:

$$\mathcal{W}_{\alpha\beta}(\mathbf{r}, \mathbf{p}, t) = \frac{1}{(2\pi\hbar)^3}\int d^3s e^{-i\mathbf{p}\cdot\mathbf{s}/\hbar}\left\langle\Psi_\beta^\dagger\left(\mathbf{r} + \frac{\mathbf{s}}{2}, t\right)\Psi_\alpha\left(\mathbf{r} - \frac{\mathbf{s}}{2}, t\right)\right\rangle$$

where the spin indices $\alpha, \beta \in \{\uparrow, \downarrow\}$. This spinor-valued Wigner function $\mathcal{W}$ can be decomposed into a scalar (spin-averaged) part and a spin vector component using the Pauli matrices:

$$\mathcal{W} = f(\mathbf{r}, \mathbf{p}, t)\mathbb{I} + \mathbf{S}(\mathbf{r}, \mathbf{p}, t)\cdot\boldsymbol{\sigma}$$

where f is the scalar distribution function and $\mathbf{S}$ represents the spin distribution vector in phase space. This decomposition is crucial for constructing moment hierarchies that separately track particle and spin observables.

The dynamical equations governing f and $\mathbf{S}$ are obtained from the Pauli equation by performing a Wigner transform and collecting terms up to $\mathcal{O}(\hbar^2)$, resulting in a set of generalized Vlasov–Pauli equations:

$$\frac{\partial f}{\partial t} + \frac{\mathbf{p}}{m} \cdot \nabla_{\mathbf{r}} f + q\left(\mathbf{E} + \frac{\mathbf{p}}{m} \times \mathbf{B}\right) \cdot \nabla_{\mathbf{p}} f + \mu_B (\nabla_{\mathbf{r}} \mathbf{B}) \cdot \nabla_{\mathbf{p}} \cdot \mathbf{S} = 0$$

$$\frac{\partial \mathbf{S}}{\partial t} + \frac{\mathbf{p}}{m} \cdot \nabla_{\mathbf{r}} \mathbf{S} + q\left(\mathbf{E} + \frac{\mathbf{p}}{m} \times \mathbf{B}\right) \cdot \nabla_{\mathbf{p}} \mathbf{S} + \mu_B (\mathbf{B} \times \mathbf{S}) = 0$$

The first equation describes the transport of charge density, while the second governs spin precession and magnetization transport in the presence of the background magnetic field. Notably, the term $\mathbf{B} \times \mathbf{S}$ induces Larmor precession, highlighting the quantum non-scalar nature of magnetized plasma evolution.

From the spin distribution $\mathbf{S}$, the macroscopic magnetization vector $\mathbf{M}(\mathbf{r}, t)$ is defined as:

$$\mathbf{M}(\mathbf{r}, t) = 2\mu_B \int \mathbf{S}(\mathbf{r}, \mathbf{p}, t) d^3 p S$$

This magnetization acts as a source term in Maxwell's equations, modifying the electromagnetic field dynamics self-consistently. The emergence of spin-modified Vlasov–Maxwell equations hence includes both spin-polarization currents and spin-magnetic pressure anisotropy. Additionally, the spin contribution to the pressure tensor is characterized by the second moment of $\mathbf{S}$, leading to a spin pressure tensor $\Pi_{ij}^{(S)}$, which modifies the fluid momentum balance and introduces magneto-spin acoustic modes in the plasma.

Spin-aligned particle ensembles can support low-frequency spin waves, with dispersion relations fundamentally distinct from conventional magnetosonic, or Langmuir modes. In quantum degenerate regimes, such waves manifest collective behaviour that interweaves Fermi statistics, magnetization, and quantum coherence.

5.1.5 Landau Damping in Quantum Plasmas

In classical plasma physics, Landau damping represents a quintessential example of a collisionless kinetic phenomenon where electrostatic waves are attenuated due to phase-resonant interactions between wave modes and particles in the velocity distribution tail. This process, discovered by Lev Landau in 1946, is a purely kinetic effect not derivable from fluid models, and it relies crucially on the analytic continuation of the plasma response function into the complex plane. In quantum plasmas, the generalization of Landau damping acquires novel features arising from quantum statistical effects, nonlocality in phase space, and modifications to the dispersion relation, due to the finite de Broglie wavelength.

Let us consider a homogeneous, collisionless, unmagnetized quantum plasma in one spatial dimension for simplicity, composed of electrons of charge $-e$, mass m, and number density n_0, immersed in a uniform ionic background. The quantum evolution of the distribution function is governed by the Wigner–Moyal equation, which serves as a quantum analogue to the Vlasov equation. For a scalar electrostatic potential $\phi(x, t)$, the Wigner function $f(x, p, t)$ evolves as:

$$\frac{\partial f}{\partial t} + \frac{p}{m}\frac{\partial f}{\partial x} - \frac{ie}{\hbar}\int \frac{d\xi}{2\pi\hbar} e^{i\xi p/\hbar}\left[\phi\left(x+\frac{\xi}{2}\right) - \phi\left(x-\frac{\xi}{2}\right)\right] f(x, p', t) = 0.$$

Linearizing about the equilibrium distribution $f_0(p)$ and Fourier transforming both spatial and temporal variables yields the perturbed Wigner equation in momentum space:

$$\left(-i\omega + \frac{ikp}{m}\right) f_1(k, p, \omega) = \frac{ie}{\hbar}\int \frac{d\xi}{2\pi\hbar} e^{i\xi p/\hbar}\left[\phi_1(k, \omega)\left(e^{ik\xi/2} - e^{-ik\xi/2}\right)\right] f_0(p).$$

Evaluating the integral, one obtains:

$$f_1(k, p, \omega) = \frac{e\phi_1(k, \omega)}{i\hbar(\omega - kp/m)}\left[f_0\left(p + \frac{\hbar k}{2}\right) - f_0\left(p - \frac{\hbar k}{2}\right)\right].$$

This expression elucidates the nonlocality in momentum space induced by quantum corrections. The dielectric response function $\varepsilon(k, \omega)$ is derived via Poisson's equation in Fourier space:

$$\varepsilon(k, \omega) = 1 + \frac{e^2}{\varepsilon_0 k}\int \frac{dp}{\hbar}\frac{1}{\omega - kp/m + i0^+}\left[f_0\left(p + \frac{\hbar k}{2}\right) - f_0\left(p - \frac{\hbar k}{2}\right)\right].$$

This defines the quantum dielectric function, whose analytic properties determine the nature of wave propagation and damping. For small $\hbar$, one may expand the numerator in powers of $\hbar k$ to recover the classical limit:

$$f_0\left(p + \frac{\hbar k}{2}\right) - f_0\left(p - \frac{\hbar k}{2}\right) \approx \hbar k\frac{\partial f_0}{\partial p} + \frac{1}{24}(\hbar k)^3\frac{\partial^3 f_0}{\partial p^3} + \cdots,$$

which reproduces the classical Landau dielectric function as the leading order term and introduces quantum corrections as higher-order gradient terms.

The complex frequency $\omega = \omega_r + i\gamma$ for an electrostatic wave corresponds to a root of the dispersion relation $\varepsilon(k, \omega) = 0$. The damping rate γ is determined from the imaginary part of the dielectric function near resonance. In the quantum case, the damping rate generalizes to:

$$\gamma \approx -\frac{\pi e^2}{2\varepsilon_0 \hbar k}\left[f_0\left(p_r + \frac{\hbar k}{2}\right) - f_0\left(p_r - \frac{\hbar k}{2}\right)\right],$$

where $p_r = m\omega_r/k$ is the resonant momentum. Notably, the finite width of the quantum resonance, characterized by the difference of f_0 across a finite $\hbar k$ interval, contrasts with the classical Landau expression:

$$\gamma_{\text{cl}} = -\frac{\pi e^2}{\varepsilon_0 k^2} \frac{\partial f_0}{\partial v}\bigg|_{v=\omega_r/k}.$$

Hence, quantum effects regularize the singularity associated with the pole in the classical dielectric function by smearing the resonance over a finite momentum width $\sim \hbar k$. This quantum smoothing suppresses damping in regimes where $\hbar k$ becomes non-negligible compared to the characteristic momentum scales of the distribution [13].

For plasmas in thermodynamic equilibrium, the equilibrium distribution $f_0(p)$ corresponds to the Fermi–Dirac distribution:

$$f_0(p) = \frac{1}{1 + \exp\left[\frac{p^2/2m-\mu}{k_B T}\right]},$$

where μ is the chemical potential. In the degenerate limit $T \ll T_F$, where T_F is the Fermi temperature, the distribution becomes a step function at the Fermi momentum. In such regimes, the damping rate is profoundly affected, phase-space availability is reduced due to Pauli blocking, and only electrons near the Fermi surface can participate in resonant interactions. As a result, quantum Landau damping in degenerate plasmas is substantially suppressed compared to classical expectations.

This suppression is particularly evident in ultradense systems such as white dwarfs, laser-compressed matter, or strongly correlated electron fluids, where the interplay of degeneracy and nonlocal effects leads to the emergence of undamped quantum modes. In some cases, wave modes with phase velocities below the Fermi velocity can remain weakly damped or even stable due to the vanishing of the resonant population gradient.

Quantum corrections to the dispersion relation emerge from the nonlocal structure of the Wigner–Moyal kernel. The full quantum dispersion relation, incorporating both collective and kinetic effects, reads:

$$1 + \frac{e^2}{\varepsilon_0 k} \int \frac{dp}{\hbar} \frac{f_0(p + \hbar k/2) - f_0(p - \hbar k/2)}{\omega - kp/m + i0^+} = 0.$$

This must typically be solved numerically, as the analytic structure is highly sensitive to the shape of f_0, temperature, degeneracy, and wavevector magnitude. In the long-wavelength limit $k \to 0$, expansions yield a modified Bohm potential and higher-order dispersive corrections. For finite k, the full quantum dispersion predicts phenomena absent in classical theory, such as quantum recoil, cutoff frequencies, and mode bifurcation near the Fermi edge [14].

These features are critical for accurately modeling high-frequency Langmuir oscillations in miniaturized devices and nanoplasmonic structures, where the de Broglie length is comparable to the system scale, and Landau damping becomes fundamentally quantum in character.

5.1.6 Kinetic Models in Ultra-Cold and Dense Plasma Environments

The formulation of kinetic models in ultra-cold and dense plasma environments presents a refined theoretical challenge, necessitating the unification of quantum statistical effects, strong inter-particle correlations, and collective dynamical behaviour on scales where classical kinetic theory becomes inadequate. In such regimes, the thermal de Broglie wavelength of the constituent particles becomes comparable to, or exceeds, the inter-particle spacing, leading to an inherent overlap of quantum wavefunctions and the manifestation of exchange interactions, Fermi degeneracy pressure, and quantum diffraction.

In contrast to classical plasmas where the phase space density is well described by the Boltzmann distribution, the ultra-cold and dense plasma requires a formulation that respects the fundamental indistinguishability and quantum statistics of particles. The appropriate starting point is the single-particle Wigner function $f_W(\mathbf{r}, \mathbf{p}, t)$, which serves as a quantum quasi-probability distribution in phase space. The Wigner function is obtained from the single-particle density matrix $\hat{\rho}\left(\mathbf{r}, \mathbf{r}', t\right)$ via the Weyl transform:

$$f_W(\mathbf{r}, \mathbf{p}, t) = \frac{1}{(2\pi\hbar)^3} \int d^3 s e^{-i\mathbf{p}\cdot\mathbf{s}/\hbar} \rho\left(\mathbf{r} + \frac{\mathbf{s}}{2}, \mathbf{r} - \frac{\mathbf{s}}{2}, t\right).$$

This function encodes both the local particle distribution and the non-local quantum coherence properties of the system. Unlike its classical analog, f_W can attain negative values, reflecting the underlying non-classical correlations and interference effects.

In the mean-field limit, where collisional effects are neglected but long-range collective fields are retained, the evolution of the Wigner function is governed by the quantum Vlasov equation:

$$\frac{\partial f_W}{\partial t} + \frac{\mathbf{p}}{m} \cdot \nabla_{\mathbf{r}} f_W - \mathcal{Q}\left[V, f_W\right] = 0,$$

where the quantum potential operator $\mathcal{Q}$ accounts for non-local field-particle interactions and is given by:

$$\mathcal{Q}\left[V, f_W\right] = \frac{i}{\hbar} \int \frac{d^3 s d^3 p'}{(2\pi\hbar)^3} e^{-i(\mathbf{p}-\mathbf{p}')\cdot s/\hbar} \left[V\left(\mathbf{r} + \frac{\mathbf{s}}{2}\right) - V\left(\mathbf{r} - \frac{\mathbf{s}}{2}\right)\right] f_W(\mathbf{r}, \mathbf{p}', t).$$

This integral form retains all orders of $\hbar$, making it exact for the single-particle quantum system in the presence of a smooth potential $V(\mathbf{r})$. In the semiclassical limit ($\hbar \to 0$), the term $\mathcal{Q}$ reduces to the classical force $-\nabla V \cdot \nabla_{\mathbf{p}} f$, recovering the classical Vlasov equation.

However, in ultra-cold and dense environments, higher-order $\hbar$ corrections cannot be neglected. These include quantum recoil, tunneling through self-consistent potentials, and wavefunction interference on the scale of the inter-particle distance [15].

When binary interactions, or many-body collisions cannot be ignored, one must extend the model to include collisional dynamics within the constraints of quantum statistics. For fermionic species, the Uehling-Uhlenbeck (U) equation serves as the quantum generalization of the Boltzmann equation:

$$\left(\frac{\partial f}{\partial t}\right)_{\text{coll}} = \int d^3p_2 \int d\Omega |\mathcal{M}|^2 \delta(E_i - E_f)$$
$$\left[f_{1'} f_{2'} (1 - f_1)(1 - f_2) - f_1 f_2 (1 - f_{1'})(1 - f_{2'})\right],$$

where $f_i \equiv f(\mathbf{p}_i, t)$, and primed quantities denote post-collision states. The Pauli exclusion principle is explicitly embedded in the phase-space blocking terms $(1 - f)$, crucial for degenerate plasmas where occupation of low-momentum states saturates.

For weakly coupled yet dense quantum plasmas, the Lenard-Balescu equation can be modified to include dynamic screening and quantum statistics through the dynamically screened dielectric function $\epsilon(\mathbf{k}, \omega)$, leading to quantum analogs of Landau damping and energy exchange between collective modes and particles. The quantum dielectric function itself must be derived from a linearized quantum kinetic response, yielding expressions that are valid in regimes where the Fermi energy and thermal energy compete, or cooperate [16].

An alternative, but complementary route to kinetic modeling invokes the decomposition of the Wigner equation into moments, leading to quantum hydrodynamics (QHD). When applied to the ultra-cold limit, this framework exposes a quantum pressure-like term known as the Bohm potential:

$$V_B(\mathbf{r}, t) = -\frac{\hbar^2}{2m} \frac{\nabla^2 \sqrt{n(\mathbf{r}, t)}}{\sqrt{n(\mathbf{r}, t)}},$$

which arises naturally from the second-order moment closure of the Wigner-Moyal expansion, or equivalently from the Madelung representation of the Schrödinger equation. Physically, this potential accounts for the nonlocal dispersive tendencies of quantum particles, playing a role akin to a quantum pressure even at zero temperature. In ultra-cold plasma clouds, especially those prepared via laser cooling and field ionization of Rydberg gases, the Bohm potential significantly alters wave propagation, stabilization of shocks, and damping rates.

Depending on the particle species involved, one must consider Fermi-Dirac or Bose-Einstein quantum kinetic models. For electrons at high density but low temperature, Fermi degeneracy dominates the distribution function, leading to the replacement of the classical Maxwell-Boltzmann distribution with:

$$f_{FD}(\mathbf{p}) = \frac{1}{e^{(\epsilon_p - \mu)/k_B T} + 1}.$$

This form modifies the particle and current densities, Debye screening lengths, and transport coefficients. In dense positron-electron plasmas, or in contexts where polaritonic or photonic modes couple to matter, bosonic kinetic formulations may be warranted. For such cases, Bose-Einstein condensation effects, superfluid-like behaviour, and coherent quantum transport must be incorporated into the kinetic formalism.

The computation of transport coefficients such as thermal conductivity, electrical resistivity, and viscosity requires integration over quantum-modified collision operators and response functions. In particular, the Chapman-Enskog expansion must be adapted to the quantum domain, with orthogonality relations between distribution function perturbations now governed by Fermi, or Bose statistics. These expansions reveal anomalous scaling behaviours in ultra-cold dense plasmas, such as enhanced conductivity due to Pauli blocking of scattering channels, or suppression of thermal diffusion owing to degeneracy-induced stiffening of the energy distribution.

Further, the linear response of such plasmas to electromagnetic perturbations yields non-trivial dielectric and magnetic susceptibilities. These are encoded in the longitudinal and transverse components of the dielectric tensor $\epsilon_{L,T}(\mathbf{k}, \omega)$, which can be derived via the quantum polarization function obtained from the Lindhard formalism, or time-dependent Hartree-Fock approximations. These functions describe the interplay between quantum recoil, Landau damping, and particle-hole excitations [17].

Finally, it is essential to recognize that many kinetic processes in ultra-cold dense plasmas cannot be captured by perturbative, or Markovian approximations. Memory effects, entanglement-induced decoherence, and non-instantaneous correlations between particles manifest as non-local kernels in the generalized master equation, or via integro-differential extensions of the quantum Boltzmann equation. The projection operator formalism and the Mori-Zwanzig approach provide rigorous bases for constructing such generalized kinetic theories. These non-Markovian extensions are indispensable when dealing with coherent excitations, such as plasmons with long lifetimes, or when systems exhibit strong coupling to background fields, as in cavity quantum electrodynamics with dense electron fluids.

5.2 Quantum Hydrodynamic Model

Quantum hydrodynamics (QHD) provides a powerful mesoscopic framework that bridges the underlying quantum kinetic behaviour of plasma particles and the emergent collective fluid-like phenomena observable at larger scales. In the context of quantum plasmas, where Fermi statistics, quantum coherence, tunneling, and dispersive corrections play critical roles, it is imperative to formulate fluid models that transcend classical assumptions and incorporate quantum corrections consistently.

A mathematically rigorous path from the many-body quantum kinetic equations to hydrodynamic equations involves taking statistical moments of the distribution function in phase space, necessitating a systematic closure scheme.

We begin with the Wigner equation, which is the quantum analogue of the classical Vlasov equation, and governs the evolution of the single-particle Wigner distribution function $f_W(\mathbf{r}, \mathbf{p}, t)$ in phase space:

$$\frac{\partial f_W}{\partial t} + \frac{\mathbf{p}}{m} \cdot \nabla_{\mathbf{r}} f_W - \Theta[V] f_W = 0.$$

Here, $\Theta[V]$ is the quantum potential operator defined as

$$\Theta[V] f_W = \frac{1}{i\hbar}(V \star f_W - f_W \star V),$$

which, upon expansion, yields

$$\Theta[V] f_W = \sum_{n=1}^{\infty} \frac{(-1)^n \hbar^{2n}}{2^{2n}(2n+1)!} \nabla_{\mathbf{r}}^{2n+1} V \cdot \nabla_{\mathbf{p}}^{2n+1} f_W.$$

This infinite series encapsulates the inherently nonlocal character of quantum dynamics and replaces the classical force term $-\nabla V \cdot \nabla_{\mathbf{p}} f$. To extract fluid-like equations, we define velocity moments of f_W as follows:

$$n(\mathbf{r}, t) = \int f_W d^3p, \quad n\mathbf{u}(\mathbf{r}, t) = \int \frac{\mathbf{p}}{m} f_W d^3p, \quad \Pi_{ij} = \int \frac{p_i p_j}{m^2} f_W d^3p,$$

and so forth, where n is the number density, $\mathbf{u}$ is the mean fluid velocity, and Π_{ij} is the pressure tensor. This hierarchy must be truncated to yield a closed system, which constitutes the essence of moment closure.

The zeroth moment of the Wigner equation yields the continuity equation:

$$\frac{\partial n}{\partial t} + \nabla \cdot (n\mathbf{u}) = 0.$$

The first-order moment provides the momentum transport equation:

$$\frac{\partial (n u_i)}{\partial t} + \partial_j \big(n u_i u_j\big) = -\partial_j \Pi_{ij} + n F_i^{\text{ext}} + Q_i,$$

where F_i^{ext} represents the external classical force (e.g., Lorentz force in electromagnetic fields), and Q_i denotes quantum corrections from the Moyal expansion. For a scalar potential $V(\mathbf{r})$, and assuming a weakly coupled quantum plasma, the leading-order quantum correction from the Moyal series is retained:

$$Q_i = -\frac{n}{m}\partial_i\left(\frac{\hbar^2}{2m}\frac{\nabla^2\sqrt{n}}{\sqrt{n}}\right),$$

which is the quantum Bohm potential:

$$V_B = -\frac{\hbar^2}{2m}\frac{\nabla^2\sqrt{n}}{\sqrt{n}}.$$

Hence, the momentum equation takes the form:

$$mn\left(\frac{\partial \mathbf{u}}{\partial t} + (\mathbf{u}\cdot\nabla)\mathbf{u}\right) = -\nabla P - n\nabla V + \frac{\hbar^2}{2m}n\nabla\left(\frac{\nabla^2\sqrt{n}}{\sqrt{n}}\right),$$

where P is the scalar pressure, closing the system at the second moment.

To complete the closure, an equation of state (EOS) is required to relate pressure to density. For a degenerate Fermi plasma at zero temperature, the pressure arises from Pauli exclusion and is given by:

$$P = \frac{\left(3\pi^2\right)^{2/3}\hbar^2}{5m}n^{5/3},$$

which corresponds to the Thomas–Fermi pressure for an ideal electron gas. For finite temperatures or non-degenerate conditions, the pressure can be derived from quantum statistical integrals using the Fermi–Dirac distribution.

In the general case, the momentum equation in conservative form with quantum corrections reads:

$$\frac{\partial(n\mathbf{u})}{\partial t} + \nabla\cdot(n\mathbf{u}\otimes\mathbf{u}) = -\nabla P + n\mathbf{F} + \frac{\hbar^2}{2m^2}n\nabla\left(\frac{\nabla^2\sqrt{n}}{\sqrt{n}}\right).$$

This is identified as the quantum Euler equation, valid in the inviscid and non-dissipative regime.

To incorporate viscosity and thermal conductivity relevant for strongly coupled, high-temperature, or non-ideal quantum plasmas higher-order moments must be retained or modeled through constitutive relations [18]. This leads to the quantum Navier–Stokes equations. In analogy with classical fluid theory, the stress tensor is decomposed as:

$$\Pi_{ij} = P\delta_{ij} + \sigma_{ij}, \quad \sigma_{ij} = \eta\left(\partial_i u_j + \partial_j u_i - \frac{2}{3}\delta_{ij}\nabla\cdot\mathbf{u}\right) + \zeta\delta_{ij}\nabla\cdot\mathbf{u},$$

where η and ζ are the shear and bulk viscosities, respectively. Quantum corrections to transport coefficients arise from inclusion of exchange–correlation effects, quantum scattering cross sections, and collective oscillations (e.g., plasmons).

The energy equation, derived from the second-order moment, includes terms such as:

$$\frac{3}{2}nk_B\left(\frac{\partial T}{\partial t}+\mathbf{u}\cdot\nabla T\right)=-P\nabla\cdot\mathbf{u}+\nabla\cdot(\kappa\nabla T)+\Phi_{\text{visc}}+Q_E,$$

where κ is the thermal conductivity, Φ_{visc} is the viscous dissipation, and Q_E contains quantum energy transport terms, including quantum heat flux derived from higher-order moments of the Wigner function.

The quantum Euler and Navier-Stokes equations presented above incorporate distinct quantum mechanical features that are essential for accurately modeling quantum plasmas. A key component is the Bohm potential, which introduces a nonlocal, dispersive correction representing quantum pressure arising from the Heisenberg uncertainty principle and wavefunction interference. Additionally, degeneracy pressure stemming from the Pauli's exclusion principle becomes dominant in regimes of high density and low temperature, such as in white dwarfs or the crusts of neutron stars. Viscous and thermal transport processes also require quantum corrections, particularly in dense laboratory plasmas, quantum wells, or systems confined at the nanoscale. These equations serve as the foundation of quantum hydrodynamic modeling in contexts where a full quantum kinetic approach is computationally infeasible, yet quantum effects are too significant to ignore. They are crucial for capturing phenomena such as electron density oscillations, quantum tunneling, collective excitations, and complex nonlinear dynamics inherent to quantum plasmas.

Modern developments incorporate nonlocal hydrodynamic models, where closure is obtained not via algebraic equations of state, but through convolution integrals over memory kernels and spatial nonlocalities derived from quantum functional methods. The inclusion of spin–orbit coupling, magnetization dynamics, and gauge fields within QHD formalism is also an active research area. Further, connections with Density Functional Theory (DFT) and time-dependent current density functional theory (TDCDFT) have motivated hybrid approaches wherein the moment closure is informed by orbital-free quantum kinetic energy functionals.

In high-energy density physics and quantum electrodynamics regimes, relativistic extensions of QHD are being explored, where closure must handle not just energy-momentum conservation but also field–current self-consistency under strong-field quantum conditions.

5.2.1 Bohm Potential and Quantum Pressure

In the context of quantum hydrodynamics, two central constructs emerge as direct manifestations of quantum mechanical principles within fluid descriptions, the Bohm potential and the quantum pressure. These elements capture inherently non-classical effects such as quantum dispersion, nonlocality, and wavefunction coherence, thereby extending the applicability of hydrodynamic models into regimes where classical approximations become inadequate.

To systematically identify the Bohm potential, and the quantum pressure term, one begins with the single-particle time-dependent Schrödinger equation for a nonrelativistic particle of mass m and charge q, in the absence of magnetic fields for clarity:

$$i\hbar\frac{\partial\psi(\mathbf{r},t)}{\partial t} = \left[-\frac{\hbar^2}{2m}\nabla^2 + V(\mathbf{r},t)\right]\psi(\mathbf{r},t),$$

where $\psi(\mathbf{r},t)$ is the wavefunction and $V(\mathbf{r},t)$ is the classical electrostatic potential. The wavefunction is expressed using the Madelung decomposition:

$$\psi(\mathbf{r},t) = \sqrt{n(\mathbf{r},t)}e^{iS(\mathbf{r},t)/\hbar},$$

where $n(\mathbf{r},t) = |\psi|^2$ is the probability density interpreted as the quantum fluid number density, and $\nabla S(\mathbf{r},t)/m$ defines the velocity field $\mathbf{u}(\mathbf{r},t)$ of the fluid-like entity. Substituting into the Schrödinger equation and separating real and imaginary parts yields two key equations. The imaginary part produces the continuity equation:

$$\frac{\partial n}{\partial t} + \nabla\cdot(n\mathbf{u}) = 0,$$

which ensures conservation of mass. The real part yields a quantum-corrected momentum equation:

$$m\left(\frac{\partial\mathbf{u}}{\partial t} + \mathbf{u}\cdot\nabla\mathbf{u}\right) = -\nabla V - \nabla Q,$$

where

$$Q(\mathbf{r},t) = -\frac{\hbar^2}{2m}\frac{\nabla^2\sqrt{n}}{\sqrt{n}},$$

is the quantum Bohm potential. This term encapsulates the nonlocal kinetic energy associated with quantum fluctuations and spatial inhomogeneities of the wavefunction amplitude. It arises purely from the quantum nature of the system and has no classical analogue.

The Bohm potential, Q, introduces a second-order spatial derivative acting on the amplitude of the wavefunction (or square root of density), which reflects the quantum kinetic contribution due to wavefunction curvature. Its form guarantees that even in the absence of classical forces, or pressure gradients, the system can exhibit dynamical evolution if density gradients are present. Mathematically, it is a consequence of the Hermitian nature of the kinetic operator in the Schrödinger equation and represents an internal force field generated by quantum dispersion [19].

Physically, this term accounts for quantum coherence, tunneling, and wavepacket spreading. It plays a particularly critical role in regimes where the de Broglie wavelength of particles is comparable to characteristic length scales, such as in high-density degenerate plasmas, cold plasmas, and in nanostructured or mesoscopic systems. In such systems, the Bohm potential mediates the response of the plasma to perturbations, leading to modified dispersion relations, altered Landau damping behaviour, and even the suppression of classical instabilities under certain parameter regimes.

A direct consequence of including Q in the fluid model is the emergence of quantum dispersive effects in the collective excitation spectrum, most notably the appearance of the quantum correction in the dispersion relation of longitudinal plasma oscillations:

$$\omega^2 = \omega_p^2 + \frac{3k^2 v_{th}^2}{1 + k^2\lambda_D^2} + \frac{\hbar^2 k^4}{4m^2},$$

where ω_p is the plasma frequency, v_{th} is the thermal velocity λ_D is the Debye length, and the last term originates from the Bohm potential, illustrating its dispersive quantum contribution.

Quantum Pressure

In quantum statistical plasmas, the pressure term receives contributions beyond the classical ideal gas form due to Fermi–Dirac statistics and quantum kinetic effects. The pressure tensor P_{ij} in quantum hydrodynamics is therefore composed of two parts:

$$P_{ij} = P_{ij}^{\mathrm{cl}} + P_{ij}^{\mathrm{q}},$$

where $P_{ij}^{\mathrm{cl}} = p(n, T)\delta_{ij}$ represents the classical pressure derived from an equation of state (ideal, adiabatic, etc.), while P_{ij}^{q} encodes purely quantum mechanical corrections. In the case of a zero-temperature degenerate Fermi gas, the quantum pressure arises from Pauli exclusion and the associated Fermi pressure:

$$p_F = \frac{\left(3\pi^2\right)^{2/3}\hbar^2}{5m} n^{5/3},$$

valid in the Thomas–Fermi limit. This relation is foundational for modeling white dwarfs, neutron star crusts, and dense laboratory plasmas, where the electron component behaves as a degenerate gas.

More generally, the quantum pressure term can be extracted from the Wigner–Moyal kinetic formalism by evaluating the second-order moment of the Wigner function $f_W(\mathbf{r}, \mathbf{p}, t)$, with gradient expansions yielding quantum stress terms involving higher-order derivatives of n. To leading order, the pressure gradient appears in the momentum equation as:

$$\frac{1}{mn}\nabla P(n) + \frac{\hbar^2}{2m^2}\nabla\left(\frac{\nabla^2\sqrt{n}}{\sqrt{n}}\right),$$

where the first term is the classical barotropic pressure contribution and the second term is precisely the Bohm potential, now seen as part of a more general pressure gradient including quantum mechanical contributions.

In the quantum Euler model, the inclusion of the Bohm potential modifies the conventional force balance and renders the fluid model dispersive and nonlocal. The resulting system becomes:

$$\frac{\partial n}{\partial t} + \nabla \cdot (n\mathbf{u}) = 0,$$

$$m\left(\frac{\partial \mathbf{u}}{\partial t} + \mathbf{u} \cdot \nabla\mathbf{u}\right) = -\nabla V - \frac{1}{n}\nabla P + \nabla Q.$$

In quantum Navier–Stokes models, additional viscous and dissipative terms are incorporated while retaining Q. These are important for capturing collisional or weakly turbulent quantum plasma dynamics and form the theoretical foundation for modeling systems such as cold plasma jets, shock waves in degenerate matter, and near-critical density laser-plasma interactions.

The introduction of the Bohm potential and quantum pressure completes the transition from a purely classical hydrodynamic description to one capable of encompassing quantum coherence, nonlocality, and degeneracy effects. Their mathematical origin lies in the quantum kinetic energy operator and the statistical mechanics of indistinguishable particles, while their physical role is to encode the dispersive and Fermi-statistical corrections that dominate at small length scales, or high densities. As such, they form indispensable components of any realistic theoretical model for quantum plasmas, from astrophysical settings to laboratory ultracold and laser-driven plasmas [20].

5.2.2 *Spin Quantum Hydrodynamics and Magnetization Current Dynamics*

In the framework of quantum plasma theory, spin quantum hydrodynamics (SQHD) introduces a significant enrichment of classical fluid models by incorporating intrinsic quantum degrees of freedom, specifically the spin angular momentum of constituent particles. The inclusion of spin dynamics and its coupling to the electromagnetic field reveals novel collective phenomena that are inaccessible through spinless formulations.

The starting point for constructing a spin quantum hydrodynamic model is the quantum mechanical description of many-body systems with spin-½ particles, typically electrons, obeying Fermi-Dirac statistics. In the semiclassical limit, the dynamics of such systems are governed by the Pauli equation for a single particle, given by

$$i\hbar\frac{\partial\psi}{\partial t} = \left[\frac{1}{2m}(-i\hbar\nabla - q\mathbf{A})^2 + q\phi - \mu_B\boldsymbol{\sigma}\cdot\mathbf{B}\right]\psi,$$

where $\psi = \psi(\mathbf{r}, t) \in \mathbf{C}^2$ is a spinor wavefunction, q and m are the charge and mass of the particle, $\mathbf{A}$ and ϕ are the electromagnetic vector and scalar potentials respectively, $\boldsymbol{\sigma} = \left(\sigma_x, \sigma_y, \sigma_z\right)$ is the triplet of Pauli matrices, $\mathbf{B} = \nabla \times \mathbf{A}$ is the magnetic field, and $\mu_B = \frac{q\hbar}{2m}$ is the Bohr magneton.

Taking the Madelung transformation and decomposing the spinor wavefunction into fluid variables allows the derivation of fluid-like equations, including continuity, momentum, and spin evolution equations. This process involves defining the density $n = \psi^\dagger\psi$, fluid velocity $\mathbf{v} = \frac{1}{m}\mathrm{Im}\left(\psi^\dagger\nabla\psi\right)/n$, and spin density $\mathbf{S} = \psi^\dagger\boldsymbol{\sigma}\psi/n$. The spin density $\mathbf{S} \in \mathbf{R}^3$ represents the local expectation value of the spin operator and satisfies the normalization $\mathbf{S}| \leq 1$, with the equality achieved in fully polarized configurations.

The SQHD framework extends the usual fluid equations by introducing the spin transport equation and accounting for the force exerted by magnetic field gradients on spin-aligned particles. The governing equations for a spin-½ quantum plasma with no collisions are:

i. *Continuity equation*:

$$\frac{\partial n}{\partial t} + \nabla\cdot(n\mathbf{v}) = 0,$$

ii. *Momentum equation*:

$$m\left(\frac{\partial\mathbf{v}}{\partial t} + \mathbf{v}\cdot\nabla\mathbf{v}\right) = q(\mathbf{E} + \mathbf{v}\times\mathbf{B}) - \frac{1}{n}\nabla P + \frac{\hbar^2}{2m}\nabla\left(\frac{\nabla^2\sqrt{n}}{\sqrt{n}}\right) + \mu_B\nabla(\mathbf{S}\cdot\mathbf{B}),$$

iii. *Spin evolution equation* (*Bloch-type equation*):

$$\frac{\partial\mathbf{S}}{\partial t} + (\mathbf{v}\cdot\nabla)\mathbf{S} = \frac{2\mu_B}{\hbar}\mathbf{S}\times\mathbf{B} + \boldsymbol{\Gamma}_S,$$

where P is the scalar pressure (which may follow Fermi-Dirac, or other quantum-statistical equations of state), and $\boldsymbol{\Gamma}_S$ represents spin relaxation or decoherence terms arising from collisions, quantum entanglement, or external interactions. The quantum potential term $\frac{\hbar^2}{2m}\nabla\left(\frac{\nabla^2\sqrt{n}}{\sqrt{n}}\right)$ embodies the Bohm-de Broglie quantum correction and encapsulates the dispersive effects due to quantum coherence.

The term $\mu_B \nabla(\mathbf{S} \cdot \mathbf{B})$ is the magnetic dipole force density and introduces the coupling between spatial spin inhomogeneities and magnetic field gradients. This term becomes dynamically dominant in spin-polarized, or partially magnetized plasmas [21].

A key outcome of spin coupling is the generation of magnetization currents arising from spatial and temporal variations of the spin density field. These currents contribute to the macroscopic electromagnetic field dynamics through the modification of Maxwell's equations. The magnetization $\mathbf{M}$ is related to the spin density by

$$\mathbf{M} = n\mu_B \mathbf{S},$$

and the corresponding Magnetization current density $\mathbf{J}_M$ appears in the modified Ampère–Maxwell law as

$$\nabla \times \mathbf{B} = \mu_0 \left(\mathbf{J}_f + \nabla \times \mathbf{M} + \varepsilon_0 \frac{\partial \mathbf{E}}{\partial t} \right),$$

where $\mathbf{J}_f = nq\mathbf{v}$ is the free current density and $\nabla \times \mathbf{M}$ corresponds to the bound current induced by the collective spin alignment. The term $\nabla \times \mathbf{M}$ encapsulates the internal spin-structure of the plasma and leads to spin-induced magnetic field self-organization.

Additionally, the continuity of must be preserved through

$$\frac{\partial \mathbf{M}}{\partial t} + \nabla \cdot \mathbf{J}_M = 0,$$

providing a constraint on the evolution of $\mathbf{S}$ consistent with total angular momentum conservation in the absence of dissipative processes.

To analyze wave propagation and instabilities in spin quantum plasmas, the SQHD equations are linearized about a homogeneous, possibly magnetized equilibrium $(n_0, \mathbf{v}_0 = 0, \mathbf{B}_0, \mathbf{S}_0)$. Perturbations of the form $\delta n, \delta \mathbf{v}, \delta \mathbf{S}, \delta \mathbf{B}, \delta \mathbf{E} \propto e^{i(\mathbf{k}\cdot\mathbf{r}-\omega t)}$ are inserted, and the linearized equations yield a coupled dispersion relation.

In the presence of an ambient magnetic field $\mathbf{B}_0$, the spin precession term $\mathbf{S} \times \mathbf{B}_0$ introduces resonant phenomena such as spin-cyclotron modes, and modifies the standard Alfvén and whistler branches. Notably, magnetization currents induce frequency shifts and mode coupling that are absent in spinless models [22]. The dispersion relation for transverse electromagnetic waves propagating along $\mathbf{B}_0$ typically acquires corrections of the form:

$$\omega^2 = \omega_p^2 + c^2k^2 + \Delta_{\text{spin}}(\omega, \mathbf{k}),$$

where Δ_{spin} is a spin-induced shift determined by the magnetization dynamics and spin polarization. In high-density, strongly magnetized regimes, these corrections can significantly affect wave damping, instability thresholds, and energy transport mechanisms.

SQHD enables the identification of instabilities unique to spin-carrying plasmas. A canonical example is the spin-flip instability, wherein collective spin precession couples to plasma flow or wave propagation, resulting in the growth of spin-polarized perturbations. The instability condition depends on the alignment between $\nabla \cdot \mathbf{M}$ and $\mathbf{B}_0$, and is influenced by quantum statistical pressure, exchange-correlation effects, and spin-orbit coupling (if included).

Additionally, quantum spin waves emerge as low-frequency, long-wavelength excitations of the spin field. These are analogues of classical magnetization waves (magnons) in ferromagnets, but here propagate in a fluid medium coupled to electromagnetic fields. The group velocity of such waves is affected by the quantum pressure and magnetization stiffness, and the dynamics are governed by a linearized form of the spin evolution equation:

$$\frac{\partial^2 \delta \mathbf{S}}{\partial t^2} + \omega_s^2 \delta \mathbf{S} = 0,$$

with ω_s being the spin resonance frequency.

Spin quantum hydrodynamics opens multiple avenues for exploring novel plasma behaviour s, especially in regimes where spin polarization is non-negligible. These include magnetized white dwarfs, neutron star atmospheres, laser-driven solid-density plasmas, and spintronic plasma devices. Extensions of SQHD incorporate spin–orbit coupling, quantum entanglement-induced transport, and kinetic generalizations via the Wigner–Pauli formalism. Further, the coupling of spin with strong electromagnetic fields invites relativistic generalizations within the Dirac–Maxwell framework and motivates the development of spin-resolved Schwinger–Keldysh formalisms for non-equilibrium phenomena.

Quantum hydrodynamic (QHD) models offer a semi-classical framework capable of interpolating between weak and moderate coupling. These models rely on moment expansions of the Wigner–Boltzmann equation and incorporate the Bohm potential $V_B = -\frac{\hbar^2}{2m}\frac{\nabla^2\sqrt{n}}{\sqrt{n}}$, which captures leading-order quantum pressure. However, in the intermediate regime, higher-order gradient corrections and finite-correlation contributions are essential.

An improved gradient expansion of the quantum pressure tensor P_{ij} includes terms such as:

$$P_{ij} = p\delta_{ij} + \frac{\hbar^2}{4m^2}\left[\partial_i\partial_j n - \frac{1}{2n}(\partial_i n)(\partial_j n)\right] + \cdots ,$$

where omitted terms account for exchange and higher-order nonlocal corrections. Closure relations for pressure in such systems may be derived from variational principles, or matched asymptotically with known limits of the partition function under quantum statistical mechanics.

The intermediate coupling regime is characterized by a competition between ordering tendencies due to interparticle interactions and dispersive tendencies from quantum delocalization. Physically, this leads to phenomena such as partial localization, increased entropy production, and suppressed long-range coherence. The limitations of mean-field and QHD models in this domain stem from their neglect of multi-particle correlations, their reliance on local closure assumptions, and their inadequate treatment of time-retarded dissipation. Quantum kinetic models, while more accurate, often become computationally prohibitive and sensitive to initial conditions due to their inherent nonlinearity and nonlocality in time [23].

Recent developments using time-dependent density functional theory (TDDFT) provide a promising bridge, offering computationally feasible schemes with exchange-correlation potentials tailored to quantum plasma environments. However, current TDDFT implementations still lack robust parametrizations for dynamically screened interactions in intermediate coupling plasmas and often rely on adiabatic approximations that neglect memory effects.

The accurate modeling of quantum plasmas in the intermediate coupling regime necessitates hybrid approaches combining kinetic, hydrodynamic, and field-theoretic perspectives. Promising directions include the use of non-equilibrium Green's functions (NEGF), real-time Schwinger–Keldysh techniques for quantum transport with dissipation, and machine-learned effective potentials trained on quantum molecular dynamics data. Further, the development of Gauge-invariant dissipative formalisms compatible with quantized electromagnetic fields remains an open and foundational challenge, especially relevant for next-generation plasma technologies and laboratory astrophysics experiments probing semi-degenerate states.

5.3 Density Functional Theory in Quantum Plasmas

In quantum plasma physics, especially in systems governed by partially or fully degenerate fermionic constituents, the treatment of many-body effects becomes paramount. Conventional field-theoretic techniques, while powerful in perturbative regimes, are computationally intractable when applied to dense quantum plasmas where Coulomb correlations, Fermi degeneracy, and kinetic inhomogeneities coexist. A transformative approach is offered by Density Functional Theory (DFT), which reformulates the quantum many-body problem using the electron density as the fundamental variable, rather than the many-electron wavefunction. The cornerstone of this formalism rests on the Hohenberg–Kohn (HK) theorems and the operational realization via the Kohn–Sham (KS) equations [24].

In the traditional Schrödinger representation of quantum mechanics, the full N-electron wavefunction $\Psi(\mathbf{r}_1, \mathbf{r}_2, \ldots, \mathbf{r}_N)$ resides in a 3N-dimensional configuration space. This exponential scaling poses a severe computational bottleneck. However, all ground-state observables are ultimately functionals of the ground-state electron density $n(\mathbf{r})$, which resides in three-dimensional real space. The shift from the many-body wavefunction to the density represents a profound conceptual simplification.

Let us consider a quantum plasma composed of electrons under the influence of a fixed ionic background and external fields, where the system is described by the Hamiltonian:

$$\hat{H} = \hat{T} + \hat{V}_{\text{ext}} + \hat{V}_{\text{int}} = -\sum_{i=1}^{N} \frac{\hbar^2}{2m}\nabla_i^2 + \sum_{i=1}^{N} v_{\text{ext}}(\mathbf{r}_i) + \frac{1}{2}\sum_{i\neq j} \frac{e^2}{|\mathbf{r}_i - \mathbf{r}_j|},$$

where $v_{\text{ext}}(\mathbf{r})$ encodes external electrostatic, confinement, or electromagnetic interactions, while the third term represents inter-electron Coulomb repulsion. In quantum plasmas, the background may further contribute an effective field due to Debye shielding, in which case $v_{\text{ext}}(\mathbf{r})$ may involve a screened Coulomb potential, or Yukawa-like corrections [25].

The Hohenberg–Kohn theorems (1964) provide the theoretical bedrock for density-based approaches. They address two fundamental questions, whether the ground-state density uniquely determines the external potential, and whether there exists a variational principle formulated in terms of the density. Both are affirmatively answered.

First Hohenberg–Kohn Theorem (HK1): Uniqueness

For a many-electron system in an external potential $v_{\text{ext}}(\mathbf{r})$, there exists a one-to-one correspondence between the ground-state electron density $n_0(\mathbf{r})$ and the external potential $v_{\text{ext}}(\mathbf{r})$, up to an additive constant. This implies that all ground-state observables including the total energy are functionals of the ground-state density. Physically, this is a nontrivial statement: the entire complexity of the quantum state is encoded in a scalar function over $\mathbf{R}^3$.

Second Hohenberg–Kohn Theorem (HK2): Variational Principle

Define the universal functional

$$F[n] = \langle \Psi[n] | \hat{T} + \hat{V}_{\text{int}} | \Psi[n] \rangle,$$

which includes the kinetic and interaction energies, but is independent of $v_{\text{ext}}(\mathbf{r})$. The total energy functional is

$$E[n] = F[n] + \int d^3 r v_{\text{ext}}(\mathbf{r}) n(\mathbf{r}).$$

Then, the ground-state density minimizes this functional, for any admissible density $n(\mathbf{r})$,

$$E[n] \geq E[n_0],$$

with equality only when $n = n_0$. In quantum plasmas, where external fields may include plasma wave excitations, or electrostatic confinement, the theorem guarantees that once $n_0(\mathbf{r})$ is known, all thermodynamic and dynamical ground-state properties can be deduced.

5.3.1 The Kohn–Sham Construction

While the HK theorems are conceptually potent, they offer no practical route to compute the universal functional F[n], as it contains the full many-body kinetic and interaction energy. Kohn and Sham (1965) devised a workaround by introducing an auxiliary non-interacting system whose ground-state density matches that of the interacting one. The Kohn–Sham scheme replaces the intractable many-body problem with a set of self-consistent single-particle equations [26]. The total energy functional is decomposed as:

$$E[n] = T_s[n] + \int d^3r v_{\text{ext}}(\mathbf{r})n(\mathbf{r}) + \frac{e^2}{2}\int d^3r d^3r' \frac{n(\mathbf{r})n(\mathbf{r}')}{|\mathbf{r}-\mathbf{r}'|} + E_{\text{xc}}[n],$$

where: $T_s[n]$ is the kinetic energy of the non-interacting system, the Hartree term represents classical Coulomb repulsion, and $E_{\text{xc}}[n]$ is the exchange-correlation functional, encapsulating all many-body effects, exchange, correlation, and the correction to the kinetic energy beyond $T_s[n]$.

The corresponding Kohn–Sham equations for the single-particle orbitals $\{\psi_i(\mathbf{r})\}$ are:

$$\left[-\frac{\hbar^2}{2m}\nabla^2 + v_{\text{eff}}(\mathbf{r})\right]\psi_i(\mathbf{r}) = \epsilon_i\psi_i(\mathbf{r}),$$

with the effective potential:

$$v_{\text{eff}}(\mathbf{r}) = v_{\text{ext}}(\mathbf{r}) + \int d^3r' \frac{e^2 n(\mathbf{r}')}{|\mathbf{r}-\mathbf{r}'|} + v_{\text{xc}}(\mathbf{r}),$$

where $v_{\text{xc}}(\mathbf{r}) = \delta E_{\text{xc}}[n]/\delta n(\mathbf{r})$ is the exchange-correlation potential. These equations must be solved self-consistently, since $v_{\text{eff}}(\mathbf{r})$ depends on $n(\mathbf{r}) = \sum_i |\psi_i(\mathbf{r})|^2$.

In the context of quantum plasmas, the KS equations allow for the inclusion of strong coupling effects, non-uniform densities, and magnetic or electric field-induced structure. The external potential may arise from background ion configurations (as in jellium approximations), Debye-screened potentials, or external lasers and magnetic fields. Moreover, in strongly coupled degenerate plasmas as encountered in warm dense matter, dense astrophysical interiors, or laser-compressed targets the electron correlation effects cannot be neglected. Here, $E_{\text{xc}}[n]$ becomes crucial, and quantum

Monte Carlo-derived parameterizations of exchange-correlation functionals are often employed.

From a physical perspective, the Kohn–Sham construction offers an effective single-particle description where collective plasma phenomena, like screening and oscillations, are absorbed into the structure of $v_{\text{eff}}(\mathbf{r})$. The response of such a plasma to external perturbations, or its linear and non-linear susceptibilities, can then be computed by applying linear response theory around the Kohn–Sham ground state.

Despite its success, standard DFT within the KS framework is a ground-state theory, inherently static and limited to equilibrium configurations. Extensions to finite-temperature DFT, as required for quantum plasmas, and time-dependent DFT (TDDFT) for analyzing non-equilibrium phenomena such as those relevant in ultra-fast plasma evolution will be developed in the subsequent sections. The incorporation of relativistic corrections, spin-polarization, and current-density effects are essential when addressing quantum plasmas in strong magnetic fields or at high energies.

Notably, the exchange-correlation functional remains the primary unknown, and considerable effort in the quantum plasma community is directed toward developing accurate functionals that reflect both strong-coupling and field-theoretic structures native to plasma environments.

The Hohenberg–Kohn theorems and the Kohn–Sham equations lay the theoretical and computational foundation for DFT-based treatments of quantum plasmas. They enable the simplification of the many-body problem to a self-consistent single-particle framework, retaining essential quantum statistical and correlation effects. As such, DFT particularly when extended beyond its ground-state formulation serves as an indispensable tool for modeling and interpreting the rich structure of strongly correlated, field-permeated quantum plasma systems [27].

5.3.2 Thomas–Fermi and Thomas–Fermi–Dirac Approximations

The Thomas–Fermi (TF) and Thomas–Fermi–Dirac (TFD) models represent cornerstone approximations in the statistical description of many-electron systems and offer a tractable semi-classical framework for analyzing the equilibrium properties of quantum plasmas. These approximations eschew the complexities of many-body wavefunctions in favour of a functional dependence on local particle densities, thereby bridging quantum statistical mechanics and macroscopic observables.

The original Thomas–Fermi model, developed independently by Thomas and Fermi in 1927, constitutes a statistical approximation to quantum many-body systems at equilibrium. It replaces the many-electron wavefunction with a position-dependent number density $n(\mathbf{r})$, defined as the number of electrons per unit volume at position $\mathbf{r}$, and assumes a locally uniform Fermi gas in which the electrons obey Fermi–Dirac statistics.

In this formulation, one begins with the semi-classical expression for the electron density in terms of the local chemical potential μ and the external potential $V(\mathbf{r})$. At zero temperature, the momentum distribution is a step function, and the number of available electron states per unit volume with momentum less than the local Fermi momentum $p_F(\mathbf{r})$ is

$$n(\mathbf{r}) = \frac{g_s}{(2\pi\hbar)^3}\int_{|\mathbf{p}|\leq p_F(\mathbf{r})} d^3p = \frac{g_s}{6\pi^2\hbar^3}p_F^3(\mathbf{r}),$$

where $g_s = 2$ is the spin degeneracy factor for electrons. The local Fermi energy is defined by

$$\varepsilon_F(\mathbf{r}) = \mu - V(\mathbf{r}) = \frac{p_F^2(\mathbf{r})}{2m},$$

yielding the Thomas–Fermi relation:

$$n(\mathbf{r}) = \frac{1}{3\pi^2}\left(\frac{2m}{\hbar^2}\right)^{3/2}[\mu - V(\mathbf{r})]_+^{3/2},$$

where $[\cdot]_+$ denotes that the expression is defined only for $\mu > V(\mathbf{r})$. This fundamental expression relates the local density to the electrostatic potential, thereby enabling the coupling of the quantum electron gas to Poisson's equation:

$$\nabla^2 V(\mathbf{r}) = -4\pi e^2[n(\mathbf{r}) - Z\delta(\mathbf{r})],$$

for a system with a single nuclear charge Ze located at the origin. In this self-consistent field approach, the total energy of the system is formulated as a functional of the electron density:

$$E[n] = T_{\mathrm{TF}}[n] + E_{\mathrm{ext}}[n] + E_{\mathrm{H}}[n],$$

where $T_{\mathrm{TF}}[n]$ is the Thomas–Fermi kinetic energy functional given by

$$T_{\mathrm{TF}}[n] = C_{\mathrm{TF}}\int d^3r n^{5/3}(\mathbf{r}), \quad C_{\mathrm{TF}} = \frac{3}{10}\left(\frac{3\pi^2}{2}\right)^{2/3}\frac{\hbar^2}{m},$$

$$E_{\mathrm{ext}}[n] = \int d^3r V_{\mathrm{ext}}(\mathbf{r})n(\mathbf{r}),$$

and $E_{\mathrm{H}}[n] = \frac{1}{2}\int d^3r d^3r' \frac{n(\mathbf{r})n(\mathbf{r}')}{|\mathbf{r}-\mathbf{r}'|}$ is the Hartree electrostatic energy.

Physically, the Thomas–Fermi model captures the leading-order effect of Pauli exclusion via the kinetic energy term and Coulomb interaction via the Hartree energy. Its central merit lies in its scalability to large particle numbers and high densities characteristic of white dwarfs, laser-compressed plasmas, and metallic states of matter where orbital resolution becomes computationally infeasible. However, the model neglects both quantum shell effects and exchange–correlation phenomena intrinsic to fermionic statistics, rendering it inadequate for low-density, or strongly inhomogeneous systems. This necessitates a refinement that accounts for the antisymmetry of the many-body wavefunction more precisely [28].

To extend the TF model beyond its mean-field approximation and incorporate quantum exchange interactions stemming from fermionic antisymmetry, Dirac introduced an approximate exchange energy functional based on the homogeneous electron gas. The Thomas–Fermi–Dirac (TFD) model thus modifies the total energy functional to

$$E_{\mathrm{TFD}}[n] = T_{\mathrm{TF}}[n] + E_{\mathrm{ext}}[n] + E_{\mathrm{H}}[n] + E_{\mathrm{x}}[n],$$

where $E_{\mathrm{x}}[n]$ represents the exchange energy functional given by

$$E_{\mathrm{x}}[n] = -C_{\mathrm{x}} \int d^3 r n^{4/3}(\mathbf{r}), \quad C_{\mathrm{x}} = \frac{3}{4}\left(\frac{3}{\pi}\right)^{1/3} e^2.$$

This expression follows from the local density approximation (LDA) for the exchange energy of a uniform electron gas, derived via Slater determinants. The exchange term lowers the total energy due to the Fermi hole around each electron, which reduces the Coulomb repulsion arising from indistinguishability and antisymmetry.

Minimization of $E_{\mathrm{TFD}}[n]$ under the constraint of particle number conservation $\int d^3 r n(\mathbf{r}) = N$ leads to a modified Thomas–Fermi equation:

$$\mu = V(\mathbf{r}) + \frac{\delta T_{\mathrm{TF}}[n]}{\delta n(\mathbf{r})} + \frac{\delta E_{\mathrm{H}}[n]}{\delta n(\mathbf{r})} + \frac{\delta E_{\mathrm{x}}[n]}{\delta n(\mathbf{r})},$$

explicitly,

$$\mu = V(\mathbf{r}) + \frac{5}{3} C_{\mathrm{TF}} n^{2/3}(\mathbf{r}) + \int d^3 r' \frac{n(\mathbf{r}')}{|\mathbf{r} - \mathbf{r}'|} - \frac{4}{3} C_{\mathrm{x}} n^{1/3}(\mathbf{r}).$$

This integro-differential equation determines the spatially varying density $n(\mathbf{r})$ self-consistently. The exchange term introduces a nonlinearity of lower power than the kinetic energy term, reflecting its subleading role at high densities, yet it becomes essential in intermediate regimes where degeneracy and exchange compete.

From a physical standpoint, the inclusion of the Dirac exchange term significantly improves the accuracy of density profiles, ionization potentials, and screening lengths in quantum plasma systems. For instance, in metallic clusters and dense laser-compressed matter, the TFD approximation better reproduces the electron spill-out and edge behaviour near boundaries, which are crucial for interpreting transport and optical properties.

In the context of quantum plasmas, the TF and TFD models serve as foundational approximations for understanding screening, electrostatic structure, and equation of state in degenerate regimes. The Debye–Hückel length in a classical plasma is supplanted by the Thomas–Fermi screening length:

$$\lambda_{\mathrm{TF}} = \left(\frac{\pi \hbar^2}{2e^2 m k_F}\right)^{1/2}, \quad k_F = \left(3\pi^2 n\right)^{1/3},$$

which governs the response of the electron gas to perturbations in the potential and encodes quantum degeneracy effects. In ultradense plasmas, such as those encountered in inertial confinement fusion cores, or the crust of neutron stars, the TF or TFD approximation often forms the base of more sophisticated density functional constructions.

Moreover, modern generalizations of the TF and TFD functionals incorporate gradient corrections, temperature dependence, and relativistic effects. The semi-classical expansion of the kinetic energy functional introduces gradient terms of the form $\propto |\nabla n|^2/n$, leading to the Thomas–Fermi–Weizsäcker model [29]. Finite-temperature extensions utilize the Fermi–Dirac integrals and allow for a smooth interpolation between quantum and classical regimes essential for modeling warm dense matter (WDM), where quantum statistical mechanics must coexist with thermally excited states.

5.3.3 *Orbital-Free Density Functional Theory*

The quantum statistical mechanics of plasmas at finite temperature presents a significant theoretical challenge due to the need to account simultaneously for thermal effects, quantum degeneracy, collective interactions, and electromagnetic field fluctuations. In this context, Orbital-Free Density Functional Theory (OFDFT) emerges as an efficient and theoretically elegant framework that circumvents the computational and conceptual limitations of orbital-based approaches when applied to extended, high-temperature, and strongly interacting plasma systems.

The essence of OFDFT lies in expressing the free energy of a many-body fermionic system solely as a functional of the local particle density $n(\mathbf{r})$, thereby obviating the need to compute Kohn–Sham orbitals. The foundation of OFDFT at finite temperature is built upon the Mermin extension of DFT, which recasts the Hohenberg–Kohn theorem in the thermodynamic context. For a system described by the grand canonical ensemble at temperature T, the equilibrium particle density $n(\mathbf{r})$ minimises the grand canonical potential functional:

$$\Omega[n] = F[n] + \int d^3 r n(\mathbf{r})(V_{\mathrm{ext}}(\mathbf{r}) - \mu),$$

where $F[n]$ is the intrinsic free energy functional, incorporating kinetic, exchange-correlation, and interaction energies; $V_{\text{ext}}(\mathbf{r})$ is the external electrostatic potential, and μ is the chemical potential. The Mermin free energy $F[n]$ is decomposed as:

$$F[n] = T_s[n] + F_{\text{xc}}[n] + F_{\text{H}}[n],$$

where $T_s[n]$ is the non-interacting kinetic free energy, $F_{\text{xc}}[n]$ denotes the exchange-correlation free energy, and $F_{\text{H}}[n]$ is the classical Hartree energy. In orbital-free theory, the pivotal challenge is constructing accurate approximations to $T_s[n]$ and $F_{\text{xc}}[n]$ without explicit reference to single-particle wavefunctions.

For an ideal Fermi gas at finite temperature, the non-interacting free energy $T_s[n]$ can be formally derived using the grand partition function $\mathcal{Z}$, leading to the local density approximation (LDA) for the kinetic contribution:

$$T_s^{\text{LDA}}[n] = \int d^3r \tau_{\text{TF}}(n(\mathbf{r}), T),$$

where τ_{TF} is the finite-temperature Thomas–Fermi kinetic energy density, expressible in terms of the polylogarithmic function $\text{Li}_{5/2}$, as:

$$\tau_{\text{TF}}(n, T) = \frac{2}{\sqrt{\pi}} \left(\frac{2m}{\hbar^2} \right)^{3/2} T^{5/2} \text{Li}_{5/2}\left(-e^{\beta(\mu_{1\text{oc}} - V_{\text{ext}})} \right).$$

Here, $\mu_{1\text{oc}}$ is the local chemical potential and $\beta = 1/(k_B T)$. Corrections beyond the LDA often involve gradient expansions of the kinetic energy functional, such as the second-order gradient correction (the von Weizsäcker term) and higher-order nonlocal kernels.

In the plasma context, the orbital-free formalism must be extended to accommodate charged particle interactions, screening phenomena, and collective electromagnetic effects. The density field $n(\mathbf{r})$ now represents not only electronic but also ionic components, and must be coupled self-consistently to the mean-field electrostatic potential $\phi(\mathbf{r})$ through the Poisson's equation:

$$\nabla^2 \phi(\mathbf{r}) = -\frac{e}{\varepsilon_0} \left(n_e(\mathbf{r}) - \sum_j Z_j n_j(\mathbf{r}) \right),$$

where $n_e(\boldsymbol{r})$ is the electron density, $n_j(\boldsymbol{r})$ are the ion species densities with respective charges $Z_j e$, and ε_0 is the vacuum permittivity. The electron density itself depends on the total electrostatic potential via the variational principle:

$$\frac{\delta \Omega[n]}{\delta n(\mathbf{r})} = 0 \Rightarrow \frac{\delta F[n]}{\delta n(\mathbf{r})} + V_{\text{ext}}(\mathbf{r}) - \mu = 0.$$

This yields a nonlinear integro-differential equation for $n(\mathbf{r})$, embedded in a self-consistent loop with the Poisson's equation.

In the presence of collective plasma oscillations and thermal fluctuations, the effective external potential $V_{\text{ext}}(\mathbf{r})$ must be augmented by the dynamically screened field, involving the dielectric response function $\varepsilon(\mathbf{r}, \mathbf{r}'; \omega)$, which itself depends on the instantaneous configuration of the plasma. At the level of mean-field theory, however, such dynamical contributions can be incorporated into an effective potential via Debye–Hückel or quantum-statistical closures.

One of the pivotal applications of orbital-free DFT in plasma physics is the derivation of quantum hydrodynamic (QHD) equations that include consistent quantum statistical pressure and Bohm-like quantum potential terms. Starting from the Euler–Lagrange equations associated with the minimisation of the free energy functional, one obtains the QHD momentum equation for the electron fluid:

$$m_e\left(\frac{\partial \mathbf{v}}{\partial t} + \mathbf{v} \cdot \nabla \mathbf{v}\right) = -\nabla(V_{\text{ext}} + V_H + V_{\text{xc}} + V_B),$$

where V_H is the Hartree potential, V_{xc} is the exchange-correlation potential derived from $F_{xc}[n]$, and $\mathrm{V_B}$ is the quantum Bohm potential. In OFDFT, V_B emerges naturally from the gradient expansion of the kinetic functional:

$$V_B[n] = -\frac{\hbar^2}{2m_e}\frac{\nabla^2\sqrt{n}}{\sqrt{n}}.$$

This representation avoids any ad hoc assumptions about the quantum pressure term and provides a firm variational basis for incorporating quantum effects into the fluid dynamics of plasmas.

One critical advantage of orbital-free DFT in the plasma setting is its built-in thermodynamic consistency. Since all quantities derive from a single free energy functional, the relationships between pressure, energy density, and chemical potential respect the fundamental Maxwell relations and thermodynamic identities. The total pressure can be computed via:

$$P = -\left(\frac{\partial \Omega}{\partial V}\right)_{T,\mu},$$

and the equation of state (EOS) naturally includes quantum statistical effects through the form of $F[n]$. This contrasts with empirical, or hybrid models where thermodynamic inconsistencies often arise due to mixing of classical and quantum contributions in an uncontrolled fashion.

Nevertheless, limitations persist. The accuracy of OFDFT relies heavily on the choice of kinetic and exchange-correlation functionals, and exact forms are only known in specific limits, such as the homogeneous electron gas. Moreover, OFDFT may underestimate shell and confinement effects in strongly inhomogeneous systems or near boundaries, where orbital-based methods may be necessary for fidelity.

Recent advances have extended OFDFT into new directions relevant to quantum plasmas. Machine-learned functionals based on neural networks or kernel ridge regression have been developed to fit high-fidelity data from quantum Monte Carlo, or path integral simulations. These approaches have proven particularly effective in capturing nonlocal kinetic energy contributions and exchange-correlation behaviour under extreme conditions. Furthermore, the integration of OFDFT with quantum kinetic theory allows for time-dependent generalisations, leading to orbital-free time-dependent DFT (OF-TDDFT) applicable to ultrafast plasma dynamics, including non-equilibrium ionisation and rapid thermalisation [30].

In parallel, extensions incorporating Gauge-covariant kinetic energy functionals have been developed to ensure consistency with electromagnetic Gauge invariance, thereby making the theory compatible with dynamically evolving magnetic fields and vector potentials a feature indispensable for high-intensity laser–plasma interactions and astrophysical magnetoplasmas.

Orbital-Free Density Functional Theory, and its extensions to finite-temperature and plasma contexts, offers a compelling and formally grounded framework for modelling dense, hot, and dynamically complex quantum plasmas. It provides a variationally consistent and computationally tractable alternative to orbital-based methods, particularly suited for high-temperature, high-density regimes where wavefunction-based approaches become infeasible. Its integration with quantum hydrodynamics, machine learning, and kinetic theory continues to expand its utility, laying the foundation for a comprehensive ab initio description of collective and quantum-statistical effects in plasma environments across laboratory and astrophysical settings.

5.3.4 Exchange–Correlation Functionals

The exchange–correlation (XC) functional constitutes the central component of Density Functional Theory (DFT), encoding the quantum mechanical effects of electron–electron interactions that are not captured by the non-interacting Kohn–Sham (KS) reference system. In the context of quantum plasmas, particularly under finite-temperature and strongly coupled conditions, the classical approximations to XC functional developed largely for ground-state condensed matter systems fail to capture the complex interplay of quantum statistical, thermal, and collective plasma effects. This necessitates the formulation of exchange–correlation functionals specifically tailored for plasma environments, where electron degeneracy, strong coupling, and quantum diffraction dominate the system's response.

A plasma-tailored XC functional must satisfy several theoretical and physical constraints that distinguish it from its zero-temperature or ground-state counterparts:

(i) *Temperature Dependence*: The inclusion of explicit temperature dependence T is indispensable in quantum plasmas. This modifies not only the occupation statistics (via the Fermi–Dirac distribution), but also the screening properties and short-range order within the electron gas.

(ii) *Strong Coupling*: In dense, or low-temperature plasma regimes where the coupling parameter $\Gamma = \frac{e^2}{ak_BT}$ exceeds unity, correlation effects become non-perturbative. This requires non-local and beyond local-density approximations (LDA) to correctly represent electron correlations.

(iii) *Quantum Diffraction and Non-locality*: Quantum diffraction, stemming from the finite de Broglie wavelength, introduces non-local exchange–correlation effects. These necessitate gradient corrections, or fully non-local functionals to capture the momentum-dependent interactions.

(iv) *Gauge and Symmetry Compliance*: For electromagnetic field-coupled plasmas, Gauge invariance and current conservation must be explicitly respected in time-dependent and magnetized extensions of XC functionals.

(v) *Hydrodynamic Consistency*: The XC functionals must be consistent with the conservation laws and hydrodynamic moment equations derived from quantum kinetic, or Wigner–Moyal formalisms.

The extension of the ground-state exchange–correlation energy $E_{xc}[n]$ to finite temperature requires a redefinition of the XC free energy functional $F_{xc}[n, T]$. This quantity enters the thermodynamic grand potential functional Ω[n] as follows:

$$\Omega[n] = \mathcal{F}_s[n, T] + \int d^3rn(\mathbf{r})(V_{\text{ext}}(\mathbf{r}) - \mu) + \mathcal{F}_{xc}[n, T],$$

where $\mathcal{F}_s[n, T]$ is the free energy of the non-interacting Kohn–Sham system and V_{ext} is the external electrostatic potential. The XC potential entering the Kohn–Sham equations is derived as a functional derivative:

$$V_{xc}(\mathbf{r}, T) = \left.\frac{\delta \mathcal{F}_{xc}[n, T]}{\delta n(\mathbf{r})}\right|_T.$$

Several models have been proposed to compute $\mathcal{F}_{xc}[n, T]$, including path integral Monte Carlo (PIMC) simulations, finite-temperature quantum Monte Carlo (QMC) parametrizations, and classical map hypernetted chain (CHNC) methods. Notably, the parametrizations by Karasiev, Dufty, and Trickey offer temperature-dependent XC functionals consistent with PIMC data for the uniform electron gas (UEG), a reference system for plasmas [27].

The homogeneous electron gas at finite temperature serves as the canonical model for developing and testing XC functionals in plasma conditions. The exchange–correlation free energy per particle $f_{xc}(r_s, T)$ depends on the Wigner–Seitz radius $r_s = \left(\frac{3}{4\pi n}\right)^{1/3}$ and the reduced temperature $\theta = k_BT/E_F$, where E_F is the Fermi energy. The coupling parameter Γ in this regime becomes a function of both r_s and θ, and the XC free energy must interpolate correctly between the high-temperature classical Debye–Hückel limit and the low-temperature quantum limit.

Accurate parametrizations of $f_{xc}(r_s, \theta)$ have been constructed using PIMC data and fitted with Padé approximants or machine learning regression models. These expressions are typically validated through the compressibility sum rule, high-density (RPA) limits, and entropy consistency conditions.

To go beyond the local density approximation(LDA), gradient corrections must be incorporated to account for inhomogeneities in the density, especially in confined or strongly non-uniform plasmas such as those found in inertial confinement fusion (ICF) targets or dense astrophysical environments. The generalized gradient approximation (GGA) modifies the XC energy density as:

$$\mathcal{F}_{xc}^{\mathrm{GGA}}[n, T] = \int d^3 r n(\mathbf{r}) f_{xc}^{\mathrm{GGA}}(n(\mathbf{r}), |\nabla n|, T),$$

where the function f_{xc}^{GGA} is constructed to satisfy known limits and constraints, including the Lieb–Oxford bound and correct linear response behaviour .

Some recent advancements introduce temperature-dependent gradient coefficients, derived either via finite-temperature linear response theory, or constrained fitting to quantum Monte Carlodata. These corrections are particularly relevant in capturing the interface regions between different plasma domains or in systems with temperature gradients.

In orbital-free DFT (OF-DFT), which is essential for large-scale simulations of dense plasmas, the challenge is to construct XC functionals that remain accurate without Kohn–Sham orbitals. This involves defining the XC contribution in terms of the density and its derivatives alone. The finite-temperature Thomas–Fermi–Dirac (TFD) model, augmented with XC corrections derived from uniform plasma data, forms the baseline. More sophisticated orbital-free models, such as those incorporating nonlocal kernel functionals (e.g., Wang–Govind–Carter-type), have been extended to include XC terms consistent with warm dense matter simulations [31].

Mathematically, such functionals take the form:

$$\mathcal{F}_{xc}^{\mathrm{OF}}[n, T] = \int d^3 r \epsilon_{xc}^{\mathrm{hom}}(n(\mathbf{r}), T) + \int d^3 r d^3 r' K_{xc}\big(\mathbf{r}, \mathbf{r}', T\big) n(\mathbf{r}) n\big(\mathbf{r}'\big),$$

where K_{xc} is a nonlocal exchange–correlation kernel encoding spatial correlations at finite temperature.

Quantum plasma systems often evolve dynamically and may reside in magnetized environments. Accordingly, the extension to time-dependent DFT (TDDFT) and current density functional theory (CDFT) is essential. In these frameworks, the exchange–correlation effects are captured through vector potentials $\mathbf{A}_{xc}$ and memory functionals that depend on the history of the current or density. The formalism involves the construction of XC vector potentials:

$$\mathbf{A}_{xc}(\mathbf{r}, t) = \frac{\delta \mathcal{A}_{xc}[\mathbf{j}]}{\delta \mathbf{j}(\mathbf{r}, t)},$$

where $\mathbf{A}_{xc}[\mathbf{j}]$ is a functional of the current density $\mathbf{j}$. This extension is crucial in accurately modeling Landau damping, magnetized screening, and collective mode modifications in quantum plasmas.

Tailored XC functionals for quantum plasmas are indispensable for predictive simulations in high-energy-density physics, astrophysical modeling of white dwarfs and neutron star crusts, and the design of inertial confinement fusion experiments. Accurate XC functionals bridge the gap between first-principles many-body physics and tractable computational models, providing consistent equations of state, transport properties, and response functions across a wide parameter space of density, temperature, and field strength [32].

The future direction includes incorporating machine-learned XC functionals trained on high-fidelity data sets, constructing XC functionals compatible with Wigner–Moyal and Kadanoff–Baym approaches, and ensuring seamless integration with quantum hydrodynamic and kinetic theories for multi-scale quantum plasma modeling.

5.3.5 Time-Dependent Density Functional Theory

The limitations of static density functional theory (DFT) in describing non-equilibrium and temporally evolving plasma environments have motivated the extension of the formalism to time-dependent systems, giving rise to time-dependent density functional theory (TDDFT). In the context of quantum plasmas, where the interplay between collective electron dynamics, electromagnetic field fluctuations, and quantum statistical correlations dictates the system's behaviour, TDDFT offers a compelling and formally rigorous route to capture both single-particle and many-body dynamics in real-time.

At its core, TDDFT rests upon the Runge-Gross theorem, which extends the Hohenberg-Kohn theorem into the time domain. The Runge-Gross theorem asserts that for a given initial many-body state Ψ_0 and a time-dependent external potential $v_{\text{ext}}(\mathbf{r}, t)$ that is Taylor-expandable in time, the one-to-one correspondence between the potential and the time-dependent density $n(\mathbf{r}, t)$ is preserved. Thus, all observables of the system can be expressed as functionals of the time-dependent density.

The quantum plasma system is governed by the time-dependent many-body Schrödinger equation:

$$i\hbar \frac{\partial \Psi(\mathbf{r}_1, \dots, \mathbf{r}_N, t)}{\partial t} = \hat{H}(t)\Psi(\mathbf{r}_1, \dots, \mathbf{r}_N, t),$$

with a Hamiltonian of the form:

$$\hat{H}(t) = \sum_{i=1}^{N} \left[-\frac{\hbar^2}{2m} \nabla_i^2 + v_{\text{ext}}(\mathbf{r}_i, t) \right] + \frac{1}{2} \sum_{i \neq j} \frac{e^2}{\left|\mathbf{r}_i - \mathbf{r}_j\right|},$$

where $v_{\text{ext}}(\mathbf{r}, t)$ incorporates external laser fields, confining potentials, background charges, and induced plasma wakefields. In dense plasmas, particularly those in extreme or relativistic regimes, this potential becomes a functional of the self-consistently determined electromagnetic field dynamics and the background dielectric response.

TDDFT bypasses the intractability of the full many-body wavefunction by mapping the problem onto a fictitious system of non-interacting Kohn–Sham particles evolving in an effective time-dependent potential. The evolution of the single-particle Kohn–Sham orbitals $\phi_j(\mathbf{r}, t)$ is governed by:

$$i\hbar \frac{\partial}{\partial t} \phi_j(\mathbf{r}, t) = \left[-\frac{\hbar^2}{2m} \nabla^2 + v_{\text{KS}}[n](\mathbf{r}, t) \right] \phi_j(\mathbf{r}, t),$$

where the Kohn–Sham potential is defined as:

$$v_{\text{KS}}[n](\mathbf{r}, t) = v_{\text{ext}}(\mathbf{r}, t) + v_{\text{H}}[n](\mathbf{r}, t) + v_{\text{xc}}[n](\mathbf{r}, t),$$

with $v_{\text{H}}[n](\mathbf{r}, t)$ being the time-dependent Hartree potential,

$$v_{\text{H}}[n](\mathbf{r}, t) = \int \frac{e^2 n\left(\mathbf{r}', t\right)}{|\mathbf{r} - \mathbf{r}'|} d^3\mathbf{r}',$$

and $v_{\text{xc}}[n](\mathbf{r}, t)$ the time-dependent exchange-correlation potential, whose construction remains the central challenge in TDDFT for strongly coupled plasmas.

The applicability of TDDFT to plasma systems is particularly powerful when one considers time-evolving fields, such as those driven by ultrashort, intense laser pulses or occurring in relativistically modulated plasma wakefields. In such systems, the time scale of the external driving field is often comparable to or faster than the characteristic response times of the plasma (e.g., plasma frequency ω_p, Fermi time τ_F), necessitating a fully time-resolved treatment.

In this dynamical regime, the time-dependent density encodes not only the instantaneous spatial distribution of the electron gas, but also the system's history through memory effects embedded within v_{xc}. The exchange-correlation potential, in general, becomes explicitly non-local in time:

$$v_{\text{xc}}[n](\mathbf{r}, t) = v_{\text{xc}}[n]\left(\mathbf{r}, t; \left\{n\left(\mathbf{r}, t'\right)\right\}_{t' \leq t}\right),$$

capturing retardation, decoherence, and dissipation arising from past interactions. This becomes especially relevant for quantum plasmas exhibiting non-Markovian electron-ion collisions, electron-hole recombination, or delayed polarization response due to vacuum or bound-state contributions.

In the context of plasma electrodynamics, TDDFT must be coupled self-consistently to Maxwell's equations, or their quantum-corrected extensions. The current density, derived from the Kohn–Sham orbitals as:

$$\mathbf{J}(\mathbf{r}, t) = -\frac{e\hbar}{2mi} \sum_{j=1}^{N} \left[\phi_j^*(\mathbf{r}, t) \nabla \phi_j(\mathbf{r}, t) - \phi_j(\mathbf{r}, t) \nabla \phi_j^*(\mathbf{r}, t) \right],$$

acts as a source term for the evolving electromagnetic fields. In practical implementations, one evolves the coupled system:

$$\nabla \times \mathbf{B} = \mu_0 \left(\mathbf{J} + \epsilon_0 \frac{\partial \mathbf{E}}{\partial t} \right), \quad \nabla \times \mathbf{E} = -\frac{\partial \mathbf{B}}{\partial t},$$

alongside the TD-Kohn–Sham equations, thereby capturing feedback between field dynamics and charge evolution a critical aspect in quantum plasma oscillations, electromagnetic pulse formation, and nonlinear mode conversion.

In extreme conditions, such as those involving relativistic intensities or vacuum polarization, Maxwell's equations may be augmented to include nonlinear terms arising from the Heisenberg-Euler Lagrangian. The induced current derived from TDDFT remains a pivotal element in these extensions, and TDDFT itself must incorporate relativistic corrections to maintain consistency.

Despite its elegance and formal rigor, the practical implementation of TDDFT in plasma systems encounters formidable challenges. Chief among these is the lack of universally accurate time-dependent exchange-correlation (XC) functionals suitable for highly excited, strongly coupled, or relativistic regimes. Most current approximations, such as the adiabatic local density approximation (ALDA), neglect Memory effects and can fail to capture dynamic correlation phenomena, such as Langmuir collapse, quantum shock formation, or turbulent density fluctuations.

Recent developments have explored orbital-dependent XC potentials, many-body perturbation theory corrections via time-dependent optimized effective potential (TDOEP) methods, and incorporation of kinetic equations for the Wigner function to back-propagate memory kernels into the XC potential. In hybridized approaches, TDDFT is supplemented by quantum Boltzmann, or Wigner kinetic equations for non-local transport and decoherence effects, especially where phase-breaking mechanisms and collisional thermalization dominate.

Time-dependent density functional theory (TDDFT) has emerged as a powerful tool for modeling ultrafast phenomena in quantum plasmas, offering insights into complex dynamical processes with high temporal and spatial resolution. In laser-driven plasma dynamics, TDDFT is employed to simulate critical events such as field ionization, attosecond-scale charge migration, and the generation of nonlinear harmonics. It also proves instrumental in studying shock-compressed matter and warm dense plasmas, enabling real-time tracking of changes in electronic structures and energy distribution under intense compressive forces. Further, TDDFT aids in understanding quantum turbulence and filamentation by capturing the spatiotemporal evolution of charge currents that drive instabilities, or lead to nonlinear wave-breaking. Additionally, it provides a robust framework for analyzing plasma surface and interface phenomena, including sheath formation, excitation of surface plasmons, and the generation of hot electrons, processes central to applications in nanoplasmonics and high-field physics.

Each of these domains benefits from TDDFT's ability to treat inhomogeneous, finite-temperature, and time-resolved electronic processes within a formally exact single-particle framework. Time-dependent Density Functional Theory constitutes a powerful formalism for modeling quantum plasmadynamics in real time, with the capacity to incorporate complex many-body and electromagnetic field effects through a density-centric viewpoint. While its full potential is yet to be unlocked, particularly in the regime of relativistic and ultra-dense plasmas, ongoing advances in functional development, coupling to electrodynamics, and hybridized computational schemes continue to establish TDDFT as a cornerstone method in the theoretical arsenal of modern quantum plasmaphysics [33, 34].

5.4 Quantum Field Theory

The development of quantum field theoretic (QFT) frameworks in plasma modeling arises from foundational limitations intrinsic to particle-based descriptions, particularly when the plasma enters regimes characterized by strong correlations, relativistic velocities, or high-frequency quantum fluctuations.

Traditional plasma descriptions rely on either kinetic theory, as embodied by the Vlasov or Boltzmann equations, or on fluid models derived from moments of the distribution function. These frameworks presuppose the validity of the quasi-particle picture: that the collective behaviour of many-body systems may be effectively modeled by treating particles as classical entities following well-defined trajectories, subject to mean-field forces and collisional perturbations. Even in quantum kinetic formulations such as the Wigner function approach particles are still conceptually treated as coherent wavepackets with quasi-classical attributes. These approximations break down in two critical scenarios: (i) when quantum entanglement between plasma constituents leads to non-factorizable many-body states, and (ii) when relativistic velocities induce frame-dependent distortions of electromagnetic fields that cannot be captured by Galilean-covariant equations [35].

To quantify the breakdown, consider the dimensionless coupling parameter

$$\Gamma = \frac{e^2}{ak_BT},$$

where $a \sim (3/4\pi n)^{1/3}$ is the inter-particle spacing, and T is the temperature. For $\Gamma \ll 1$, the plasma is weakly coupled and amenable to perturbative kinetic theory. However, in dense plasmas such as those found in inertial confinement fusion, or neutron star crusts, $\Gamma \gtrsim 1$, leading to non-perturbative correlations. Similarly, the relativistic parameter

$$\chi = \frac{eE\lambda_C}{m_e c^2}$$

where $\lambda_C = \hbar/m_e c$ is the Compton wavelength, characterizes quantum radiation dominance; $\chi \sim 1$ signals the onset of strong-field QED effects where classical trajectories are ill-defined.

In such regimes, particle-based models fail to incorporate spontaneous emission, vacuum polarization, and the back-reaction of quantum fluctuations, on the plasma medium. These effects require a reconstitution of plasma theory wherein fields not particles serve as the fundamental dynamical variables.

The field-theoretic formulation replaces the many-body Schrödinger or density-matrix dynamics with second-quantized fields. For instance, a fully quantized electron–photon plasma system is described by the Lagrangian density:

$$\mathcal{L} = \overline{\psi}\left(i\gamma^\mu D_\mu - m\right)\psi - \frac{1}{4}F^{\mu\nu}F_{\mu\nu} + \mathcal{L}_{\text{int}},$$

where $D_\mu = \partial_\mu + ieA_\mu$ is the Gauge covariant derivative, ψ is the Dirac field, and $F^{\mu\nu}$ is the electromagnetic field strength tensor. The interaction Lagrangian $\mathcal{L}_{\text{int}}$ can be modified to include Podolsky-type higher-derivative terms, or effective current–current interactions mediated by virtual exchange fields in strongly correlated media.

This framework naturally incorporates:

- *Vacuum fluctuations*, via loop corrections in Feynman diagrams,
- *Gauge invariance*, ensuring local charge conservation and compatibility with relativistic covariance, and
- *Statistical ensemble averaging*, through the path-integral formalism which encodes thermal and quantum effects on equal footing.

A central object is the generating functional:

$$Z[J] = \int \mathcal{D}\overline{\psi}\mathcal{D}\psi\mathcal{D}A_\mu e^{i\int d^4x(\mathcal{L}+J^\mu A_\mu)},$$

which encapsulates all correlation functions and response properties. Its functional derivatives with respect to the source J^μ yield n-point Green's functions that describe the plasma's linear and nonlinear susceptibilities, photon emission rates, and transport coefficients.

In strongly coupled plasmas, the single-particle spectral function $A(\omega, \mathbf{k})$, defined via the imaginary part of the retarded Green's function $G^R(\omega, \mathbf{k})$, ceases to exhibit a sharp quasi-particle peak. Mathematically,

$$A(\omega, \mathbf{k}) = -2\ \mathrm{Im}\ G^R(\omega, \mathbf{k}),$$

develops broad features when the mean free path becomes comparable to the de Broglie wavelength, or inter-particle separation. This implies the breakdown of the Fermi-liquid picture, and demands the treatment of collective modes such as plasmon, phonon, or polariton excitations as fundamental degrees of freedom [36].

Similarly, in ultra-relativistic plasmas encountered in laser–plasma interactions,or astrophysical jets, the field self-energy acquires non-trivial tensorial structure:

$$\Pi^{\mu\nu}(k) = \left(g^{\mu\nu} - \frac{k^\mu k^\nu}{k^2}\right)\Pi_T(k^2) + \frac{k^\mu k^\nu}{k^2}\Pi_L(k^2),$$

where $\Pi_{T,L}(k^2)$ encode the dynamical screening and Landau damping. These effects cannot be captured by a Boltzmann description without artificially introducing field-dependent collision integrals.

Moreover, stochastic radiation reaction, spin–orbit coupling, and quantum recoil necessitate resumption of infinite-order diagrams, a task that naturally falls within the remit of Schwinger–Dyson, or 2PI (two-particle irreducible) formalisms.

From a physical standpoint, the core advantage of field-theoretic approaches lies in their capacity to encode fluctuations and correlations beyond mean-field. In a particle picture, collective effects such as wakefield generation, bremsstrahlung, or virtual pair creation must be superimposed as external effects, or ad hoc modifications to the collision term. In contrast, in a QFT model, these phenomena arise naturally from interaction vertices and loop diagrams.

Additionally, the relativistic covariance of QFT allows one to consistently describe phenomena across different frames critical when considering plasmas moving at relativistic speeds or embedded in expanding spacetimes (e.g., early universe QED plasmas). The observer-dependence of particle content (as in the Unruh effect) underscores the inadequacy of defining plasma merely as a collection of particles; instead, one must analyze the vacuum structure and field modes appropriate to the observer's frame.

5.4.1 Coherent States of Plasma Modes

The coherent-state formalism and the quantized field description offer a powerful framework for modeling collective plasma excitations in a manner that bridges classical wave behaviour and the underlying quantum dynamics. This approach becomes especially critical in regimes where quantum coherence, phase correlations, and field quantization effects manifest strongly such as in degenerate plasmas, photon-permeated quantum cavities, or high-intensity laser-plasma interactions. The quantized field treatment not only enables a unified description of electromagnetic and plasma modes, but also offers a rigorous operator-based formulation for exploring nonlinear, dispersive, and dissipative phenomena from first principles [37].

The starting point for a field-theoretic description of plasma oscillations is the identification of the normal modes of the system in the linear regime. For a cold, collisionless, isotropic electron plasma embedded in a neutralizing background and coupled to the electromagnetic field, the linearized Maxwell–Vlasov or Maxwell–fluid system yields longitudinal (Langmuir) and transverse (electromagnetic) collective excitations. These modes are solutions to the linearized dielectric dispersion relation:

$$\epsilon_L(\omega, \mathbf{k}) = 1 - \frac{\omega_p^2}{\omega^2} \quad \text{(longitudinal)}, \quad \epsilon_T(\omega, \mathbf{k}) = 1 - \frac{\omega_p^2}{\omega^2 - c^2k^2} \quad \text{(transverse)}.$$

In the quantized description, these normal modes are elevated to field operators bosonic excitations that satisfy canonical commutation relations. The quantized field operator for longitudinal plasma modes (e.g., Langmuir waves) takes the generic form:

$$\hat{\phi}(\mathbf{r}, t) = \sum_{\mathbf{k}} \left[\phi_{\mathbf{k}}(\mathbf{r})\hat{a}_{\mathbf{k}}(t) + \phi_{\mathbf{k}}^*(\mathbf{r})\hat{a}_{\mathbf{k}}^\dagger(t) \right],$$

where $\hat{a}_{\mathbf{k}}$ and $\hat{a}_{\mathbf{k}}^\dagger$ are the annihilation and creation operators for the $\mathbf{k}$-th mode satisfying

$$\left[\hat{a}_{\mathbf{k}}, \hat{a}_{\mathbf{k}'}^\dagger \right] = \delta_{\mathbf{k},\mathbf{k}'}.$$

The mode functions $\phi_{\mathbf{k}}(\mathbf{r})$ are determined by the dielectric response and the spatial geometry of the plasma. In the transverse case, the electromagnetic vector potential $\widehat{\mathbf{A}}(\mathbf{r}, t)$ is quantized similarly under the Coulomb or Lorenz Gauge, and plasma effects modify the photon dispersion relation via a refractive index $n(\omega, \mathbf{k})$ derived from the dielectric tensor $\epsilon_{ij}(\omega, \mathbf{k})$.

A coherent state $|\alpha_{\mathbf{k}}\rangle$ for the mode $\mathbf{k}$ is an eigenstate of the annihilation operator:

$$\hat{a}_{\mathbf{k}}|\alpha_{\mathbf{k}}\rangle = \alpha_{\mathbf{k}}|\alpha_{\mathbf{k}}\rangle,$$

where $\alpha_{\mathbf{k}} \in \mathbf{C}$ encodes the complex amplitude of the excitation. In the many-mode case, the global coherent state is a tensor product:

$$\{\alpha_{\mathbf{k}}\}\rangle = \bigotimes_{\mathbf{k}} |\alpha_{\mathbf{k}}\rangle.$$

Physically, coherent states represent the quantum analogue of classical field configurations with well-defined amplitude and phase. They exhibit minimal uncertainty and Poissonian number statistics, rendering them ideal for describing macroscopic plasma waves such as driven Langmuir oscillations, electromagnetic pulses in plasmas, or wakefields in laser-plasma systems.

The expectation value of the field operator in a coherent state yields the classical wave:

$$\langle\alpha|\hat{\phi}(\mathbf{r},t)|\alpha\rangle = \sum_{\mathbf{k}} \left[\phi_{\mathbf{k}}(\mathbf{r})\alpha_{\mathbf{k}}(t) + \phi^*_{\mathbf{k}}(\mathbf{r})\alpha^*_{\mathbf{k}}(t)\right].$$

This classicalization mechanism allows the field-theoretic formalism to recover fluid, or kinetic plasma equations in appropriate limits, with additional insight into quantum fluctuations and coherence decay.

Plasmon modes collective quantized oscillations of electron density can be described via bosonic operators $\hat{b}_{\mathbf{k}}$, similar to phonons in solid-state systems. These obey a Plasmon Hamiltonian:

$$\hat{H}_{\text{pl}} = \sum_{\mathbf{k}} \hbar\omega_p(\mathbf{k})\left(\hat{b}^\dagger_{\mathbf{k}}\hat{b}_{\mathbf{k}} + \frac{1}{2}\right),$$

with dispersion $\omega_p(\mathbf{k})$ determined by the plasma dielectric function and any background magnetic field (e.g., Bernstein modes in magnetized plasma). Interaction with photons, electrons, or ions introduces coupling terms of the form:

$$\hat{H}_{\text{int}} = \sum_{\mathbf{k},\mathbf{q}} g_{\mathbf{k},\mathbf{q}}\left(\hat{b}^\dagger_{\mathbf{k}}\hat{a}_{\mathbf{q}} + \hat{b}_{\mathbf{k}}\hat{a}^\dagger_{\mathbf{q}}\right),$$

where $\hat{a}_{\mathbf{q}}$ are photon annihilation operators and $g_{\mathbf{k},\mathbf{q}}$ encapsulates mode coupling via current fluctuations or nonlinear polarization. Such Hamiltonians underlie effective models of photon–plasmon interactions in warm dense matter, quantum cavities, and laser-excited plasmas.

When coherent states are adopted for $\hat{b}_{\mathbf{k}}$, the plasmon field acquires a macroscopic amplitude, leading to phenomena such as parametric instabilities, wakefield formation, or radiation pressure—induced ponderomotive effects in the quantum regime. Further, decoherence of these statesvia damping, mode-mode interactions, or entanglement with other subsystems can be systematically treated using quantum master equations, Wigner functions, or 2PI Schwinger-Keldysh techniques [38].

The phase-space representation via Wigner functions complements the coherent state approach by allowing a quasi-probabilistic analysis of quantum field evolution. For the field operator $\hat{\phi}$, one defines the Wigner distribution:

$$W(\phi,\pi;\mathbf{r},t) = \frac{1}{\pi\hbar}\int d\xi e^{-2i\pi\xi/\hbar}\langle\phi + \frac{\xi}{2}|\hat{\rho}(\mathbf{r},t)\Big|\phi - \frac{\xi}{2}\Big\rangle,$$

where $\hat{\rho}$ is the density operator and π is the field momentum conjugate to ϕ. The Moyal bracket governs time evolution:

$$\partial_t W = \{H, W\}_{\text{MB}},$$

with the Moyal bracket reducing to the Poisson bracket in the classical limit. This formulation captures quantum dispersion, phase-space folding, and tunneling between coherent configurations, thus serving as a diagnostic for coherence loss and entropy growth in evolving plasma systems.

The use of coherent states in quantized field theories of plasma provides a systematic correspondence principle, classical plasma waves emerge as coherent superpositions of quantized excitations. This correspondence permits the rigorous derivation of nonlinear Schrödinger equations for envelope dynamics, Gross–Pitaevskii-like models for dense plasma systems, and mean-field approximations with quantum corrections. It also facilitates the inclusion of phase decoherence, fluctuation-driven instabilities, and spontaneous emission processes in otherwise classical treatments [39].

From the physical perspective, coherent states allow the modeling of laser pulses as quantum wavepackets, the analysis of stimulated and spontaneous scattering (e.g., Raman, Compton), and the interpretation of plasma turbulence as a dephasing of multimode coherent backgrounds. Importantly, they also serve as initial states in non-equilibrium quantum field evolution, particularly relevant for Schwinger–Keldysh treatments of quantum quench dynamics and plasma prethermalization in ultra-fast processes.

5.4.2 *Canonical and Path-Integral Quantization*

The quantization of plasma systems, particularly those incorporating electromagnetic interactions and collective field dynamics, mandates a coherent formulation that systematically promotes classical observables to quantum operators and accounts for their inherent statistical nature. The quantum plasma, composed of electrons, ions, and an electromagnetic field often subject to constraints like Gauge invariance, finite temperature, or external boundaries requires a formalism capable of capturing both the dynamical and statistical correlations.

A. *Canonical Second Quantization of Plasma Constituents*

The canonical quantization of charged matter fields begins with the promotion of classical fields describing particles typically scalar, or spinor fields to operator-valued distributions on a Hilbert space. For non-relativistic fermionic constituents (e.g., electrons), the starting point is the field-theoretic Hamiltonian of a many-body system minimally coupled to a Gauge field.

Let the field operator for electrons be denoted by $\hat{\psi}(\mathbf{r}, t)$, satisfying equal-time anti-commutation relations:

$$\left\{\hat{\psi}(\mathbf{r}, t), \hat{\psi}^{\dagger}\left(\mathbf{r}', t\right)\right\} = \delta^{(3)}\left(\mathbf{r} - \mathbf{r}'\right), \quad \left\{\hat{\psi}(\mathbf{r}, t), \hat{\psi}\left(\mathbf{r}', t\right)\right\} = 0.$$

The Hamiltonian density for the fermionic plasma in the presence of a Gauge field $A_{\mu} = (\phi, \mathbf{A})$ is given by:

$$\mathcal{H}_{\text{fermion}} = \hat{\psi}^{\dagger}(\mathbf{r})\left[\frac{1}{2m}(-i\hbar\nabla - q\mathbf{A}(\mathbf{r}))^2 + q\phi(\mathbf{r})\right]\hat{\psi}(\mathbf{r}).$$

This form captures both the kinetic energy of particles and their coupling to electromagnetic potentials. The canonical approach quantizes $\hat{\psi}(\mathbf{r}, t)$ and evolves the system using the Heisenberg equation of motion derived from the total Hamiltonian.

For completeness, a bosonic species such as collective ion modes, or phonons would instead obey canonical commutation relations:

$$\left[\hat{\phi}(\mathbf{r}, t), \hat{\phi}^{\dagger}\left(\mathbf{r}', t\right)\right] = \delta^{(3)}\left(\mathbf{r} - \mathbf{r}'\right), \quad \left[\hat{\phi}(\mathbf{r}, t), \hat{\phi}\left(\mathbf{r}', t\right)\right] = 0.$$

The extension to relativistic or multi-component plasmas requires a Lagrangian formalism, from which the Hamiltonian and canonical momenta are derived by Legendre transformation. This is particularly important for high-temperature or high-field regimes, where relativistic effects become non-negligible, necessitating the Dirac or Klein-Gordon field representation [40].

B. *Canonical Quantization of Gauge Fields*

The electromagnetic field, being a gauge field, demands careful treatment under quantization due to the redundant degrees of freedom introduced by gauge symmetry. The canonical quantization of the gauge field proceeds by identifying conjugate momenta and imposing equal-time commutation relations. Working in the Coulomb gauge ($\nabla\cdot\mathbf{A} = 0$), the transverse components of the vector potential $\mathbf{A}_{\perp}$ are quantized as:

$$\left[\hat{A}_i^{\perp}(\mathbf{r}, t), \hat{\Pi}_j^{\perp}\left(\mathbf{r}', t\right)\right] = i\hbar\delta_{ij}^{\perp}\left(\mathbf{r} - \mathbf{r}'\right), \quad \hat{\Pi}_i^{\perp} = -\epsilon_0\partial_t\hat{A}_i^{\perp},$$

where $\delta_{ij}^{\perp}\left(\mathbf{r} - \mathbf{r}'\right)$ is the transverse delta function ensuring the constraint consistency. The longitudinal modes are eliminated via the Gauss constraint, and only physical (transverse) photon polarizations remain.

This formalism leads to a photon field operator expansion:

$$\widehat{\mathbf{A}}(\mathbf{r}, t) = \sum_{\lambda=1,2}\int\frac{d^3k}{(2\pi)^3}\sqrt{\frac{\hbar}{2\varepsilon_0\omega_k}}\left[\hat{a}_{\mathbf{k},\lambda}\epsilon_{\mathbf{k},\lambda}e^{i\mathbf{k}\cdot\mathbf{r}} + \text{H.c.}\right],$$

where $\hat{a}_{\mathbf{k},\lambda}$ and $\hat{a}^{\dagger}_{\mathbf{k},\lambda}$ are annihilation and creation operators for the photon with wavevector $\mathbf{k}$ and polarization λ, and $\epsilon_{\mathbf{k},\lambda}$ are polarization vectors satisfying $\mathbf{k}\cdot\epsilon_{\mathbf{k},\lambda} = 0$.

C. *Path Integral Quantization: Matter and Gauge Fields*

Canonical quantization provides operator-based formulations that are particularly useful for systems with discrete symmetries or small Hilbert spaces. However, in dealing with strongly coupled, non-equilibrium, or topologically non-trivial quantum plasmas, the path integral approach offers broader utility [41].

The generating functional for a quantum plasma composed of fermions (electrons), bosons (ions or collective excitations), and gauge fields is written as:

$$Z[J,\overline{\eta},\eta] = \int \mathcal{D}[A_\mu]\mathcal{D}[\overline{\psi},\psi]\exp\left\{\frac{i}{\hbar}\int d^4x[\mathcal{L}[\overline{\psi},\psi,A_\mu] + J^\mu A_\mu + \overline{\eta}\psi + \overline{\psi}\eta]\right\},$$

where J^μ, $\overline{\eta}$, η are external sources for the gauge and matter fields, respectively. The action functional is constructed from the gauge-invariant Lagrangian:

$$\mathcal{L} = \overline{\psi}(i\hbar\gamma^\mu D_\mu - mc)\psi - \frac{1}{4\mu_0}F^{\mu\nu}F_{\mu\nu},$$

with $D_\mu = \partial_\mu + i\frac{q}{\hbar}A_\mu$, and $F_{\mu\nu} = \partial_\mu A_\nu - \partial_\nu A_\mu$.

Gauge invariance imposes the need for Gauge fixing, typically implemented via the Faddeev-Popov procedure, introducing ghost fields (in covariant Gauges) and ensuring a well-defined functional measure. For non-Abelian extensions or in non-perturbative regimes (e.g., confinement analogues in quantum plasma), the inclusion of topological sectors becomes essential.

D. *Statistical Ensembles and Quantum Kinetics*

To capture thermodynamic, or real-time evolution of quantum plasmas, the formalism must incorporate statistical averaging over quantum states. In the canonical picture, this corresponds to a density matrix evolution governed by the von Neumann equation, while in the path-integral formalism, one evaluates expectation values using the Keldysh-Schwinger contour for non-equilibrium dynamics, or the Matsubara formalism for finite-temperature equilibrium:

- Matsubara formalism (imaginary time): fields are periodic (bosons) or antiperiodic (fermions) in imaginary time, with discrete Matsubara frequencies ω_n leading to temperature-dependent propagators.
- Schwinger-Keldysh formalism (real time): fields are doubled and defined along a closed-time contour $C = C_+ \cup C_-$, enabling computation of causal (retarded/advanced) correlation functions and real-time spectral dynamics.

In this context, propagators acquire matrix structure in the contour space, and self-energies describe quantum damping, decoherence, and memory effects crucial to plasma equilibration.

E. *Physical Interpretation and Theoretical Synthesis*

The transition from single-particle quantum mechanics to quantum field theory marks the recognition that particles are excitations of underlying fields. In the plasma context, this recognition translates to a dynamical interplay between quantized matter fields (electrons, ions) and quantized gauge fields (photons), with second quantization enabling particle creation and annihilation processes essential for modeling ionization, recombination, and bremsstrahlung in high-energy plasmas.

The path-integral formulation emphasizes the statistical and topological aspects of the system, accommodating non-perturbative phenomena such as quantum turbulence, plasma instabilities, and field-induced transitions. These methods collectively build the theoretical backbone for quantum kinetic theory, effective field theories of plasma modes, and real-time simulation frameworks, forming a consistent description of quantum plasmas in both equilibrium and dynamical scenarios.

5.4.3 Derivation of QED Plasmas via Functional Integrals

The theoretical description of quantum electrodynamic(QED) plasmas within the path-integral framework offers a formally elegant and computationally versatile avenue for exploring collective behaviours of relativistic, quantum-degenerate, and strongly interacting systems of charged particles coupled to gauge fields [42].

We begin by constructing the generating functional for a QED system that comprises relativistic electrons and positrons interacting with quantized electromagnetic fields. In the absence of external sources, the partition function at finite temperature $T = \beta^{-1}$ and chemical potential μ, in the grand canonical ensemble, is written as a Euclidean functional integral over fermionic spinor fields $\psi(x)$, $\overline{\psi}(x)$ and gauge fields $A_\mu(x)$:

$$\mathcal{Z}_{\text{QED}} = \int \mathcal{D}\overline{\psi}\mathcal{D}\psi\mathcal{D}A_\mu \, \exp\left(-S_E\left[\overline{\psi}, \psi, A_\mu\right]\right),$$

where the Euclidean action S_E reads:

$$S_E\left[\bar{\psi}, \psi, A_\mu\right] = \int_0^\beta d\tau \int d^3x \Big\{ \bar{\psi}\left[\gamma_0(\partial_\tau - \mu) + i\gamma^i(\partial_i + ieA_i)\right.$$
$$\left. + m\right]\psi + \frac{1}{4}F_{\mu\nu}F^{\mu\nu} + \frac{1}{2\xi}\left(\partial^\mu A_\mu\right)^2 \Big\}.$$

Here, γ^μ are the Dirac gamma matrices in Euclidean space, $F_{\mu\nu} = \partial_\mu A_\nu - \partial_\nu A_\mu$ is the field strength tensor, and ξ is the gauge-fixing parameter (e.g., $\xi = 1$ for Feynman Gauge). The temporal coordinate τ is compactified on a circle of circumference β, implying periodic boundary conditions for bosonic fields and anti-periodic ones for fermions, in accordance with the Matsubara formalism. The chemical potential enters naturally as a shift in the temporal derivative, ensuring grand canonical ensemble consistency [43].

The next step involves integrating out the fermionic fields to obtain an effective action for the gauge field. Because the action is quadratic in ψ and $\overline{\psi}$, the Grassmann integration yields a fermion determinant:

$$\mathcal{Z}_{\text{QED}} = \int \mathcal{D}A_\mu \, \det\big[\gamma_0(\partial_\tau - \mu) + i\gamma^i(\partial_i + ieA_i) + m\big]$$
$$\exp\left(- \int d^4x \left[\frac{1}{4} F_{\mu\nu}F^{\mu\nu} + \frac{1}{2\xi}\left(\partial^\mu A_\mu\right)^2 \right] \right).$$

This determinant encodes the quantum fluctuations of the fermionic sector and introduces nonlocal, nonlinear corrections to the gauge sector, accounting for phenomena such as vacuum polarization, screening, and pair production. The determinant is typically expanded in powers of A_μ when interactions are weak, or resummed non-perturbatively using large-N, or Schwinger-Dyson techniques in strongly coupled regimes. This gives rise to an effective Gauge field action:

$$S_{\text{eff}}\big[A_\mu\big] = \int d^4x \left[\frac{1}{4} F_{\mu\nu}F^{\mu\nu} + \frac{1}{2\xi}\left(\partial^\mu A_\mu\right)^2 \right] - \ln \det\big[\not{D}_E + m - \mu\gamma_0\big],$$

where$\not{D}_E = \gamma_\mu(\partial^\mu + ieA^\mu)$ is the Euclidean covariant Dirac operator.

A central object of interest in the QED plasma context is the polarization tensor $\Pi_{\mu\nu}(x, y)$, which arises from the functional differentiation of the fermionic determinant:

$$\Pi_{\mu\nu}(x, y) = - \left. \frac{\delta^2 \ln \det\big[\not{D}_E + m - \mu\gamma_0\big]}{\delta A_\mu(x)\delta A_\nu(y)} \right|_{A=0}.$$

In momentum space and at finite temperature and density, this tensor depends explicitly on the Matsubara frequencies and chemical potential, encoding both vacuum and medium modifications to the photon propagator. Physically, the real part of $\Pi_{\mu\nu}(k)$ governs dispersion relations of plasma modes (e.g., longitudinal and transverse excitations), while its imaginary part describes damping, Landau dissipation, and pair production.

The one-loop expression in momentum space reads:

$$\Pi_{\mu\nu}(k) = -e^2 \int \frac{d^4p}{(2\pi)^4} \text{Tr}\big[\gamma_\mu S(p)\gamma_\nu S(p + k)\big],$$

where $S(p)$ is the fermionic propagator with temperature and chemical potential dependence. The analytic continuation of Matsubara frequencies to real time permits the derivation of dielectric functions, refractive indices, and response tensors characteristic of QED plasmas.

It is imperative to ensure that the functional formalism respects local Gauge invariance. This requirement manifests through Ward-Takahashi identities, which constrain the structure of the polarization tensor and ensure the transversality of the photon self-energy:

$$k^{\mu}\Pi_{\mu\nu}(k) = 0,$$

reflecting current conservation. These identities also govern the consistency of vertex corrections and guarantee the renormalizability of the theory.Gauge fixing, as introduced via the term $\frac{1}{2\xi}\left(\partial_{\mu}A^{\mu}\right)^{2}$, breaks Gauge invariance explicitly but allows for well-defined propagators. However, the full generating functional remains gauge-independent when ghost contributions are accounted for (if non-Abelian generalizations are considered), or when appropriate BRST symmetry is preserved.

The presence of a finite chemical potential μ introduces asymmetries between particles and antiparticles, which is central in astrophysical and laboratory plasma settings. The corresponding fermionic propagator is modified as:

$$S(p_0, \vec{p}) = \frac{1}{i\omega_n + \mu - \vec{\gamma}\cdot\vec{p} - m},$$

where $\omega_n = (2n+1)\pi/\beta$ are fermionic Matsubara frequencies. The effective photon propagator in such a medium is no longer Lorentz invariant, and dielectric tensors derived from $\Pi_{\mu\nu}$ exhibit anisotropic behaviour and complex permittivities. This framework allows one to derive plasma dispersion relations, damping rates, and collective excitation spectra, all directly from the underlying microscopic QED action.

While the above formalism applies strictly to systems in thermal equilibrium, many realistic QED plasma systems such as those in laser–plasma interactions, astrophysical transients, or heavy-ion collisions are inherently out of equilibrium. The Schwinger-Keldysh (closed-time path) formalism extends the functional integral into the complex-time plane, providing real-time correlation functions, causal propagators, and quantum kinetic equations.

The generating functional in the Schwinger-Keldysh framework becomes:

$$\mathcal{Z}_{\text{CTP}} = \int \mathcal{D}\overline{\psi}_{\pm}\mathcal{D}\psi_{\pm}\mathcal{D}A^{\mu}_{\pm}\exp\left(iS_{\text{QED}}\left[\overline{\psi}_{+}, \psi_{+}, A^{\mu}_{+}\right] - iS_{\text{QED}}\left[\overline{\psi}_{-}, \psi_{-}, A^{\mu}_{-}\right]\right),$$

where $(+)$ and $(-)$ denote the forward and backward time branches, respectively. This formulation is indispensable for studying decoherence, quantum noise, photon production, and entropy generation in dynamical quantum plasma environments [44].

The derivation of QED plasma dynamics through functional integrals not only furnishes a nonperturbative definition of quantum-statistical electrodynamics, but also enables systematic derivations of macroscopic observables from thermodynamic potentials to collective excitations. This framework underlies the computation of photon self-energies, electron screening lengths, Debye masses, and photon emission rates, which are fundamental in diverse contexts such as early-universe cosmology, neutron staratmospheres, relativistic laser–plasma experiments, and strongly magnetized plasmas.

In particular, medium-modified photon propagators inform the behaviour of soft electromagnetic modes, governing phenomena like photon trapping, anomalous dispersion, and radiation transfer in hot and dense media. Additionally, the effective action approach provides a foundation for developing resummed theories of transport, including generalized Boltzmann and Kadanoff-Baym equations.

5.5 Comparative Analysis

In modeling quantum plasma systems, the choice of theoretical framework hydrodynamic, kinetic, or field-theoretic depends fundamentally on the interplay of spatiotemporal scales, statistical regimes, and the degree of collective versus microscopic resolution required.

5.5.1 *Hydrodynamic Models*

Hydrodynamic models of quantum plasmas, also referred to as quantum fluid models (QFMs), emerge from moments of the quantum kinetic equations, typically the Wigner–Moyal equation or quantum Vlasov–Poisson–Maxwell system. They are most appropriate when the system exhibits strong collective behaviour, weak mean-free-paths, and near-equilibrium local thermodynamic variables. The zeroth-, first-, and second-order moments of the distribution function yield the number density $n(\mathbf{x}, t)$, current density $\mathbf{j}(\mathbf{x}, t)$, and energy density $\mathcal{E}(\mathbf{x}, t)$, which evolve according to continuity, momentum, and energy conservation equations respectively.

However, hydrodynamic models suffer from the necessity of closure assumptions, which truncate the hierarchy of moments and may fail to capture kinetic resonances, non-Maxwellian features, and anisotropies. Therefore, they are best employed in low-temperature, high-density degenerate plasmas such as white dwarfs, laser-compressed solids, and metallic nanoplasmas, where fast collective motions dominate and individual particle kinetics are subdominant.

5.5.2 *Kinetic Models*

Quantum kinetic theories retain the phase-space structure of the plasma and allow for a detailed resolution of non-equilibrium, dispersive, and collisional processes. The quantum kinetic equation takes the form of the Wigner equation,

$$\frac{\partial f_W}{\partial t} + \frac{\mathbf{p}}{m} \cdot \nabla_{\mathbf{x}} f_W - \mathcal{Q}\big[V, f_W\big] = C\big[f_W\big],$$

where $f_W(\mathbf{x}, \mathbf{p}, t)$ is the Wigner function, $\mathcal{Q}\big[V, f_W\big]$ encodes quantum nonlocality via the Moyal bracket expansion,

$$\mathcal{Q}\big[V, f_W\big] = \sum_{n=1}^{\infty} \frac{(-1)^n (\hbar/2)^{2n}}{(2n+1)!} \nabla_x^{2n+1} V \cdot \nabla_p^{2n+1} f_W,$$

and $C\big[f_W\big]$ is a quantum collision integral, often approximated by relaxation-time, or Lindblad-type operators in open systems.

Kinetic models are indispensable in capturing Landau damping, velocity-space diffusion, Debye shielding, and kinetic instabilities. In the quantum regime, phenomena such as quantum recoil, zitterbewegung, and quantum tunneling in momentum space can only be accurately represented via kinetic theory. Moreover, quantum kinetic models can describe both coherent effects (such as superposition and interference) and incoherent processes (such as quantum decoherence and entanglement erosion due to environment-induced scattering).

From a parametric standpoint, kinetic models are necessary when Kn $\sim \mathcal{O}(1)$, and when the mean de Broglie wavelength $\lambda_{\mathrm{dB}} \sim L$, rendering the quantum statistical fluctuations and phase coherence across phase-space cells dynamically significant. These models are best suited for semi-relativistic or non-equilibrium plasmas in laser-plasma interactions, inertial confinement fusion cores during ignition phases, and high-intensity beam-plasma systems.

5.5.3 Field-Theoretic Models

Field-theoretic descriptions form the most general and fundamental level of modeling, in which plasma constituents and Gauge fields are treated as quantum fields. The standard formulation involves second quantization in the path-integral, or operator formalism, with the plasma described by a Lagrangian density $\mathcal{L}\big[\psi, A_\mu\big]$ that incorporates spinor fields ψ for charged particles and Gauge fields A_μ for photons. The dynamics follow from the action principle:

$$S\big[\psi, A_\mu\big] = \int d^4x \left[\overline{\psi}\left(i\gamma^\mu D_\mu - m\right)\psi - \frac{1}{4} F_{\mu\nu} F^{\mu\nu} + \mathcal{L}_{\mathrm{int}} \right],$$

where $D_\mu = \partial_\mu + iqA_\mu$ is the Gauge-covariant derivative, and $\mathcal{L}_{\mathrm{int}}$ includes terms such as Yukawa couplings, anomalous magnetic moments, or higher-derivative Podolsky-type corrections.

Field-theoretic modeling becomes essential in several advanced plasma regimes where conventional approaches fail to capture the underlying complexity. In the strong coupling regime, where the plasma parameter is high, the system exhibits non-perturbative correlations and collective quantum coherence. These features necessitate sophisticated tools such as Dyson–Schwinger equations, two-particle-irreducible (2PI) effective actions, and lattice gauge theory techniques to accurately model the plasma behavior. In ultrarelativistic and quantum radiative regimes, such as those encountered in pulsar magnetospheres, gamma-ray bursts, or QED cascades triggered by ultra-intense laser-plasma interactions, radiative corrections become significant. Here, quantum electrodynamics (QED) methods are required, including Feynman diagrammatic expansions, loop corrections, and the renormalization group evolution of effective couplings. Further, for plasmas involving non-Abelian charges, such as quark–gluon plasmas or axion–plasma systems, the governing gauge symmetry transitions from the Abelian U(1) to non-Abelian groups. This shift dramatically influences the vacuum structure, dispersion relations, and can lead to anomaly-induced transport phenomena. Thus, field-theoretic frameworks are indispensable for exploring these complex plasma systems with strong coupling, relativistic effects, and rich gauge symmetries.

Field theory also accommodates quantum entanglement, decoherence, and photon–photon scattering through effective action techniques. The Schwinger–Keldysh closed-time-path (CTP) formalism is particularly well suited for non-equilibrium quantum plasmas, as it permits the computation of real-time correlators, spectral functions, and fluctuation–dissipation relations beyond the linear regime [45].

Ultimately, the selection among hydrodynamic, kinetic, or field-theoretic descriptions must be dictated by the physical scales of interest, the level of resolution required, and the nature of the observables under investigation. In contemporary research, hybrid models and multiscale matching (e.g., kinetic–hydrodynamic coupling or mean-field–quantum-loop matching) are becoming increasingly relevant, especially in complex systems such as quantum radiative shocks, turbulent quantum plasmas, and warm dense matter under strong fields.

References

1. Srokowski, T. (2007). Non-Markovian Lévy diffusion in nonhomogeneous media. *Physical Review E—Statistical, Nonlinear, and Soft Matter Physics, 75*(5), 051105.
2. Koshelev, A. E., Bulaevskii, L. N., & Maley, M. P. (2000). Josephson coupling, phase correlations, and Josephson plasma resonance in vortex liquid phase. *Physical Review B, 62*(21), 14403.
3. Hochstuhl, D. (2008). *Nonequilibrium Green functions approach to ionization processes* (Doctoral dissertation, Inst. für Theoretische Physik und Astrophysik der Christian-Albrechts-Univ.).
4. Balzer, K. (2007). *Nonequilibrium Green's function approach to artificial atoms* (Doctoral dissertation, Inst. für Theoretische Physik und Astrophysik der Christian-Albrechts-Univ.).
5. Maresca, E. (2025). Noncommutative geometry and space time: A historical reconstruction. In *Journal of Physics: Conference Series* (Vol. 2948, No. 1, p. 012011). IOP Publishing.
6. Ota, Y., Machida, M., & Koyama, T. (2011). Macroscopic quantum tunneling in multigap superconducting Josephson junctions: Enhancement of escape rate via quantum fluctuations of the Josephson-Leggett mode. *Physical Review B—Condensed Matter and Materials Physics, 83*(6), 060503.
7. Thoma, M. H. (2009). Colloquium: Field theoretic description of ultrarelativistic electron–positron plasmas. *Reviews of Modern Physics, 81*(3), 959–968.
8. Matsumoto, H. (1985). Coherent nonlinear effects on electromagnetic wave-particle interactions. *Space Science Reviews, 42*(3), 429–448.
9. Liu, D. (2021). All-optical modulation of linear and nonlinear emissions in plasmonic nanogap systems.
10. Lee, J., Jeon, D. J., & Yeo, J. S. (2021). Quantum plasmonics: energy transport through plasmonic gap. *Advanced Materials, 33*(47), 2006606.
11. Villard, L., Angelino, P., Bottino, A., Allfrey, S. J., Hatzky, R., Idomura, Y., Sauter, O., & Tran, T. M. (2004). First principles based simulations of instabilities and turbulence. *Plasma physics and controlled fusion, 46*(12B), B51.
12. Asenjo, F. A., Zamanian, J., Marklund, M., Brodin, G., & Johansson, P. (2012). Semi-relativistic effects in spin-1/2 quantum plasmas. *New Journal of Physics, 14*(7), Article 073042.
13. Zhu, J., Ji, P., & Lu, N. (2009). Quantum correction to Landau damping of electron plasma waves. *Physics of Plasmas, 16*(3).
14. Saha, A., Pradhan, B., & Banerjee, S. (2020). Bifurcation analysis of quantum ion-acoustic kink, anti-kink and periodic waves of the Burgers equation in a dense quantum plasma. *The European Physical Journal Plus, 135*(2), 216.
15. Sharma, K., & Deka, U. (2024). Effect of spin quantum force on lower and upper hybrid waves' instability under the influence of an electron beam in a magnetized semiconductor plasma. *The European Physical Journal B, 97*(9), 135.
16. Benedict, L. X., Surh, M. P., Castor, J. I., Khairallah, S. A., Whitley, H. D., Richards, D. F., Glosli, J. N., Murillo, M. S., Scullard, C. R., Grabowski, P. E., & Graziani, F. R. (2012). Molecular dynamics simulations and generalized Lenard-Balescu calculations of electron-ion temperature equilibration in plasmas. *Physical Review E—Statistical, Nonlinear, and Soft Matter Physics, 86*(4), 046406.
17. Seoanez, C., Weick, G., Jalabert, R. A., & Weinmann, D. (2007). Friction of the surface plasmon by high-energy particle-hole pairs: Are memory effects important? *The European Physical Journal D, 44*, 351–358.
18. Most, E. R., Noronha, J., & Philippov, A. A. (2022). Modelling general-relativistic plasmas with collisionless moments and dissipative two-fluid magnetohydrodynamics. *Monthly Notices of the Royal Astronomical Society, 514*(4), 4989–5003.
19. Brodin, G., Ekman, R., & Zamanian, J. (2016). Quantum kinetic theories in degenerate plasmas. *Plasma Physics and Controlled Fusion, 59*(1), Article 014043.
20. Umstadter, D. (2001). Review of physics and applications of relativistic plasmas driven by ultra-intense lasers. *Physics of Plasmas, 8*(5), 1774–1785.

21. Trukhanova, M. I. (2015). Spin current evolution in the separated spin-up and spin-down quantum hydrodynamics. *Physics Letters A, 379*(42), 2777–2782.
22. Morandi, O., Zamanian, J., Manfredi, G., & Hervieux, P. A. (2014). Quantum-relativistic hydrodynamic model for a spin-polarized electron gas interacting with light. *Physical Review E, 90*(1), Article 013103.
23. Rybalka, D. O., Gorbar, E. V., & Shovkovy, I. A. (2019). Hydrodynamic modes in a magnetized chiral plasma with vorticity. *Physical Review D, 99*(1), Article 016017.
24. Nalewajski, R. F., & Nalewajski, R. F. (2012). *Density functional theory* (pp. 255–368). Springer.
25. Dharma-Wardana, M. W. C., Stanek, L. J., & Murillo, M. S. (2022). Yukawa-Friedel-tail pair potentials for warm dense matter applications. *Physical Review E, 106*(6), Article 065208.
26. Zhang, S., Wang, H., Kang, W., Zhang, P., & He, X. T. (2016). Extended application of Kohn-Sham first-principles molecular dynamics method with plane wave approximation at high energy—From cold materials to hot dense plasmas. *Physics of Plasmas, 23*(4).
27. Bonitz, M., Filinov, A., Abraham, J. W., & Loffhagen, D. (2018). Extending first principle plasma-surface simulations to experimentally relevant scales. *Plasma Sources Science and Technology, 27*(6), Article 064005.
28. Danel, J. F., Kazandjian, L., & Zérah, G. (2008). Equation of state of a dense boron plasma by Thomas–Fermi–Dirac–Weizsäcker molecular dynamics. *Physics of Plasmas, 15*(7).
29. Rahim, Z., Adnan, M., Qamar, A., & Saha, A. (2018). Nonplanar dust-acoustic waves and chaotic motions in Thomas Fermi dusty plasmas. *Physics of Plasmas, 25*(8).
30. Aleksandrov, N. L., Kindysheva, S. V., Nudnova, M. M., & Starikovskiy, A. Y. (2010). Mechanism of ultra-fast heating in a non-equilibrium weakly ionized air discharge plasma in high electric fields. *Journal of Physics D: Applied Physics, 43*(25), Article 255201.
31. Sarcinella, F., Fabiano, E., Constantin, L. A., & Della Sala, F. (2021). Nonlocal kinetic energy functionals in real space using a Yukawa-potential kernel: Properties, linear response, and model functionals. *Physical Review B, 103*(15), Article 155127.
32. Kwong, N. H., & Bonitz, M. (2000). Real-time Kadanoff-Baym approach to plasma oscillations in a correlated electron gas. *Physical review letters, 84*(8), 1768.
33. Deuchler, L. (2022). *TDDFT study of the charge and energy transfer of ions interacting with solid surfaces* (Doctoral dissertation).
34. Hollebon, P. (2019). *Density functional theory studies of solid density plasmas* (Doctoral dissertation, University of Oxford).
35. Brown, L. S., & Yaffe, L. G. (2001). Effective field theory for highly ionized plasmas. *Physics Reports, 340*(1–2), 1–164.
36. Pines, D., & Schrieffer, J. R. (1962). Approach to equilibrium of electrons, plasmons, and phonons in quantum and classical plasmas. *Physical Review, 125*(3), 804.
37. Perepelkin, E. E., Sadovnikov, B. I., Inozemtseva, N. G., & Suchkov, D. A. (2019). First principles-based applications of the Vlasov equation to dissipative systems. *Plasma Research Express, 1*(2), Article 025003.
38. Ivanov, Y. B., & Voskresensky, D. N. (2014). Non-equilibrium 2PI potential and its possible application to evaluation of bulk viscosity. *Physics of Atomic Nuclei, 77*, 1115–1129.
39. Kuzelev, M. V., & Rukhadze, A. A. (2008). Spontaneous and stimulated emission induced by an electron, electron bunch, and electron beam in a plasma. *Physics-Uspekhi, 51*(10), 989.
40. Varró, S. (2014). New exact solutions of the Dirac and Klein-Gordon equations of a charged particle propagating in a strong laser field in an underdense plasma. *Nuclear Instruments and Methods in Physics Research Section A: Accelerators, Spectrometers, Detectors and Associated Equipment, 740*, 280–283.
41. Shevkunov, S. V. (2002). A dense hydrogen plasma modeled by the path integral-Monte Carlo method. *Journal of Experimental and Theoretical Physics, 94*, 943–965.
42. Colonna, G., Pintassilgo, C. D., Pegoraro, F., Cristofolini, A., Popoli, A., Neretti, G., Gicquel, G., Duigou, O., Bieber, T., Hassouni, K., & Laguardia, L. (2021). Theoretical and experimental aspects of non-equilibrium plasmas in different regimes: Fundamentals and selected applications. *The European Physical Journal D, 75*(6), 183.

43. Engelhardt, M. (1995). Finite-sized plasmas. *Physical Review D, 52*(2), 1267.
44. Thimsen, E. (2021). Entropy production and chemical reactions in nonequilibrium plasma. *AIChE Journal, 67*(9), Article e17291.
45. Kanekar, A., Schekochihin, A. A., Dorland, W., & Loureiro, N. F. (2015). Fluctuation-dissipation relations for a plasma-kinetic Langevin equation. *Journal of Plasma Physics, 81*(1), 305810104.

Chapter 6
Magnetized Quantum Plasma

The confluence of magnetic fields and quantum plasma phenomena represents a frontier regime in plasma physics that unveils novel collective behaviors, transport dynamics, and field-particle interactions inaccessible through classical, or weak-field approximations. Magnetized quantum plasma (MQP), defined as a plasma system in which both quantum mechanical effects and magnetic field influences are non-negligible, becomes essential for understanding matter under extreme densities, low temperatures, or high field strengths regimes ubiquitously encountered in compact astrophysical objects and modern high-energy laboratory experiments.

From a theoretical standpoint, MQP presents a unique domain where canonical fluid descriptions, kinetic theories, and field-theoretic approaches must be rigorously generalized to include spin effects, quantum statistics, non-local correlations, and quantized Landau-level dynamics under intense magnetic backgrounds. Unlike unmagnetized quantum plasmas, the presence of a magnetic field introduces anisotropy in transport, modifies dispersion relations, alters degeneracy pressure via density-of-states quantization, and mediates novel instabilities and collective excitations whose nature is profoundly quantum.

In astrophysical environments, MQP governs the thermodynamic and dynamic behaviour of matter in white dwarfs, neutron stars, and magnetars. For instance, in the crustal layers of neutron stars, electrons form a degenerate, magnetized Fermi gas that experiences Landau quantization, while ions arrange into a Coulomb lattice. Magnetic field strengths ranging from 10^9 to 10^{15} G render classical magnetohydrodynamics (MHD) inadequate, necessitating the inclusion of quantum spin–magnetic coupling, relativistic degeneracy, and quantized conductivity tensors. Observational consequences such as polarization-dependent radiative transport, anisotropic cooling, and quantum-modified Alfvénoid oscillations in magnetars are direct signatures of MQP phenomena [1].

In laboratory settings, particularly within ultra-intense laser–solid interactions, magnetized inertial confinement fusion (ICF) targets, and quantum Hall devices, MQP-like states are increasingly relevant. High-density plasmas formed at solid

P. Kumar, *Quantum Plasma*, https://doi.org/10.1007/978-981-95-0481-7_6

densities exhibit quantum degeneracy, and external magnetic fields in the multi-tesla range, or self-generated fields via Biermann battery effects, impose quantization constraints on particle dynamics. Moreover, recent advances in laser wakefield acceleration, Magnetized plasma mirrors, and quantum spintronics involve plasma states where magnetic and quantum effects are intertwined. These applications stimulate the development of accurate MQP models that transcend traditional MHD and cold plasma approximations.

From a conceptual perspective, MQP serves as a theoretical laboratory for exploring fundamental quantum-field–plasma interactions. It offers insight into the interplay between spin, gauge fields, and collective excitations, akin to phenomena in condensed matter (e.g., topological insulators), yet at plasma densities and field intensities. In this context, MQP bridges quantum field theory in curved spacetime, transport in anisotropic Fermi systems, and nonequilibrium many-body dynamics. For instance, the emergence of chiral magnetic effects, spin–vorticity coupling, and quantum reconnection layers represents deep interconnections between quantum anomalies and plasma instabilities in magnetized domains [2, 3].

6.1 Overview of Regimes

The coexistence of magnetic fields, quantum effects, and collective plasma interactions gives rise to a spectrum of physical regimes that deviate markedly from both classical plasma dynamics and conventional quantum many-body systems. The classification of these regimes requires a careful delineation of the dominant energy scales, relevant quantum-statistical parameters, and magnetization thresholds that determine the plasma's macroscopic and microscopic behaviour.

The dynamics of MQP systems are determined by the interrelation among several key parameters. Let n denote the number density, T the temperature, B the magnetic field strength, $\hbar$ the reduced Planck's constant, e the elementary charge, m the particle mass (typically electron), and k_B the Boltzmann constant. The following characteristic scales provide a diagnostic foundation:

i. Thermal de Broglie wavelength

$$\lambda_{\mathrm{dB}} = \frac{h}{\sqrt{2\pi m k_B T}}$$

Quantum effects become non-negligible when $\lambda_{\mathrm{dB}} \gtrsim n^{-1/3}$, which marks the breakdown of classical Boltzmann statistics and the emergence of Fermi–Dirac or Bose–Einstein behaviour.

ii. Cyclotron frequency and magnetic quantization energy

$$\omega_c = \frac{eB}{mc}, \quad \hbar\omega_c = \textit{Landau level spacing}$$

When $\hbar\omega_c \gtrsim K_BT$ *or* $\hbar\omega_c \gtrsim E_F$ (Fermi energy), Landau quantization of orbital motion significantly alters the kinetic and thermodynamic properties of the plasma.

iii. Quantum plasma parameter (degeneracy factor)

$$\chi_Q = \frac{\lambda_{\text{dB}}^3}{n^{-1}} = n\lambda_{\text{dB}}^3$$

This dimensionless parameter quantifies the degree of wavefunction overlap. For $\chi_Q \gg 1$, the plasma is strongly quantum-degenerate.

iv. Magnetization parameter

$$\beta_m = \frac{\omega_c}{\omega_p}, \quad \text{where } \omega_p = \sqrt{\frac{4\pi ne^2}{m}}$$

Here, ω_p is the electron plasma frequency. The regime $\beta_m \gg 1$ corresponds to strongly magnetized plasmas where particle motion is tightly constrained along field lines.

v. Landau quantization condition

The number of occupied Landau levels in a degenerate plasma at zero temperature is given by:

$$N_L = \frac{E_F^2 - m^2c^2}{2\hbar eBc}$$

Systems with $N_L \lesssim 1$ are in the extreme quantum limit, where only the lowest Landau level is populated, leading to effectively one-dimensional motion.

6.1.1 Regime Map and Classification

Let us now identify the primary physical regimes that arise from different asymptotic combinations of the above parameters. The interplay of field strength, density, and temperature naturally defines the following categories:

a. Classical Magnetized plasma (CMP)

This regime arises when $\lambda_{\text{dB}} \ll n^{-1/3}$ and $\hbar\omega_c \ll K_BT$, implying both quantum statistical and Landau quantization effects are negligible. However, if $\beta_m \gtrsim 1$, magnetization remains dynamically important. Examples include solar coronal loops and magnetized fusion plasmas at moderate densities and high temperatures.

b. Unmagnetized Quantum Plasma (UQP)

This domain is characterized by strong quantum degeneracy, i.e., $\chi_Q \gg 1$, but weak magnetization $\beta_m \ll 1$ and negligible Landau quantization. It describes systems such as metallic conduction electrons, dense laser-compressed targets, and certain stages of inertial confinement. In this regime, Fermi pressure, Bohm potential, and exchange effects dominate, but magnetic field lines do not guide particle dynamics.

c. Strongly Magnetized Quantum Plasma (SMQP)

This is the principal regime of interest in MQP theory. It occurs when:
$\lambda_{\mathrm{dB}} \gtrsim n^{-1/3}$(quantum degeneracy),
$\hbar\omega_c \gtrsim k_B T$ (Landau quantization),
$\beta_m \gg 1$ (strong magnetization).
In this regime, particles undergo discrete cyclotron motion, transport becomes highly anisotropic, and novel magnetization effects emerge due to spin and orbital quantization. Matter in this category is typified by neutron star crusts, magnetized white dwarfs, and ultra-dense laboratory plasmas under intense fields.

The electron pressure tensor becomes anisotropic:

$$P_{ij} = P_\perp\left(\delta_{ij} - b_i b_j\right) + P_\parallel b_i b_j,$$

where $b_i = B_i \Big/ \left|\vec{B}\right|$, and $P_\perp \neq P_\parallel$ due to Landau quantization. Thermodynamic response functions such as heat capacity and magnetic susceptibility also become tensorial and oscillatory (de Haas–van Alphen effect), especially in low-temperature and high-field limits.

d. Quantum Magneto-Hydrodynamic Regime (QMHD)

In this framework, plasma behaviour is governed by collective fluid dynamics that are refined through the inclusion of quantum potentials and magnetic field interactions. The foundational equations comprise the quantum Euler equation, which incorporates the Bohm potential to account for quantum pressure effects; a generalized form of Ohm's law that includes spin-related contributions, additional magnetization current terms that emerge from spin–orbit coupling phenomena; and a modified version of Faraday's law, adjusted to integrate quantum corrections. Together, these equations offer a comprehensive description of plasma dynamics in regimes where quantum effects cannot be ignored.

The quantum magneto-acoustic waves in this regime display dispersion of the form:

$$\omega^2 = k^2 v_A^2 + k^2 v_Q^2 + k^4 \frac{\hbar^2}{4m^2}$$

where v_A is the Alfvén speed, and v_Q encodes quantum pressure contributions.

e. Ultra-Relativistic Magnetized Degenerate Plasma

When $E_F \gtrsim mc^2$, relativistic effects must be included. The energy levels become:

$$E_n(p_z) = \sqrt{p_z^2 c^2 + m^2 c^4 + 2n\hbar\omega_c mc^2}$$

The equation of state is modified, and pressure anisotropy becomes more pronounced. In magnetars, such plasmas determine the star's stability, crustal fracture mechanics, and neutrino emissivity.

In practical systems, especially in laboratory experiments or astrophysical shocks, plasmas may traverse different regimes during their evolution. For instance, laser-solid interactions may begin in a UQP domain but transition into SMQP as compression and field amplification proceed. Similarly, in post-merger neutron star crusts, non-equilibrium cooling dynamics may shift the plasma from ultra-relativistic to partially degenerate SMQP over millisecond timescales.

The regime boundaries are not sharp, but depend on competing time scales such as cyclotron period, collision time, and quantum coherence length. This necessitates multiscale modeling techniques and hybrid kinetic–fluid formalisms to adequately capture such transitions.

The identification of regimes where quantum, magnetic, and collective plasma effects coexist is essential for the accurate modeling and experimental interpretation of high-density, high-field plasma systems. The domain of MQP is thus not a singular regime but a multidimensional landscape of overlapping asymptotics. The precise classification depends not only on equilibrium parameters but also on the dynamical and geometrical structure of the system. Subsequent sections of this chapter will develop the governing equations, dispersion relations, and field-theoretic formulations appropriate to each of these MQP domains [4].

6.1.2 Quantum Versus Classical Magnetization Responses (Pauli, Landau, Diamagnetic Effects)

In the context of magnetized quantum plasma, the response of a many-body electron system to an applied magnetic field manifests through rich and physically distinct magnetization mechanisms. These responses are categorized broadly into classical and quantum contributions, with the quantum responses further subdivided into Pauli paramagnetism, Landau diamagnetism, and magnetization due to orbital quantization

Classical Diamagnetism: Larmor Precession and the Bohr–van Leeuwen Theorem

In classical kinetic theory, an ensemble of charged particles subjected to a magnetic field undergoes gyromotion about field lines with a Larmor radius and frequency. This orbital motion induces a circulating current that opposes the applied magnetic field a manifestation of classical diamagnetism.

The magnetization **M** in classical physics, defined as magnetic moment per unit volume, is given by the ensemble average of microscopic currents:

$$\mathbf{M}_{\text{classical}} = \frac{1}{2V} \sum_i q_i \langle \mathbf{r}_i \times \mathbf{v}_i \rangle .$$

However, when treated rigorously in thermodynamic equilibrium, the Bohr–van Leeuwen theorem asserts that classical statistical mechanics predicts zero magnetization in the canonical ensemble:

$$\mathbf{M}_{\text{classical}} = 0.$$

This result implies that magnetization is inherently a quantum phenomenon, requiring quantization of orbital angular momentum and spin degrees of freedom for a non-zero response. In essence, classical plasmas can carry currents and exhibit magnetohydrodynamic behaviour but do not exhibit intrinsic magnetic ordering or equilibrium magnetization [5].

Quantum Magnetization: Origin and Formalism

Quantum magnetization arises from two principal contributions:

i. Spin magnetization (Pauli paramagnetism) due to the alignment of intrinsic spin magnetic moments in a magnetic field.
ii. Orbital Magnetization (Landau diamagnetism) due to quantized cyclotron motion in the plane orthogonal to the magnetic field.

These effects become significant when the thermal de Broglie wavelength is large compared to interparticle separation, i.e., in degenerate or partially degenerate systems such as white dwarfs, laser-compressed plasmas, and electron gases in condensed matter systems.

Pauli Paramagnetism: Spin-Dependent Magnetic Susceptibility

The spin of electrons gives rise to a magnetic moment $\boldsymbol{\mu}_s = -g\mu_B S/\hbar$, where $\mu_B = \frac{e\hbar}{2m}$ is the Bohr magneton and $g \approx 2$ for free electrons. In the presence of a weak magnetic field $\mathbf{B}$, the energy levels of spin-up and spin-down states split due to the Zeeman effect:

$$\Delta E = \pm\mu_B B.$$

In a degenerate electron gas at temperature $T \ll T_{F,}$ the difference in population between the two spin states yields a net magnetization. The Pauli susceptibility χ_P, quantifying the linear response $\mathbf{M} = \chi_P \mathbf{B}$, is given by:

$$\chi_P = \mu_0 \mu_B^2 D(E_F) = \frac{3\mu_0 n \mu_B^2}{2E_F},$$

where $D(E_F)$ is the density of states at the Fermi energy, μ_0 is the permeability of free space, and n is the electron number density.

This susceptibility is temperature independent at $T \ll T_F$ and reflects a quantum statistical feature of fermions. Pauli paramagnetism is significant in high-density plasmas where Fermi–Dirac statistics prevail.

Landau Diamagnetism: Orbital Quantization in Magnetic Fields

The second key quantum contribution arises from the quantization of electron orbits perpendicular to a uniform magnetic field **B**. In quantum mechanics, this leads to the formation of discrete Landau levels with energy spectrum:

$$E_n = \left(n + \frac{1}{2}\right)\hbar\omega_c, \quad n \in \mathbf{N}_0, \ \omega_c = \frac{eB}{m},$$

where ω_c is the Cyclotron frequency. The density of states becomes highly peaked at these discrete levels, and the partition function must be recalculated with summation over n.

The magnetization from Landau levels can be computed from the thermodynamic potential Ω, yielding for weak fields and non-relativistic degenerate gases:

$$\chi_L = -\frac{1}{3}\chi_P,$$

which is a purely quantum mechanical diamagnetic response. Landau diamagnetism is thus weaker than Pauli paramagnetism by a factor of 3 and opposes the applied magnetic field, in agreement with Lenz's law.

Combined Quantum Magnetic Response: Total Susceptibility

In degenerate quantum plasmas, both spin and orbital effects coexist. The total magnetic susceptibility is given by:

$$\chi_{total} = \chi_P + \chi_L = \frac{2}{3}\chi_P$$

This result highlights a net paramagnetic response, dominated by spin effects. However, this picture is altered under relativistic conditions, strong magnetic fields, or low dimensionality, where the orbital contribution can be enhanced or become nonlinear.

Quantum Oscillations and De Haas–van Alphen Effect

When the magnetic field strength is large enough such that only a few Landau levels are occupied (i.e., $\hbar\omega_c \gtrsim k_B T$), the magnetization becomes an oscillatory function of $1/B$, a phenomenon known as the de Haas–van Alphen effect

$$M(B) \sim \cos\left(\frac{2\pi E_F}{\hbar\omega_c}\right)$$

These oscillations are indicative of quantum coherence over cyclotron orbits and are crucial diagnostics of Fermi surfaces in solid-state systems. In MQPs, similar oscillations may arise in high-density, low-temperature conditions and have implications for quantum turbulence and magneto-quantum instabilities.

Physical Regimes and Dominance Hierarchies

The dominance of various magnetization mechanisms depends sensitively on plasma parameters:

High Density, Low Temperature (e.g., white dwarfs): Pauli paramagnetism dominates.

Low Density, High Magnetic field (e.g., neutron star crusts, ultra-intense laser plasmas): Landau quantization and de Haas–van Alphen oscillations become prominent.

Intermediate Regimes (laboratory MQPs): Mixed contributions, requiring full quantum kinetic treatments.

Further, these quantum magnetization responses feed back into the plasma collective behavior, modifying dispersion relations, instabilities, and transport coefficients, as will be addressed in subsequent sections of this chapter.

6.1.3 Role of External Versus Self-generated Magnetic Fields

In a quantum plasma environment, magnetic fields may either be externally imposed (e.g., from laboratory magnets, neutron stars, or ambient galactic fields), or arise intrinsically from the plasma's own currents, instabilities, and dynamical configurations. The distinction between externally applied and self-generated magnetic fields has profound implications for the equilibrium structure, wave dynamics, transport coefficients, and quantum statistical effects in magnetized plasmas.

In classical electromagnetism, the total magnetic field B is defined via Maxwell's equations as:

$$\nabla \cdot \mathbf{B} = 0, \ \nabla \times \mathbf{B} = \mu_0 \mathbf{J}_{\text{tot}},$$

where $\mathbf{J}_{\text{tot}} = \mathbf{J}_{\text{ext}} + \mathbf{J}_{\text{plasma}}$. The former represents imposed external sources (e.g., coils or astrophysical magnetic fields), while the latter is the current density due to plasma charges.

In quantum plasma, one must generalize this expression to include the quantum current operator $\hat{\mathbf{J}}(\mathbf{r}, t)$, arising from the expectation values of quantum field operators:

$$\hat{\mathbf{J}}(\mathbf{r}, t) = \sum_s \frac{q_s}{2m_s} \left[\hat{\psi}_s^\dagger (\mathbf{p} - q_s \mathbf{A}) \hat{\psi}_s + \text{ H.C.} \right],$$

where q_s and m_s are the charge and mass of species s, and $\hat{\psi}_s$ is the second-quantized field operator.

The magnetic field is thus decomposed as:

$$\mathbf{B}(\mathbf{r}, t) = \mathbf{B}_{\text{ext}}(\mathbf{r}, t) + \mathbf{B}_{\text{self}}(\mathbf{r}, t),$$

with the self-field derived from $\mathbf{J}_{\text{plasma}}$ via Biot–Savart law or via full dynamical evolution from Maxwell's equations coupled with quantum plasma currents.

External magnetic fields are pivotal for plasma confinement (e.g., in magnetic fusion devices) and define quantization axes for spin and orbital degrees of freedom. Their spatial uniformity, or imposed gradients crucially influence Landau quantization, magnetic moment precession, and the coupling of quantum charged particles to gauge fields.

For instance, the quantum Hamiltonian of a charged particle in a uniform external magnetic field $\mathbf{B}_{\text{ext}} = B_0\hat{z}$ is:

$$\hat{H} = \frac{1}{2m}\left(\hat{\mathbf{p}} - q\mathbf{A}_{\text{ext}}\right)^2 + q\phi,$$

with the symmetric gauge $\mathbf{A}_{\text{ext}} = \frac{1}{2}\mathbf{B}_0 \times \mathbf{r}$. The eigenstates are quantized into Landau levels, with energy spectrum:

$$E_{n,k_z} = \hbar\omega_c\left(n + \frac{1}{2}\right) + \frac{\hbar^2 k_z^2}{2m},$$

where $\omega_c = \frac{qB_0}{m}$ is the Cyclotron frequency. These quantized orbits impact thermodynamic quantities, conductivity tensors, and spinor alignment.

In relativistic settings, this structure generalizes via the Dirac equation in an external field, leading to discrete spin-resolved energy branches and anomalous magnetic moment corrections, which are crucial in astrophysical quantum plasmas, such as magnetar atmospheres, or pulsar magnetospheres.

Self-generated magnetic fields, in contrast, originate from microscopic plasma currents and collective instabilities. Such magnetic fields are dynamic, spatially complex, and evolve nonlinearly in tandem with the plasma distribution functions.

In a quantum kinetic framework, the evolution of the self-field $\mathbf{B}_{\text{self}}$ must be obtained self-consistently by solving:

$$\nabla \times \mathbf{B}_{\text{self}} = \mu_0\left\langle\hat{\mathbf{J}}_{\text{plasma}}\right\rangle,$$

with the current expectation values derived from Wigner distribution functions or path-integral formalisms. For example, within the quantum Vlasov–Maxwell system, the magnetic field evolution is coupled to the Wigner function $f(\mathbf{r}, \mathbf{p}, t)$ via:

$$\frac{\partial f}{\partial t} + \frac{\mathbf{p}}{m}\cdot\nabla_{\mathbf{r}} f + q\left(\mathbf{E} + \frac{\mathbf{p}}{m}\times\mathbf{B}_{\text{self}}\right)\cdot\nabla_{\mathbf{p}} f + \mathcal{Q}[f] = 0$$

where $\mathcal{Q}[f]$ represents quantum corrections such as the Moyal bracket expansion.

Mechanisms of self-field generation include, Weibel instabilities driven by anisotropic quantum momentum distributions, Magnetization currents arising from spin polarization and spin–orbit coupling, Biermann battery effects, enhanced in quantum plasmas with strong density, or temperature gradients at the Debye scale, Photon-polarization effects, such as vacuum birefringence in nonlinear QED formulations, contributing to magnetogenesis at ultrahigh field strengths [6].

A crucial distinction lies in the control and predictability of external fields versus the emergent, collective character of self-generated fields. External fields serve as tunable parameters and boundary conditions, while self-fields encapsulate internal plasma state evolution.

In magnetized quantum plasmas, the interplay of these fields determines the dispersion relations of collective modes (e.g., Bernstein modes, spin waves, quantum magnetosonic waves), where effective field strengths include both external and self-generated contributions.

The effective magnetic susceptibility, χ_{eff}, becomes a nonlinear functional of the total field, modifying the response functions:

$$\chi_{\text{eff}}(B) = \chi^{\text{ext}}(B_{\text{ext}}) + \chi^{\text{self}}[f(\mathbf{r}, \mathbf{p})],$$

where the latter depends on non-equilibrium Wigner functions

The quantum Bohm potential and spin quantum pressure become explicitly dependent on the magnetic field gradient:

$$V_{\text{Bohm}} \propto \frac{\hbar^2}{2m}\nabla\left(\frac{\nabla^2\sqrt{n}}{\sqrt{n}}\right), \ \mathbf{F}_{\text{spin}} \propto \mu_B\nabla(\mathbf{B}\cdot\mathbf{S}),$$

where $\mathbf{B} = \mathbf{B}_{\text{ext}} + \mathbf{B}_{\text{self}}$, and $\mathbf{S}$ is the spin density vector.

The topological structuring of fields, including quantum magnetic reconnection and knotted field configurations, becomes sensitive to both imposed and plasma-induced fields, especially under conditions of helicity conservation and quantum Hall-MHD regimes.

From a modern quantum field-theoretic viewpoint, external and self-generated fields are unified within the effective action formalis. In systems where the electromagnetic field is quantized, one computes an effective action $S_{eff}\left[A_\mu\right]$ incorporating both background (external) fields and quantum fluctuations (self-fields) [7] via:

$$e^{iS_{eff}[A]} = \int \mathcal{D}\left[\overline{\psi}, \psi\right] e^{iS\left[\overline{\psi}, \psi, A\right]},$$

and identifies emergent terms such as magnetic polarization, anomalous magnetic moments, and self-energy corrections, bridging the distinction between imposed and emergent field contributions.

In such cases, magnetic field dynamics include higher-order corrections from the photon self-energy (vacuum polarization), yielding modified Maxwell equations with quantum corrections:

$$\nabla \times \mathbf{B} = \mu_0\left(\mathbf{J}_{\text{ext}} + \langle\hat{\mathbf{J}}\rangle\right) + \nabla \times \left(\frac{\partial \mathbf{P}_{\text{vac}}}{\partial t}\right),$$

where $\mathbf{P}_{\text{vac}}$ denotes vacuum magnetization from field fluctuations.

The dichotomy of external versus self-generated magnetic fields in magnetized quantum plasma is foundational to understanding its structure, stability, and response. While external fields are often precisely specified and serve to anchor the system's symmetry properties and energy scales, self-generated magnetic fields embody the nonlinear, collective, and quantum-statistical essence of the plasma state. A comprehensive treatment thus requires a dual approach, specifying external conditions while dynamically solving for self-consistent field evolution, ideally within a quantum kinetic, or field-theoretic framework. The superposition and competition of these fields underlie key phenomena such as quantum magnetohydrodynamic stability, spin alignment, magnetic turbulence, and photon-driven plasma responses, forming the backbone of contemporary and future theoretical modeling of magnetized quantum plasmas [8].

6.2 Quantum Effects in Magnetized Plasma: Theoretical Preliminaries

6.2.1 Landau Quantization

In the presence of a strong magnetic field, the quantum behaviour of charged particles is profoundly altered due to the discretization of their transverse motion into Landau levels. This phenomenon, known as Landau quantization, introduces anisotropy into the phase space and modifies the thermodynamic and transport properties of plasma. In magnetized quantum plasmas, such quantization plays a pivotal role not only in determining the microscopic energy spectra of electrons and positrons, but also in dictating macroscopic observables such as pressure, conductivity, and magnetization. The formal treatment of this quantization is not merely a quantum correction to classical cyclotron dynamics; it necessitates a full quantum field theoretic or many-body quantum mechanical framework.

Consider a non-relativistic charged particle of charge q and mass m subjected to a uniform magnetic field $\mathbf{B} = B\hat{z}$. In the symmetric gauge, the vector potential is chosen as $\mathbf{A} = \frac{1}{2}\mathbf{B} \times \mathbf{r}$, which leads to $\mathbf{A} = \frac{B}{2}(-y, x, 0)$. The Hamiltonian of the particle is

$$\hat{H} = \frac{1}{2m}(\hat{\mathbf{p}} - q\mathbf{A})^2$$

The motion parallel to $\mathbf{B}$ remains free, while the transverse degrees of freedom experience quantization. The energy eigenvalues split into a discrete ladder of Landau levels,

$$E_{n,k_z} = \hbar\omega_c\left(n + \frac{1}{2}\right) + \frac{\hbar^2 k_z^2}{2m}, \quad n = 0, 1, 2, ...,$$

where $\omega_c = \frac{|q|B}{m}$ is the Cyclotron frequency, and k_z is the momentum component along the magnetic field. Each Landau level has a macroscopic degeneracy due to the gauge freedom in the x–y plane. The quantum mechanical degeneracy per unit area in the transverse plane is given by

$$g = \frac{|q|B}{2\pi\hbar}$$

This expression follows from the flux quantization condition, where the number of quantum states per unit area corresponds to the number of flux quanta $\Phi_0 = \frac{h}{|q|}$ threading the system.

In relativistic quantum theory, particularly for electrons in extreme magnetic environments such as those found in magnetars, or high-intensity laser-matter interactions, the Landau quantization must be described using the Dirac equation. In this context, the energy levels become

$$E_{n,k_z,s} = \sqrt{m^2c^4 + \hbar^2c^2k_z^2 + 2n\hbar|q|Bc^2}, n = 0, 1, 2, ...,$$

where s denotes the spin orientation and the term $2n\hbar|q|Bc^2$ represents the magnetic quantization. The lowest Landau level (LLL)) with $n = 0$ is special, it is spin-polarized and non-degenerate, while higher levels are doubly degenerate due to spin.

The suppression of higher Landau levels becomes apparent when $\hbar\omega_c \gg k_BT$, or equivalently when $B \gg B_Q = \frac{m^2c^3}{|q|\hbar} \approx 4.4 \times 10^{13}$ G for electrons, where B_Q denotes the quantum critical magnetic field. In such cases, the system becomes effectively one-dimensional in momentum space, confined predominantly to the LLL.

The density of states (DOS) is profoundly modified by Landau quantization. In contrast to the continuous free-particle DOS $D(E) \propto \sqrt{E}$ in three dimensions, the magnetic field induces a staircase structure. For each Landau level n, the partial DOS becomes

$$D_n(E) = \frac{g}{L_z}\delta\left(E - \hbar\omega_c\left(n + \frac{1}{2}\right) - \frac{\hbar^2k_z^2}{2m}\right),$$

where L_z is the system size along Z. Upon integrating over k_z, the total DOS is

$$D(E) = \sum_{n=0}^{\infty} \frac{g}{\sqrt{2m\left(E - \hbar\omega_c\left(n + \frac{1}{2}\right)\right)}}\Theta\left(E - \hbar\omega_c\left(n + \frac{1}{2}\right)\right),$$

where Θ is the Heaviside step function ensuring energy conservation. In the high-field limit (quantum limit), only the LLL contributes significantly, reducing the DOS to a quasi-one-dimensional form [9],

$$D(E) \sim \frac{g}{\sqrt{2m\left(E - \frac{1}{2}\hbar\omega_c\right)}}.$$

This has profound consequences for thermodynamic quantities and transport coefficients.

At zero temperature, the Fermi energy in a quantizing magnetic field must accommodate discrete Landau levels. For a fixed electron number density n_e, the occupation condition is

$$n_e = \frac{|q|B}{2\pi\hbar} \sum_{n=0}^{n_{\max}} \int_{-\infty}^{\infty} \frac{dk_z}{2\pi} f\left(E_{n,k_z}\right),$$

where $f(E)$ is the Fermi–Dirac distribution. At zero temperature, the integration yields the Fermi surface for each level. In the quantum limit where only $n = 0$ is populated,

$$n_e = \frac{|q|B}{2\pi^2\hbar} k_F,$$

where k_F is the Fermi momentum along the field direction. This yields

$$k_F = \frac{2\pi^2\hbar n_e}{|q|B}.$$

Consequently, the Fermi energy and chemical potential become magnetic-field dependent, and the system undergoes a topological transition in its Fermi surface structure as the field strength varies. At finite temperature, thermal smearing introduces Landau-level broadening, which can be handled via semiclassical quantization or Green's function methods.

One of the most striking implications of Landau quantization in quantum plasmas is magnetic catalysis the enhancement of symmetry breaking, or condensation phenomena due to dimensional reduction in strong magnetic fields. In QED plasmas, this phenomenon manifests in enhanced chiral condensates. The reduction from 3 to 1D density of states near the LLL effectively enhances interaction effects and leads to phenomena like mass generation and pairing.

Additionally, the oscillatory behaviour of thermodynamic observables such as magnetization and susceptibility in weak fields is encapsulated by the de Haas–van Alphen and Shubnikov–de Haas effects [5]. These quantum oscillations provide signatures of Landau quantization and are vital in diagnosing field strength and carrier properties in laboratory and astrophysical plasmas.

In quantum field theory, the spectral density of states is embedded in the one-loop effective action of charged fermions in a magnetic background. Schwinger's proper-time formalism reveals the modified vacuum polarization tensor and effective Lagrangian,

$$\mathcal{L}_{\text{eff}} = -\frac{1}{8\pi^2}\int_0^\infty \frac{ds}{s^3}e^{-m^2 s}\left[(eBs)\cot(eBs) - 1 - \frac{(eBs)^2}{3}\right],$$

which encodes the full effect of Landau level summation. The poles of the cotangent function correspond to the discrete Landau levels, while the asymptotic behaviour captures vacuum birefringence and nonlinear QED effects.

In quantum plasmas, the generalized propagators in the presence of a background B -field can be used to compute the modified spectral function, which determines scattering amplitudes, energy loss rates, and radiative corrections in high-field environments.

Landau quantization introduces a fundamental discretization in the phase space of charged particles in strong magnetic fields, yielding altered transport properties, equation of state, and response functions in quantum plasmas. The associated changes in density of states, occupation statistics, and symmetry-breaking phenomena play an indispensable role in both laboratory-generated high-field plasmas and astrophysical environments such as neutron stars and magnetars. The consistent incorporation of these effects requires a synthesis of quantum mechanics, statistical physics, and quantum field theory, forming the cornerstone of the theoretical framework for magnetized quantum plasmas.

6.2.2 Spin Magnetization and Spin–Orbit Coupling

In the presence of intense magnetic fields, quantum plasma systems exhibit phenomena that are profoundly shaped by the intrinsic spin degrees of freedom of the constituent fermions and their coupling to electromagnetic fields. In quantum plasma systems, electrons and ions carry intrinsic spin angular momentum characterized by the Pauli matrices $\boldsymbol{\sigma} = (\sigma_x, \sigma_y, \sigma_z)$, with spin operators satisfying the SU(2) [10] algebra $[\sigma_i, \sigma_j] = 2i\epsilon_{ijk}\sigma_k$. In a magnetized plasma, the spin couples to the magnetic field via the Zeeman term:

$$\mathcal{H}_{\text{Zeeman}} = -\boldsymbol{\mu}\cdot\mathbf{B} = -g\mu_B\mathbf{S}\cdot\mathbf{B}$$

where $\mu_B = \frac{e\hbar}{2m}$ is the Bohr magneton, g is the Landé g-factor, and $\mathbf{S} = \frac{\hbar}{2}\boldsymbol{\sigma}$ is the spin operator. In strongly magnetized quantum plasmas, such as those found in neutron star crusts, laser-compressed fusion plasmas, or magnetized quantum wells, this term cannot be neglected, and the spin polarization density contributes to the overall magnetization of the plasma.

The spin-magnetization density $\mathbf{M}_s$ arising from the collective alignment of microscopic magnetic moments is given by:

$$\mathbf{M}_s = -g\mu_B\langle\psi^\dagger\mathbf{S}\psi\rangle$$

where ψ is the second-quantized spinor field operator for the electron (or any spin-1/2 species). In thermodynamic equilibrium, this leads to a net magnetization that modifies the total magnetic field via $\mathbf{B}_{\text{eff}} = \mathbf{B} + \mu_0 \mathbf{M}_s$, thereby influencing the Lorentz force and the Maxwell equations.

To account for Spin magnetization dynamically, one must extend the Wigner formalism to include spin degrees of freedom. The spin-dependent Wigner function $f(\mathbf{r}, \mathbf{p}, t)$ becomes a 2×2 matrix in spin space:

$$f(\mathbf{r}, \mathbf{p}, t) = f_0(\mathbf{r}, \mathbf{p}, t)\mathbb{I} + \mathbf{f}_s(\mathbf{r}, \mathbf{p}, t) \cdot \boldsymbol{\sigma}$$

where f_0 denotes the scalar (spin-averaged) part and $\mathbf{f}_s$ represents the spin-polarization vector distribution. The evolution of $N_L = \frac{E_F^2 - m^2c^4}{2\hbar eBc}$ follows from the quantum Liouville equation or quantum Vlasov–Pauli equation with spin-field interactions:

$$\frac{\partial f}{\partial t} + \frac{\mathbf{p}}{m} \cdot \nabla_{\mathbf{r}} f - e\left(\mathbf{E} + \frac{\mathbf{p}}{m} \times \mathbf{B}\right) \cdot \nabla_{\mathbf{p}} f + \frac{g\mu_B}{\hbar}(\mathbf{B} \times \mathbf{f}_s) \cdot \boldsymbol{\sigma} = 0$$

The spin precession term $\frac{g\mu_B}{\hbar}(\mathbf{B} \times \mathbf{f}_s)$ manifests the Larmor precession of spin in the magnetic field, generating spin torque and dynamic magnetization. The magnetization current induced by spatial variations in spin polarization modifies Ampère's law and affects the electromagnetic response of the medium.

Spin–orbit coupling (SOC) arises from relativistic corrections to the electron motion in electric and magnetic fields. In a quantum plasma, SOC originates from the Foldy–Wouthuysen transformation of the Dirac Hamiltonian, yielding a Pauli-type Hamiltonian with the SOC term:

$$\mathcal{H}_{\text{SOC}} = \frac{\hbar}{4m^2c^2}(\boldsymbol{\sigma} \cdot (\mathbf{E} \times \mathbf{p}))$$

Physically, SOC represents the interaction of the electron's magnetic moment in its rest frame with the magnetic field generated by its motion through an external electric field. In magnetized quantum plasmas, such fields may emerge from collective charge displacements, wave propagation, or boundary-induced electric gradients. The presence of SOC leads to spin-dependent force terms in the kinetic and fluid descriptions and can induce anomalous Hall effect, spin-polarized currents, and topologically protected excitations [11].

The inclusion of SOC in the quantum hydrodynamic (QHD) framework modifies the momentum equation. For the spin-polarized electron fluid, the equation of motion becomes:

$$m\left(\frac{\partial \mathbf{u}}{\partial t} + \mathbf{u} \cdot \nabla \mathbf{u}\right) = -\nabla P + e(\mathbf{E} + \mathbf{u} \times \mathbf{B}) + \mathbf{F}_Q + \mathbf{F}_{\text{spin}} + \mathbf{F}_{\text{SOC}}$$

where $\mathbf{F}_Q$ is the quantum Bohm force, $\mathbf{F}_{\text{spin}}$ arises from spin magnetization gradients, and the spin–orbit force is:

$$\mathbf{F}_{\text{SOC}} = \nabla\left(\frac{\hbar}{4m^2c^2}\boldsymbol{\sigma}\cdot(\mathbf{E}\times\mathbf{p})\right)$$

This term couples spatial inhomogeneities in electric fields with the spin polarization and generates non-trivial current distributions even in the absence of external magnetic fields. In turbulent or structured plasma configurations, SOC may give rise to internal spin Hall currents and feedback into electromagnetic modes.

The incorporation of spin magnetization and spin–orbit coupling alters the linear response functions of quantum plasmas. For magnetized spin-1/2 systems, the longitudinal and transverse susceptibilities $\chi_{\parallel}(k,\omega)$, $\chi_{\perp}(k,\omega)$ acquire corrections due to spin alignment and precession, modifying both the dispersion relations and the damping of plasma modes. In particular, spin-aligned degenerate plasmas support spin-polarized Langmuir-like modes, spin-cyclotron hybrid modes, and spin-magnetosonic waves.

The total magnetization contributes to the susceptibility tensor through:

$$\mathbf{M}_{\text{tot}} = \mathbf{M}_{\text{orbital}} + \mathbf{M}_{\text{spin}} = \chi\mathbf{B}$$

where $\chi = \chi_L + \chi_P$ includes the Landau diamagnetic and Pauli paramagnetic components. For degenerate fermions, the Pauli susceptibility dominates:

$$\chi_P = \frac{3\mu_0\mu_B^2 n}{2E_F}$$

where E_F is the Fermi energy. However, in relativistic or non-equilibrium conditions, the interplay of spin–orbit interaction and quantum correlations leads to corrections that must be computed from the two-particle Green's function, or kinetic equations incorporating spinor dynamics.

An effective fluid model incorporating spin effects can be constructed using the Madelung decomposition of the spinor field and ensemble averaging over spin-polarized wavefunctions. The resulting quantum spin hydrodynamics (QSHD) model includes evolution equations for the fluid density n, velocity $\mathbf{u}$, and spin density $\mathbf{S}$, coupled via generalized Maxwell and Bloch-type equations:

$$\frac{\partial\mathbf{S}}{\partial t} + (\mathbf{u}\cdot\nabla)\mathbf{S} = \frac{2\mu_B}{\hbar}\mathbf{S}\times\mathbf{B}_{\text{eff}} + D_s\nabla^2\mathbf{S}$$

where D_s is the spin diffusion coefficient and $\mathbf{B}_{\text{eff}}$ includes external and internal magnetic fields. This formalism enables analysis of spin waves, relaxation dynamics, and spin–vorticity coupling in magnetized quantum plasmas.

The physical ramifications of spin magnetization and spin–orbit interactions in quantum plasmas are profound. In laboratory laser–plasma interactions, the polarization of high-density relativistic electron beams can be influenced by spin–orbit resonances. In magnetars and pulsar magnetospheres, spin-polarized pair plasmas generate anisotropic emission spectra and modify the dielectric tensor significantly.

Further, modern experimental techniques in ultracold plasma and solid-state plasma analogues (e.g., 2D electron gas systems under quantum Hall conditions) have revealed signatures of spin–orbit coupling in collective excitation spectra and non-linear magneto-transport.

Spin magnetization and spin–orbit coupling in quantum plasmas introduce inherently quantum–mechanical effects that strongly influence the macroscopic behaviour of strongly magnetized systems. Their proper inclusion necessitates a multi-layered theoretical approach spanning quantum kinetic theory, relativistic field-theoretic corrections, and fluid models with internal degrees of freedom. The synthesis of these frameworks permits a unified and predictive description of phenomena ranging from spin waves in solid-state plasmas to spin-aligned degenerate matter in astrophysical contexts, thereby reinforcing the necessity of a spin-resolved analysis in any complete theory of magnetized quantum plasma.

6.2.3 Relativistic Landau Levels and Quantum Hall Features

The interplay between relativistic quantization and collective plasma behaviour in strong magnetic fields presents a domain of quantum plasma physics replete with both profound theoretical structure and diverse physical phenomena. In highly magnetized quantum electron fluids typical in environments such as pulsar magnetospheres, laser-compressed dense matter, or relativistically degenerate laboratory plasmas the electron dynamics transverse to the magnetic field becomes quantized into relativistic Landau levels. This quantization, arising from the cyclotron motion of charged fermions in a magnetic field, underpins a series of collective and topological features, including the emergence of relativistic quantum Hall conductivity and magneto-oscillatory behaviors that influence transport and radiation processes. To derive the relativistic Landau levels for an electron, one considers the dynamics governed by the Dirac equation in the presence of an external magnetic field $\mathbf{B} = B_0\hat{z}$, assumed to be spatially uniform and temporally constant [12]. The corresponding four-vector potential can be chosen in the Landau gauge as:

$$A^\mu = (0, 0, B_0 x, 0),$$

yielding a magnetic field $\mathbf{B} = \nabla \times \mathbf{A} = B_0\hat{z}$. The Dirac equation for a spin-1/2 electron of mass m and charge -e (with $e > 0$) in an electromagnetic field reads:

$$\left(i\gamma^\mu D_\mu - m\right)\psi = 0,$$

where $D_\mu = \partial_\mu + ieA_\mu$ is the gauge-covariant derivative and γ^μ are the Dirac gamma matrices. The physical content of this equation is most transparently revealed through squaring the Dirac operator, leading to the Klein-Gordon-like operator enriched with spinor structure:

$$\left[\left(iD_\mu\right)^2 - \frac{e}{2}\sigma^{\mu\nu}F_{\mu\nu} + m^2\right]\psi = 0,$$

where $\sigma^{\mu\nu} = \frac{i}{2}\left[\gamma^\mu, \gamma^\nu\right]$ and $F_{\mu\nu} = \partial_\mu A_\nu - \partial_\nu A_\mu$ is the electromagnetic field tensor. This decomposition explicitly separates the orbital Landau quantization from the spin coupling via the magnetic moment term $\sigma^{\mu\nu}F_{\mu\nu}$.

Solving the Dirac equation in the chosen gauge, one finds that the energy spectrum of an electron in a uniform magnetic field is quantized into relativistic Landau levels given by:

$$E_n(p_z, s) = \pm\sqrt{m^2 + p_z^2 + 2eB_0 n + seB_0},$$

$$n \in \mathbf{N}_0,\ s = \pm 1,$$

where p_z is the conserved momentum along the direction of the magnetic field and s labels the spin projection. The term seB_0 arises due to the spin–magnetic field interaction and leads to Zeeman-like splitting in the relativistic regime. Importantly, the lowest Landau level (LLL)with n $= 0$ and spin antiparallel to $\mathbf{B}$ is non-degenerate, a feature of critical importance in understanding magnetic catalysis and chiral symmetry breaking in quantum plasmas.

In the ultra-relativistic regime ($p_z \gg m, eB_0$), the spacing between the Landau levels becomes less significant, but for $eB_0 \gtrsim m^2$, the quantum discreteness of the transverse motion dominates, strongly modifying the density of states and response functions of the plasma.

The quantization of transverse motion modifies the density of single-particle states from the classical expression to a discretized sum over Landau levels:

$$g(E) = \frac{eB_0}{2\pi^2}\sum_{n=0}^{n_{\max}}\int dp_z \delta(E - E_n(p_z)),$$

with degeneracy per unit area per level $eB_0/(2\pi)$. This Landau quantization imposes quantum oscillations most notably de Haas–van Alphen and Shubnikov–de Haas effects on thermodynamic and transport observables in magnetized electron fluids.

For instance, the magnetization per unit volume $\mathcal{M}$ of a degenerate relativistic electron gas exhibits oscillatory behaviour as a function of $1/B_0$, stemming from the sharp Fermi edge intersecting the discretized Landau levels. This is captured via the Lifshitz-Kosevich formalism [8] extended to relativistic regimes:

$$\mathcal{M}(B_0) \sim \sum_n \cos\left(2\pi\frac{E_F^2 - m^2}{2eB_0}\right),$$

where E_F denotes the relativistic Fermi energy. These oscillations become increasingly pronounced at lower temperatures and larger magnetic fields, and they influence the collective modes in magnetized quantum plasmas.

In two-dimensional (2D), or quasi-2D magnetized electron systems relevant, for example, in quantum wells, graphene-like materials, or surface states in dense plasma interfaces the relativistic Landau quantization lays the foundation for quantum Hall behavior. The quantized Hall conductivity in the relativistic regime is given by:

$$\sigma_{xy} = \nu \frac{e^2}{h}, \quad \nu = 4\left(n + \frac{1}{2}\right),$$

where the factor 4 accounts for spin and valley degeneracies (as in graphene) and n is the Landau level index. This half-integer quantization is a manifestation of the underlying relativistic Dirac structure and is indicative of topologically protected edge currents.

In a strongly magnetized quantum plasma, particularly one near the relativistic degeneracy limit, such Hall plateaus can manifest not just in low-dimensional electronic systems, but also in bulk transport properties through effective dimensional reduction induced by magnetic confinement. This reflects in the nonlocal conductivity tensor $\sigma^{\mu V}$, where Hall-like off-diagonal components arise due to quantum anomalies and Berry-phase curvature [13] in momentum space.

The robustness of the Hall quantization is physically rooted in the energy gap between successive Landau levels, which inhibits longitudinal scattering and allows for dissipationless transverse transport. In magnetized quantum plasmas, such mechanisms may play critical roles in charge separation, magnetovortical instabilities [23], and current filamentation processes.

Landau quantization not only alters the single-particle spectra, but also modifies the collective response of the plasma. The dielectric function $\varepsilon(\omega, \mathbf{k})$ in the presence of a strong magnetic field acquires discrete contributions due to transitions between Landau levels. In particular, the polarization tensor $\Pi^{\mu\nu}$ for a relativistic magnetized electron fluid exhibits tensorial anisotropies and resonance structures associated with inter-level transitions.

These transitions manifest as quantum Bernstein modes discrete harmonics of the Cyclotron frequency shifted by relativistic and quantum effects:

$$\omega_{n,m} \approx \sqrt{m^2 + p_z^2 + 2eB_0(n + m)},$$

where m labels the transition between different levels. These quantized excitations influence wave propagation, damping, and mode conversion in magnetized quantum plasmas and are essential in understanding radiative energy losses and dispersion characteristics under extreme magnetic conditions.

Further, in the strong field limit $\left(eB_0 \gg m^2\right)$, the system effectively undergoes a dimensional reduction from $(3 + 1)$D to $(1 + 1)$D, with only longitudinal dynamics remaining gapless. This leads to chiral transport phenomena and anomaly-induced currents, such as the chiral magnetic effect (CME), which have been proposed to occur in highly magnetized quark–gluon plasmas and dense astrophysical matter.

The study of relativistic Landau quantization and its associated quantum Hall features opens a rich window into the emergent behaviour of quantum plasmas in strong magnetic fields. In laboratory contexts, ultra-intense laser–plasma interactions and magnetized inertial confinement fusion scenarios can access regimes where Landau levels and Hall-like transport dominate the electron kinetics. In astrophysical systems, such as magnetars and the inner crusts of neutron stars, the quantized nature of electron motion under supercritical magnetic fields (up to 10^{15} G) shapes the equation of state, cooling rates, and emission spectra.

Modern theoretical approaches are extending the formalism through techniques such as the worldline formalism for quantum field propagation in background fields, Wigner function transport in quantized phase space, and holographic duality for magnetized strongly coupled plasmas. These provide deeper insights into the topological, nonperturbative, and quantum coherence aspects that define relativistic quantum Hall physics in plasmas.

Ultimately, the relativistic Landau levels form a quantum backbone upon which a myriad of complex collective phenomena in magnetized plasmas are scaffolded, establishing them as a cornerstone of high-field quantum plasma theory.

6.2.4 Quantum-Statistical Pressure and Degeneracy Pressure

The inclusion of magnetic fields in quantum-statistical systems composed of fermions, such as degenerate electrons or neutrons, modifies the quantum equation of state, alters the phase-space structure, and leads to a fundamentally anisotropic pressure tensor. For a non-interacting ensemble of fermions of mass m and charge q in thermal equilibrium at temperature T, the quantum-statistical pressure in the absence of a magnetic field follows from the grand canonical potential density,

$$\Omega = -\frac{g_s}{\beta}\int \frac{d^3p}{(2\pi)^3}\ln\left[1+e^{-\beta(\varepsilon(p)-\mu)}\right],$$

where g_s is the spin degeneracy factor, $\beta = 1/(k_BT)$, $\varepsilon(p) = \sqrt{p^2+m^2}$ (in natural units $\hbar = c = 1$), and μ is the chemical potential. The pressure is obtained via $P = -\Omega$, and the energy density from

$$\mathcal{E} = \frac{g_s}{(2\pi)^3}\int d^3p\varepsilon(p)f(p),\; f(p) = \frac{1}{e^{\beta(\varepsilon(p)-\mu)}+1}.$$

In the zero-temperature limit ($T \to 0$), the distribution function reduces to a step function: $f(p) = \Theta(p_F - p)$, where p_F is the Fermi momentum. The degeneracy pressure becomes

$$P_{\text{deg}} = \frac{g_s}{6\pi^2} \int_0^{p_F} \frac{p^4 dp}{\sqrt{p^2 + m^2}}$$

This pressure resists compression and persists even at $T = 0$, representing a purely quantum mechanical manifestation of the exclusion principle.

When a uniform magnetic field $\mathbf{B} = B\hat{z}$ is applied, the motion of charged fermions transverse to the field becomes quantized into discrete Landau levels, while the longitudinal motion remains continuous. For a relativistic fermion, the single-particle energy eigenvalues become

$$\varepsilon_{n,s}(p_z) = \sqrt{p_z^2 + m^2 + 2n|qB| - sqB},\ n \in \mathbf{Z}_{\geq 0},\ s = \pm 1,$$

where n labels the Landau level and s denotes the spin orientation. In quantum electrodynamics (QED), the spin interaction is treated via coupling to the magnetic moment: $\mu_B = q/2m$.

The transverse momentum phase space becomes discretized, and the integration over d^3p is replaced by a sum over Landau levels and integration over P_z:

$$\int \frac{d^3p}{(2\pi)^3} \rightarrow \frac{|qB|}{2\pi} \sum_{n=0}^{\infty} g_n \int \frac{dp_z}{2\pi},\ g_n = \begin{cases} 1,\ n = 0 \\ 2,\ n \geq 1 \end{cases},$$

where $\mathrm{g_n}$ accounts for spin degeneracy and the non-degenerate nature of the lowest Landau level (LLL). The quantum-statistical pressure is now anisotropic, split into longitudinal $\left(P_{\shortparallel}\right)$ and transverse $(P_{\perp})$ components.

For the longitudinal component of pressure, which reflects the dynamics along the field lines, the expression becomes

$$P_{\shortparallel} = \frac{|qB|}{2\pi} \sum_{n=0}^{\infty} g_n \int \frac{dp_z}{2\pi} \frac{p_z^2}{\varepsilon_n} f(\varepsilon_n),$$

while the transverse component is derived from the thermodynamic identity $P_{\perp} = -\Omega - B\mathcal{M}$, where $\mathcal{M}$ is the magnetization density:

$$\mathcal{M} = -\left(\frac{\partial \Omega}{\partial B}\right)_{T,\mu}$$

The anisotropy $\Delta P = P_{\shortparallel} - P_{\perp}$ arises due to the magnetic field imposing a preferred direction in phase space and leads to deformation in the Fermi surface. In the extreme quantum limit (EQL) [26], where only the LLL is occupied $n = 0$, the pressure becomes purely one-dimensional, and the transverse dynamics are frozen, yielding:

$$P_{\shortparallel}^{\mathrm{EQL}} = \frac{|qB|}{4\pi^2}\int_0^{p_{F0}} \frac{p_z^2 dp_z}{\sqrt{p_z^2 + m^2}}, \quad P_{\perp}^{\mathrm{EQL}} = 0.$$

This is a direct consequence of dimensional reduction in strong magnetic fields, where quantum confinement leads to the suppression of transverse kinetic contributions.

In ultra-relativistic conditions ($p_F \gg m$), relevant in neutron stars, magnetars, and high-intensity laser–plasma interactions, the energy dispersion simplifies to $\varepsilon_n \approx \sqrt{p_z^2 + 2n|qB|}$, and the pressure expressions scale non-linearly with field strength. In the EQL and relativistic degenerate limit,

$$P_{\shortparallel}^{\mathrm{rel}} \approx \frac{(|qB|)^2}{8\pi^2}\left(\frac{\mu^2}{|qB|} - m^2 \ln\left(\frac{\mu + \sqrt{\mu^2 - m^2}}{m}\right)\right).$$

Such strong field degeneracy pressures play a crucial role in stabilizing white dwarfs and neutron stars against gravitational collapse. They also affect the propagation of collective modes and the dispersion of waves in quantum magnetized plasmas, necessitating a modified dielectric tensor and equation of state.

The anisotropic pressure in magnetized fermion gases requires a tensorial generalization of the thermodynamic relations. The energy–momentum tensor $T^{\mu\nu}$ for a magnetized quantum fluid takes the form:

$$T^{\mu\nu} = (\mathcal{E} + P_{\perp})u^{\mu}u^{\nu} - P_{\perp}g^{\mu\nu} + \left(P_{\shortparallel} - P_{\perp}\right)b^{\mu}b^{\nu},$$

where u^{μ} is the fluid four-velocity and b^{μ} is a unit vector along the magnetic field direction. The pressure anisotropy can trigger magnetoelastic instabilities if not properly balanced by boundary conditions, or external confinement. Thermodynamic consistency demands that the entropy and compressibility tensors also reflect the anisotropic character of the quantum statistics under magnetization.

In white dwarfs, the degeneracy pressure counteracts gravitational collapse, and the inclusion of strong magnetic fields introduces quantized corrections to the mass–radius relation. For field strengths approaching, or exceeding the Schwinger critical field $\left(B_c \approx 4.41 \times 10^{15}\ \mathrm{G}\right)$, QED effects such as vacuum polarization and magnetic catalysis further modify the pressure. In laboratory contexts, strongly magnetized quantum plasmas in ultracold traps, or high-intensity laser environments can exhibit field-induced anisotropy in pressure, leading to novel transport phenomena such as magneto-thermal oscillations and anisotropic thermal conductivity.

Quantum-statistical and degeneracy pressures in magnetized fermion systems form a cornerstone of the theory of magnetized quantum plasmas. The quantization of energy levels through Landau states, the emergence of anisotropic pressure components, and the thermodynamic implications thereof are essential for a correct theoretical formulation of dense, magnetized fermionic matter. The framework developed herein not only generalizes classical plasma pressure models, but also connects the statistical behaviour of relativistic quantum gases to observable consequences in astrophysical and laboratory plasmas. These effects set the stage for subsequent sections on collective excitations, wave dynamics, and transport properties in Magnetized quantum plasma systems [14, 15].

6.2.5 *Effective Quantum Potential in the Presence of Magnetic Fields*

The concept of an effective quantum potential serves as a pivotal bridge between classical fluid descriptions and quantum–mechanical phenomena in plasmas. The central role is played by the quantum Bohm potential, an intrinsically non-classical term arising in the Madelung representation of the Schrödinger equation and more generally in quantum hydrodynamic (QHD) formulations. In the absence of magnetic fields, the one-body time-dependent Schrödinger equation for a non-relativistic, spinless particle of mass m and charge q, in a scalar potential $V(\mathbf{r}, t)$, yields the QHD system via the polar decomposition of the wavefunction $\psi = \sqrt{n}e^{iS/\hbar}$, where n is the probability density and $\nabla S/m$ is interpreted as the velocity field. The resulting momentum conservation equation reads:

$$m\left(\frac{\partial \mathbf{u}}{\partial t} + \mathbf{u} \cdot \nabla \mathbf{u}\right) = -\nabla V - \nabla Q,$$

where the quantum potential Q takes the form:

$$Q = -\frac{\hbar^2}{2m}\frac{\nabla^2\sqrt{n}}{\sqrt{n}}.$$

This term accounts for quantum pressure originating from wavefunction dispersion and reflects spatial correlations absent in classical descriptions. When generalized to many-body systems through closure approximations, or Wigner–Moyal expansions, the Bohm potential persists as an effective term encapsulating the leading-order quantum correction to classical fluid forces.

To extend this framework to magnetized systems, one must introduce the electromagnetic gauge fields $\mathbf{A}(\mathbf{r}, t)$ and $\phi(\mathbf{r}, t)$ such that the Hamiltonian becomes:

$$\hat{H} = \frac{1}{2m}(-i\hbar\nabla - q\mathbf{A})^2 + q\phi$$

Upon applying the Madelung transformation to the resulting Schrödinger equation with minimal coupling, and decomposing the current density $\mathbf{J} = n\mathbf{u}$, the quantum Euler equation in a Magnetized quantum plasma can be expressed as:

$$m\left(\frac{\partial \mathbf{u}}{\partial t} + \mathbf{u} \cdot \nabla \mathbf{u}\right) = q(\mathbf{E} + \mathbf{u} \times \mathbf{B}) - \nabla Q + \mathbf{F}_{\text{int}},$$

where $\mathbf{F}_{\text{int}}$ includes pressure gradient forces and exchange–correlation contributions. Here, $\mathbf{B} = \nabla \times \mathbf{A}$ and $\mathbf{E} = -\nabla\phi - \partial\mathbf{A}/\partial t$ represent the magnetic and electric fields, respectively.

The form of the quantum potential Q remains formally identical, but its operational role now intertwines with the magnetization effects. The presence of a magnetic field induces anisotropies in the quantum pressure distribution, particularly evident in Landau-quantized systems and degenerate magnetized fermion gases. To capture these features, one must consider quantum kinetic derivations and effective actions that incorporate magnetic gauge fields explicitly.

A rigorous extension of the Bohm potential in magnetized systems requires constructing a gauge-invariant expression compatible with electromagnetic covariance. The conventional Bohm term, being a scalar derived from the probability density, does not account for vector potential dependencies. Recent developments suggest a generalized form incorporating magnetic length scales and spin–orbit effects [16].

Consider a quantum system characterized by a Wigner distribution function $f_W(\mathbf{r}, \mathbf{p}, t)$ evolving under the magnetized Wigner–Moyal equation. Upon gradient expansion and appropriate closure, the quantum force density can be written as:

$$\mathbf{F}_{\text{quantum}} = -\nabla Q + \frac{q\hbar^2}{4m^2}\nabla\left(\frac{(\nabla \cdot \mathbf{B})}{n}\right) - \frac{q\hbar^2}{8m^2}\nabla\left(\frac{|\mathbf{B}|^2}{n}\right) + \cdots .$$

These corrections arise from the magnetic field dependence of the higher-order moments in the Wigner distribution and represent dispersive analogues of magnetization effects. The effective potential now acquires terms proportional to the spatial gradients of $\mathbf{B}$, modulated by the density. While the leading-order Bohm term remains scalar, the higher-order terms embody vectorial and tensorial interactions, necessitating covariant reformulations in strongly magnetized regimes.

An alternative formalism employs the use of a magnetized quantum pressure tensor $\Pi_{ij}^{\text{quantum}}$, constructed from the spinor decomposition of the Pauli–Schrödinger equation. For a spin-1/2 magnetized electron fluid, the effective quantum potential includes spin coupling and magnetization energy terms such as:

$$Q_{\text{eff}} = -\frac{\hbar^2}{2m}\frac{\nabla^2\sqrt{n}}{\sqrt{n}} - \mu_B \mathbf{B} \cdot \mathbf{S} - \frac{\mu_B^2}{2m}(\nabla \cdot \mathbf{B})^2/n,$$

where $\mu_B = \frac{e\hbar}{2m}$ is the Bohr magneton, and S denotes the spin vector density. This expression underscores how spin polarization and magnetic inhomogeneity deform the local quantum pressure landscape, modifying the effective dynamics even in the absence of explicit collisions.

6.2.6 Physical Interpretation and Regime Classification

From the observer's perspective, the modified quantum potential in magnetized quantum plasmas embodies the influence of spatial coherence and quantum interference modulated by magnetic geometry. The standard Bohm potential governs zero-field quantum dispersion, while its extensions encode the response of de Broglie wave interference patterns to magnetic quantization and magneto-kinetic effects.

Three characteristic regimes can be distinguished:

1. Weak Magnetic fields $(\omega_c \ll \omega_p)$: The magnetic correction terms are subdominant. The Bohm potential primarily governs quantum corrections to pressure and dispersion, preserving isotropy in the absence of magnetic anisotropy.
2. Moderate Fields $(\omega_c \sim \omega_p)$: Interference between cyclotron motion and quantum dispersion gives rise to anisotropic modifications in the Bohm pressure. These include density-aligned field modulations and modified acoustic modes in quantum magnetosonic waves.
3. Strong Fields $(\omega_c \gg \omega_p)$: Landau quantization dominates. The density of states becomes quantized, and the Bohm potential must be reinterpreted through averaged contributions over discrete Landau levels. Spin polarization, Zeeman splitting, and anisotropic wavefunction confinement significantly alter the quantum potential structure.

The extension of the Bohm potential to magnetized quantum plasmas provides a mathematically consistent and physically rich tool for understanding non-classical transport and structural dynamics. It enables the hydrodynamic modeling of systems where quantum coherence and magnetic field topology intersect such as in quantum Hall systems, magnetized white dwarfs, high-field laser-plasma interactions, and neutron star crusts.

These extensions also connect with semiclassical approaches such as the Thomas–Fermi–Dirac–Weizsäcker models [17] in magnetized configurations, while offering a real-time, field-theoretic insight into quantum pressure contributions. The framework is also foundational for numerical modeling of magnetized quantum systems, including extended quantum magnetohydrodynamics (QMHD), quantum kinetic simulations with magnetic corrections, and fully spin-resolved quantum fluid codes.

Future directions may include path-integral generalizations, stochastic Bohm potential derivations in open magnetized quantum systems, and non-abelian generalizations in systems with spin–orbit coupling, or synthetic gauge fields. These approaches promise to deepen our understanding of quantum coherence in strongly magnetized regimes across plasma, condensed matter, and astrophysical contexts.

6.3 Quantum Magnetohydrodynamics (QMHD)

Quantum Magnetohydrodynamics (QMHD) extends classical magnetohydrodynamics by incorporating quantum mechanical corrections to fluid and electromagnetic equations, thereby enabling the study of collective phenomena in quantum plasmas under the influence of magnetic fields. While classical MHD is well-suited for large-scale, macroscopic descriptions of conducting fluids, its failure to account for quantum statistical and coherence effects necessitates a formulation that incorporates the principles of quantum mechanics. The QMHD formalism is rigorously derived from the coupled Euler–Maxwell system augmented by quantum corrections, particularly those originating from the Madelung–Bohm potential, Fermi pressure, and spin contributions.

The Euler–Maxwell system for a single-species charged quantum fluid (e.g., electrons) in the presence of an external magnetic field $\mathbf{B}$ and self-consistent electric field $\mathbf{E}$ is formulated by coupling the quantum hydrodynamic (QHD) equations with Maxwell's equations. The continuity and momentum balance equations form the quantum Euler backbone:

$$\frac{\partial n}{\partial t} + \nabla \cdot (n\mathbf{v}) = 0,$$

$$m\left(\frac{\partial \mathbf{v}}{\partial t} + (\mathbf{v} \cdot \nabla)\mathbf{v}\right) = -\nabla P + q(\mathbf{E} + \mathbf{v} \times \mathbf{B}) + \mathbf{F}_Q,$$

where:

n is the number density,
$\mathbf{v}$ is the fluid velocity,
P is the scalar pressure,
q and m are the charge and mass of the species,
$\mathbf{F}_Q$ is the quantum force density.

The quantum force $\mathbf{F}_Q$ encompasses the Bohm potential correction:

$$\mathbf{F}_Q = -\frac{\hbar^2}{2m}\nabla\left(\frac{\nabla^2\sqrt{n}}{\sqrt{n}}\right),$$

which arises from the Madelung decomposition of the wavefunction in the Schrödinger–Poisson formalism and accounts for quantum diffraction effects. Additional corrections from Fermi pressure and spin can be appended where necessary, particularly in degenerate or spin-polarized systems.

The electromagnetic fields obey the Maxwell equations:

$$\nabla \cdot \mathbf{E} = \frac{q}{\varepsilon_0}(n - n_0), \;\; \nabla \cdot \mathbf{B} = 0,$$

$$\nabla \times \mathbf{E} = -\frac{\partial \mathbf{B}}{\partial t}, \quad \nabla \times \mathbf{B} = \mu_0 q n \mathbf{v} + \mu_0 \varepsilon_0 \frac{\partial \mathbf{E}}{\partial t},$$

where n_0 is a fixed neutralizing background density ensuring quasi-neutrality in the equilibrium configuration.

To derive a consistent QMHD system, one introduces appropriate magnetization constraints that simplify the Maxwell sector and ensure compatibility with magnetized plasma limits. A frequent assumption is the neglect of displacement currents ($\partial_t \mathbf{E} \approx 0$) under the quasi-neutral approximation and low-frequency conditions $(\omega \ll \omega_{pe})$ a limit justified in many astrophysical and laboratory plasmas with high conductivity. Under such conditions, Ampère's law simplifies to:

$$\nabla \times \mathbf{B} = \mu_0 q n \mathbf{v},$$

while Faraday's law and Gauss's law retain their standard form. Additionally, the electric field may be obtained from Ohm's law, or derived from the electron momentum equation in regimes where inertial effects are suppressed.

To close the hydrodynamic system, the pressure P is specified through an equation of state, frequently chosen to be isothermal or adiabatic, or more realistically derived from the Fermi–Dirac distribution for degenerate plasmas:

$$P = \frac{\left(3\pi^2\right)^{2/3} \hbar^2}{5m} n^{5/3}.$$

This pressure term accounts for the Pauli's exclusion principle, and introduces quantum statistical effects not present in classical fluids.

Combining the quantum-corrected Euler equations with the magnetized Maxwell set, and enforcing the approximations described above, one obtains the basic set of QMHD equations for a single-species fluid:

(i) Continuity Equation:

$$\frac{\partial n}{\partial t} + \nabla \cdot (n\mathbf{v}) = 0,$$

(ii) Momentum Equation (Quantum Magnetohydrodynamic Form):

$$m\left(\frac{\partial \mathbf{v}}{\partial t} + (\mathbf{v} \cdot \nabla)\mathbf{v}\right) = -\nabla P + q\mathbf{v} \times \mathbf{B} - \nabla V_B,$$

with the quantum Bohm potential:

$$V_B = -\frac{\hbar^2}{2m} \frac{\nabla^2 \sqrt{n}}{\sqrt{n}}.$$

(iii) Faraday's Law:

$$\frac{\partial \mathbf{B}}{\partial t} = -\nabla \times \mathbf{E},$$

(iv) Generalized Ohm's Law (from electron momentum balance):

In the presence of strong magnetic fields, electron inertia can be neglected, leading to:

$$\mathbf{E} + \mathbf{v} \times \mathbf{B} = -\frac{1}{qn}\nabla P + \frac{1}{qn}\nabla V_B.$$

Substituting this into Faraday's law, one obtains the induction equation:

$$\frac{\partial \mathbf{B}}{\partial t} = \nabla \times (\mathbf{v} \times \mathbf{B}) + \nabla \times \left[\frac{1}{qn}\nabla(P - V_B)\right].$$

The last term represents the quantum correction to the magnetic field evolution due to pressure and quantum potential gradients. In strongly magnetized regimes, where the Cyclotron frequency dominates over fluid vorticity, this term may introduce dispersive effects in Alfvén and magnetosonic waves.

The QMHD system encapsulates a wide array of quantum and electromagnetic interactions. Each term encodes a specific physical process. The Lorentz force $q\mathbf{v} \times \mathbf{B}$ governs the coupling between the charged fluid and magnetic field lines, leading to classical MHD effects such as flux freezing. The Bohm potential gradient ∇V_B introduces quantum pressure due to wavefunction curvature, resulting in phenomena such as quantum dispersion and tunneling-induced density oscillations. The quantum corrected induction equation reflects how quantum coherence and Fermi pressure gradients influence magnetic field dynamics, potentially modifying magnetic reconnection, turbulent cascades, and wave propagation. These corrections become particularly significant in regimes where, the thermal de Broglie wavelength is comparable to, or exceeds the inter-particle spacing, i.e., $\lambda_{\mathrm{dB}} \gtrsim n^{-1/3}$, the magnetic energy density $B^2/2\mu_0$ is of the same order, or larger than kinetic, or Fermi energies, strong degeneracy pressure dominates over classical thermodynamic pressure, the Cyclotron frequency $\omega_c = qB/m$ dominates the fluid dynamics, enforcing magnetization constraints. Such conditions are met in several astrophysical environments (white dwarfs, magnetars), in solid-density laser-compressed plasmas, and in next-generation quantum plasma devices operating under ultra-intense magnetic fields.

The QMHD model presented above is primarily applicable to single-component quantum fluids. For realistic magnetized quantum plasmas, particularly those involving electrons, ions, and positrons or spin-polarized species, one must generalize to a multi-fluid description. Each species s then obeys its own continuity and momentum equations:

$$\frac{\partial n_s}{\partial t} + \nabla \cdot (n_s \mathbf{v}_s) = 0,$$

$$m_s\left(\frac{\partial \mathbf{v}_s}{\partial t} + (\mathbf{v}_s \cdot \nabla)\mathbf{v}_s\right) = -\nabla P_s + q_s \mathbf{v}_s \times \mathbf{B} - \nabla V_{B,s},$$

with Maxwell's equations modified by the total current density $\mathbf{J} = \sum_s q_s n_s \mathbf{v}_s$. Inter-species collision terms, spin-magnetic interactions, and quantum statistical corrections must be included for high-accuracy modeling in ultra-dense regimes. The derivation of QMHD from quantum Euler–Maxwell systems under magnetic constraints reveals a consistent and physically rich framework for describing strongly magnetized quantum plasmas. The resulting equations bridge classical MHD with quantum hydrodynamics, incorporating dispersive and nonlocal quantum effects essential in high-density, low-temperature, or high-field environments. As quantum plasmas find increasing relevance in laboratory, astrophysical, and high-energy applications, QMHD models offer predictive tools to analyze transport, stability, and wave phenomena with unprecedented precision. The formalism also serves as a fertile ground for numerical simulation and experimental diagnostics of quantum magnetofluid dynamics in next-generation plasma systems.

6.3.1 Single-Fluid and Multi-species QMHD Extensions

In the macroscopic treatment of plasma, the classical magnetohydrodynamic (MHD) model provides a well-established framework that unifies Maxwell's equations with the Navier–Stokes-type fluid equations for a conducting medium. However, in quantum plasmas, especially those under high density, strong confinement, or extreme magnetization, such a formulation fails to incorporate crucial effects such as quantum diffraction, spin polarization, and Fermi pressure. The development of Quantum Magnetohydrodynamics (QMHD) represents a pivotal generalization, encompassing the microscopic quantum structure into the macroscopic evolution of plasma variables.

Single-Fluid Quantum Magnetohydrodynamics

The single-fluid QMHD model assumes a quasi-neutral, isothermal plasma with a common velocity field and thermodynamic closure. It is valid when electrons and ions are tightly coupled by electrostatic forces and thermal velocities are equilibrated. The dynamical variables are the total mass density ρ, center-of-mass velocity $\mathbf{u}$, and magnetic field $\mathbf{B}$, supplemented by quantum terms in the momentum equation.

Let $\rho = \sum_\alpha m_\alpha n_\alpha$, where m_α and n_α are the mass and number density of species α. The single-fluid QMHD equations take the form:

$$\frac{\partial \rho}{\partial t} + \nabla \cdot (\rho \mathbf{u}) = 0$$

$$\rho\left(\frac{\partial \mathbf{u}}{\partial t}+(\mathbf{u}\cdot\nabla)\mathbf{u}\right)=-\nabla P+\mathbf{J}\times\mathbf{B}+\rho\mathbf{g}+\mathbf{F}_Q+\mathbf{F}_S$$

$$\frac{\partial \mathbf{B}}{\partial t}=\nabla\times(\mathbf{u}\times\mathbf{B})-\nabla\times(\eta\mathbf{J})$$

where P is the total scalar pressure including Fermi degeneracy pressure,
$\mathbf{F}_Q=\frac{\hbar^2}{2m_e m_i}\nabla\left(\frac{\nabla^2\sqrt{\rho}}{\sqrt{\rho}}\right)$ represents the collective Bohm force,
$\mathbf{F}_S=\mu_0(\mathbf{M}\cdot\nabla)\mathbf{B}$ is the magnetization force due to spin alignment,
$\mathbf{J}=\nabla\times\mathbf{B}/\mu_0$ is the total current.
This framework captures both classical MHD features and quantum mechanical effects through $\mathbf{F}_Q$ and $\mathbf{F}_S$. For degenerate electrons, $P\sim P_F=\frac{(3\pi^2)^{2/3}\hbar^2}{5m_e}n_e^{5/3}$, reflecting Fermi pressure dominance.

The single-fluid QMHD model provides a self-consistent macroscopic description of quantum plasmas under magnetization. The quantum Bohm term, reflecting collective coherence and wavefunction interference, governs the dispersive corrections to fluid motion. The spin force, coupled to $\nabla\mathbf{B}$, allows for magnetization-induced pressure anisotropies and spin-wave excitations. In strong magnetic fields, these terms induce new magnetoacoustic modes, modify Alfvén wave dispersion, and contribute to quantum magnetosonic turbulence [18].

Multi-species Quantum Magnetohydrodynamics

The limitations of the single-fluid model particularly its neglect of species separation, charge density fluctuations, and mass asymmetry necessitate a multi-fluid generalization. This is indispensable in environments where inertial effects, charge separation, and inter-species drift currents are significant, such as in white dwarfs, laser-plasma interfaces, or QED-scale astrophysical magnetospheres.

For each species $\alpha\in\{e,i,\ldots\}$, the fluid variables n_α, $\mathbf{u}_\alpha$, and P_α evolve under:

$$\frac{\partial n_\alpha}{\partial t}+\nabla\cdot(n_\alpha\mathbf{u}_\alpha)=0$$

$$m_\alpha n_\alpha\left(\frac{\partial \mathbf{u}_\alpha}{\partial t}+(\mathbf{u}_\alpha\cdot\nabla)\mathbf{u}_\alpha\right)=-\nabla P_\alpha+q_\alpha n_\alpha(\mathbf{E}+\mathbf{u}_\alpha\times\mathbf{B})+\mathbf{F}_{Q,\alpha}+\mathbf{F}_{s,\alpha}$$

$$\nabla\cdot\mathbf{E}=\frac{1}{\epsilon_0}\sum_\alpha q_\alpha n_\alpha,\ \nabla\times\mathbf{B}=\mu_0\sum_\alpha q_\alpha n_\alpha\mathbf{u}_\alpha+\epsilon_0\mu_0\frac{\partial \mathbf{E}}{\partial t}$$

with:

$$\mathbf{F}_{Q,\alpha}=\frac{\hbar^2}{2m_\alpha}\nabla\left(\frac{\nabla^2\sqrt{n_\alpha}}{\sqrt{n_\alpha}}\right),$$

$$\mathbf{F}_{s,\alpha}=\mu_\alpha(\mathbf{M}_\alpha\cdot\nabla)\mathbf{B},$$

$\mu_\alpha$is the magnetic moment of species α.

This multi-fluid QMHD formalism distinguishes interpenetrating fluid species, each responding differently to electromagnetic fields and quantum pressures. The inclusion of separate continuity and momentum equations ensures that charge separation, streaming instabilities, and differential transport can be dynamically resolved.

Coupled QMHD Modes and Interactions

The dispersion properties of the system can be extracted by linearizing the multi-species equations around a homogeneous equilibrium and introducing plane-wave perturbations $\propto e^{i(\mathbf{k}\cdot\mathbf{r}-\omega t)}$. The quantum corrections modify the dispersion relations for classical MHD waves. For example:

Quantum-Modified Alfvén Wave:

$$\omega^2 = k^2 v_A^2 + \frac{\hbar^2 k^4}{4 m_e m_i}$$

where $v_A = B_0/\sqrt{\mu_0 \rho_0}$, showing quantum stiffening at short wavelengths.

Spin-Alfvén coupling leads to anisotropic dispersion for spin-aligned plasmas under strong magnetic field, introducing a spin resonance frequency scale. The coupling of multiple species further results in rich collective modes such as quantum ion-acoustic waves, quantum Buneman-type instabilities, and magnetized Bernstein modes with quantum broadening.

Magnetization, Pressure Closure, and Relativistic Extensions

A crucial ingredient in both single- and multi-fluid QMHD models is the closure relation. The pressure P_α is often approximated using the equation of state derived from equilibrium Fermi–Dirac distributions. However, under strong magnetization, the Landau quantization must be accounted for, leading to:

$$P_\alpha(B) = \sum_{n=0}^{\infty} g_n \int \frac{dp_z}{2\pi\hbar} \frac{\left(p_z^2 + 2n|q_\alpha|\hbar B\right)^{1/2}}{e^{\beta(E_n-\mu_\alpha)} + 1}$$

with $g_n = 2 - \delta_{n0}$ as spin degeneracy, manifesting the quantum magnetic pressure anisotropy. For ultra-strong magnetic fields (e.g., $B \gtrsim 10^{12} G$), this anisotropy becomes dominant, necessitating tensorial pressure formulations and anisotropic QMHD.

Further, relativistic generalizations of multi-species QMHD can be constructed via the covariant formulation of quantum fluid dynamics, employing the quantum generalization of the energy–momentum tensor and incorporating the Dirac spin current and the field strength tensor into a unified stress-energy framework. These extensions are pivotal in modeling quantum plasma near compact astrophysical objects and in ultra-intense laser–plasma interactions.

The QMHD framework, in both single-fluid and multi-fluid configurations, offers a powerful and self-consistent formalism for modeling the collective dynamics of magnetized quantum plasmas. The single-fluid model suffices in many laboratory and semi-classical regimes, whereas multi-species formulations are essential for capturing species-specific quantum responses, interpenetrating current systems, and magnetization-induced anisotropies. Through rigorous derivation and careful closure prescriptions incorporating quantum statistics, Bohm potential, and spin-magnetization couplings, the QMHD extensions forge an essential bridge between kinetic quantum field descriptions and fluid-scale dynamics, making them indispensable in both theoretical and applied research on quantum plasmas under strong fields.

6.4 Spin-1/2 and Spin-3/2 Fluid Models Under Magnetization

Magnetized quantum plasmas with significant spin contributions present a rich interplay between quantum statistics, spin dynamics, and electromagnetic fields. Traditional fluid models, which neglect spin degrees of freedom, fail to capture a multitude of phenomena such as spin-polarization transport, magnetization pressure anisotropy, and quantum spin torque effects.

Quantum Fluid Formalism with Spin-1/2: Pauli–Madelung Decomposition and Magnetization

The simplest extension of fluid models to include intrinsic spin arises from considering the single-particle Pauli equation in an external electromagnetic field:

$$i\hbar\frac{\partial\psi}{\partial t}=\left[\frac{1}{2m}(-i\hbar\nabla-q\mathbf{A})^2+q\phi-\mu_B\boldsymbol{\sigma}\cdot\mathbf{B}\right]\psi$$

where $\psi(\mathbf{r},t)\in\mathbf{C}^2$ is a spinor wavefunction, $\boldsymbol{\sigma}$ are the Pauli matrices, $\mu_B=\frac{q\hbar}{2m}$ is the Bohr magneton, and $\mathbf{B}=\nabla\times\mathbf{A}$. Introducing the Madelung decomposition:

$$\psi=\sqrt{n}\chi e^{iS/\hbar},\quad \chi^\dagger\chi=1$$

yields a spinor hydrodynamic formulation, where the scalar density $n(\mathbf{r},t)$, phase S, and spin direction vector $\mathbf{s}=\chi^\dagger\boldsymbol{\sigma}\chi$ become dynamical variables. One then derives the following continuity and momentum equations:

$$\frac{\partial n}{\partial t}+\nabla\cdot(n\mathbf{v})=0$$

$$m\left(\frac{\partial}{\partial t}+\mathbf{v}\cdot\nabla\right)\mathbf{v}=q(\mathbf{E}+\mathbf{v}\times\mathbf{B})-\nabla Q+\mathbf{F}_{\text{spin}}\,.$$

Here, $Q = -\frac{\hbar^2}{2m}\frac{\nabla^2\sqrt{n}}{\sqrt{n}}$ is the Bohm quantum potential and $\mathbf{F}_{\text{spin}} = \mu_B\nabla(\mathbf{s}\cdot\mathbf{B})$ is the spin-magnetic coupling force. The evolution of the spin field is governed by a precession equation:

$$\frac{d\mathbf{s}}{dt} = \frac{2\mu_B}{\hbar}\mathbf{s}\times\mathbf{B}$$

Thus, the spin-1/2 quantum fluid under magnetization exhibits three-fold coupling: Lorentz dynamics of charged carriers, spin precession due to magnetic torque, and a spin-induced pressure tensor that contributes anisotropically to the momentum flux.

In dense magnetized plasmas, such as those in white dwarfs or solid-density laser-compressed targets, the spin polarization $\langle\mathbf{s}\rangle$ becomes macroscopic. This gives rise to spin-dependent pressure anisotropy, modified Alfvénic modes, and spin alignment instabilities. A notable feature is the magnetization current $\mathbf{J}_M = \nabla\times\mathbf{M}$, where $\mathbf{M} = n\mu_B\mathbf{s}$ behaves as an intrinsic magnetization vector and contributes to the total current entering Maxwell's equations [19].

Spin-1/2 Kinetic Closure and Magneto-Spin Transport Tensor

A more refined description arises from quantum kinetic theory. The spin-1/2 Wigner function $W(\mathbf{r},\mathbf{p},t)\in\mathbf{C}^{2\times 2}$ encodes both position-momentum and spin correlations. It can be decomposed as:

$$W = f_0 I + \mathbf{f}_s\cdot\boldsymbol{\sigma}$$

where f_0 is the scalar phase-space distribution and $\mathbf{f}_s$ is the spin polarization distribution function. The quantum Vlasov equation, linearized around a magnetized background, leads to coupled transport equations:

$$\left(\frac{\partial}{\partial t} + \frac{\mathbf{p}}{m}\cdot\nabla + q\left(\mathbf{E} + \frac{\mathbf{p}}{m}\times\mathbf{B}\right)\cdot\nabla_p\right)f_0 = \mu_B\left(\mathbf{B}\cdot\nabla_p\right)\cdot\mathbf{f}_s$$

$$\left(\frac{\partial}{\partial t} + \frac{\mathbf{p}}{m}\cdot\nabla + q\left(\mathbf{E} + \frac{\mathbf{p}}{m}\times\mathbf{B}\right)\cdot\nabla_p\right)\mathbf{f}_s = \mu_B\left(\mathbf{B}\times\mathbf{f}_s - \left(\nabla_p f_0\right)\times\mathbf{B}\right)$$

These equations describe spin–orbit coupling, magnetic moment precession, and the spin Hall-like transport behavior. Integrating over momentum yields macroscopic fluid moments, including the spin-aligned magnetization tensor Π^{ij} and spin-vorticity coupling terms, which are essential for modeling spin-polarized flows in rotating and sheared plasma media.

Spin-3/2 Quantum Fluids: The Rarita–Schwinger Approach

While spin-1/2 quantum fluids are directly associated with electrons and positrons, spin-3/2 systems become relevant for particles such as delta baryons $\left(\Delta^{\pm}\right)$, gravitinos in supersymmetric models, or emergent collective quasi-particles in spin-polarized nuclear plasmas. The field-theoretic description of spin-3/2 particles is given by the

Rarita–Schwinger equation [20]:

$$(i\gamma^{\mu}\partial_{\mu} - m)\psi_{\nu} = 0,\ \gamma^{\mu}\psi_{\mu} = 0$$

where $\psi_{\mu} \in \mathbf{C}^4 \otimes \mathbf{R}^4$ is a spinor-vector field, satisfying the auxiliary transversality and gamma-trace constraints. The propagation of such fields in electromagnetic backgrounds requires consistent coupling that avoids superluminal propagation and maintains causality. In a magnetized plasma, the minimal-coupling prescription modifies the Dirac operator:

$$(i\gamma^{\mu}D_{\mu} - m)\psi_{\nu} = 0,\ D_{\mu} = \partial_{\mu} + iqA_{\mu}$$

Hydrodynamically, a spin-3/2 plasma is described by a set of generalized moments: the particle density n, four-velocity u^{μ}, and a third-rank polarization tensor $S^{\mu\nu\rho}$, symmetric in its last two indices. The evolution of $S^{\mu\nu\rho}$ under magnetization involves spin precession driven by the Lorentz torque and higher-order magnetic multipoles. The covariant conservation equation takes the schematic form:

$$\partial_{\lambda}T^{\lambda\mu} = qF^{\mu\nu}J_{V} + \nabla_{V}M^{\mu\nu},\ \partial_{\lambda}S^{\lambda\mu\nu} = \mathcal{T}^{\mu\nu}$$

where $M^{\mu\nu}$ is the magnetization-polarization tensor, and $\mathcal{T}^{\mu\nu}$ is the spin torque density. These equations are essential to model transport in spin-3/2-dominated environments, such as nuclear spin-polarized matter, or quark-gluon plasmas with high isospin chemical potentials [23].

Physically, spin-3/2 systems exhibit richer coupling structures due to their extended magnetic moment tensors and possible quadrupole interactions. Under strong magnetization, such particles can align in multiple spin configurations, resulting in non-trivial collective excitations and anisotropic stress contributions. In particular, wave dispersion relations in spin-3/2 plasmas show additional branches corresponding to multipolar spin-flip modes, modified by magneto-geometric effects [24].

The inclusion of spin degrees of freedom under magnetization modifies the plasma dielectric response tensor, leading to emergent collective modes. For spin-1/2 plasmas, spin-Alfvén and spin-cyclotron modes arise due to the coupling between magnetization currents and electromagnetic perturbations. The generalized dielectric tensor $\varepsilon_{ij}(\omega, \mathbf{k})$ acquires corrections of the form:

$$\delta\varepsilon_{ij}^{(\text{spin})} \propto \frac{n\mu_B^2 B_0}{m\omega^2}\left(\delta_{ij} - \frac{k_i k_j}{k^2}\right)$$

which leads to spin-polarization dependent splitting of wave dispersion branches. For spin-3/2 fluids, the additional internal degrees of freedom allow for higher-order magnetic resonance phenomena and spin-induced anisotropic pressure evolution, affecting plasma stability.

In ultra-relativistic regimes, the spin dynamics must be treated within the framework of the Bargmann–Michel–Telegdi (BMT) equation [21] for spin precession, or its generalization for higher-spin fields. The effective spin-polarization density couples to the vorticity and leads to spin-vorticity-induced current generation an effect closely related to the chiral vortical effect in quantum field theory.

The formulation of spin-1/2 and spin-3/2 fluid models in magnetized quantum plasmas reveals a complex and highly structured dynamical landscape. For spin-1/2 systems, magnetization induces intrinsic coupling between charge, velocity, and spin degrees of freedom, giving rise to collective spin dynamics and anisotropic stresses. For spin-3/2 systems, additional tensor degrees of freedom emerge, modifying the conservation laws and the spectrum of excitations. Both classes of models underscore the necessity of incorporating spin dynamics into high-field plasma modeling, particularly in astrophysical and high-energy density laboratory settings. These frameworks enrich the theoretical foundations of quantum plasma physics and serve as indispensable tools for analyzing next-generation plasma experiments and compact object environments.

6.5 Quantum Instabilities in Magnetized Plasmas

In magnetized quantum plasmas, the presence of strong external magnetic fields, quantum degeneracy, and intrinsic particle correlations leads to a complex interplay between kinetic fluctuations and collective modes. This interplay can give rise to a rich spectrum of plasma instabilities, where small perturbations in the distribution function, current density, or field configuration grow exponentially in time. Unlike classical plasmas, quantum instabilities in such systems are significantly modified by effects including Fermi pressure, Bohm potential, spin polarization, and Landau quantization.

6.5.1 The Weibel Instability: Quantum Kinetic Anisotropy

The Weibel Instability arises when the plasma distribution function is anisotropic in momentum space, typically due to a temperature imbalance between directions parallel and perpendicular to the ambient magnetic field. Classically, this instability generates spontaneous magnetic fields via current filamentation. In quantum plasmas, Fermi degeneracy and quantum diffraction modify the dispersion relations and growth rates.

Let the equilibrium distribution be an anisotropic Fermi–Dirac distribution $f_0(\mathrm{p})$ with $T_\perp \neq T_\parallel$. The quantum dielectric tensor for transverse electromagnetic modes is derived from the Wigner–Maxwell system, yielding the dispersion relation:

$$\det\left[\epsilon_{ij}^{(\varrho)}(\omega, \mathbf{k}) - \left(\frac{c^2k^2}{\omega^2}\delta_{ij} - \frac{c^2k_ik_j}{\omega^2}\right)\right] = 0,$$

where $\epsilon_{ij}^{(\varrho)}$ incorporates quantum corrections, including the Wigner–Vlasov term and the Bohm potential. Linearizing the Wigner equation with quantum closure, the modified Weibel growth rate in the long-wavelength limit becomes:

$$\gamma_W^2 \approx \frac{\omega_p^2 v_{\mathrm{tb},\perp}^2}{c^2}\left(\frac{T_\perp}{T_\parallel} - 1\right) - \frac{\hbar^2k^4}{4m^2},$$

highlighting the stabilizing role of quantum dispersion $\left(\sim \hbar^2\right)$ that suppresses short-wavelength filamentation. In ultra-relativistic degenerate plasmas, the anisotropy may emerge not from temperature, but from phase-space deformation due to spin–orbit interaction or confinement geometry, further complicating the Weibel growth mechanism.

6.5.2 *Firehose Instability: Quantum Anisotropic Pressure Balance*

The firehose instability is associated with negative pressure anisotropy where $P_\parallel > P_\perp$, violating the force balance along magnetic field lines. This leads to the growth of Alfvénic perturbations in a magnetized plasma. In quantum settings, the momentum transport equations acquire additional contributions from the quantum pressure tensor and spin-aligned magnetic susceptibility. The modified fluid momentum equation along field-aligned direction reads:

$$\rho\frac{d\mathbf{v}}{dt} = -\nabla P + \frac{1}{\mu_0}(\nabla \times \mathbf{B}) \times \mathbf{B} + \mathbf{F}_{\mathrm{Bohm}} + \mathbf{F}_{\mathrm{spin}},$$

with:

$\mathbf{F}_{\mathrm{Bohm}} = \frac{\hbar^2}{2m}\nabla\left(\frac{\nabla^2\sqrt{n}}{\sqrt{n}}\right)$ representing quantum dispersion, and

$\mathbf{F}_{\mathrm{spin}} \sim \nabla(\chi_m\mathbf{B})$ accounting for spin-induced paramagnetism.

Linearizing about a homogeneous background and introducing pressure anisotropy $\Delta = P_\parallel - P_\perp$, the quantum firehose criterion becomes:

$$\Delta > \frac{B^2}{\mu_0}\left(1 + \delta_Q\right), \ \delta_Q \sim \frac{\hbar^2k^2}{m^2v_\parallel^2},$$

indicating that quantum corrections increase the threshold for instability by enhancing the effective magnetic tension.

In high-density degenerate plasmas such as those in white dwarfs or magnetars, Landau quantization may further stabilize the firehose mode by discretizing the transverse pressure response and modifying the magnetization.

6.5.3 Magnetoacoustic Instabilities: Quantum Magnetosonic Modes

Magnetoacoustic Instabilities arise from coupling between compressional plasma waves and magnetic perturbations. The basic modes slow and fast magnetosonic waves are modified in quantum plasmas by the inclusion of quantum pressure, magnetization, and spin–magnetic couplings. The quantum MHD equations yield a linearized dispersion relation for magnetosonic waves propagating at an angle θ to $\mathbf{B}_0$:

$$\omega^2 = k^2\left(c_s^2 + v_A^2\right) \pm k^2\sqrt{c_s^4 + v_A^4 - 2c_s^2 v_A^2 \cos 2\theta + Q(k)},$$

where: c_s is the quantum-modified sound speed,
$v_A = \frac{B_0}{\sqrt{\mu_0 \rho}}$ is the Alfvén speed, $Q(k)$ includes quantum corrections: $Q(k) \sim \frac{\hbar^2 k^2}{m^2}\left(\frac{1}{n}\frac{\partial^2 P}{\partial n^2} + \frac{\chi_m}{\mu_0}\right)$.

When dissipative or anisotropic effects are introduced such as differential streaming or spin polarization the slow branch can develop an imaginary component, signifying a quantum magnetoacoustic instability. These modes are particularly important in astrophysical magnetoplasmas where strong gradients and nonuniform magnetic fields couple to degeneracy pressure and spin alignment.

6.5.4 Streaming Instabilities: Quantum Counterstreaming and Anisotropic Drift

Quantum streaming instabilities arise in two or more interpenetrating plasma populations with a relative drift velocity. In classical plasmas, this manifests as two-stream or filamentation instabilities. In quantum plasmas, the Pauli's exclusion principle restricts phase-space overlap, and quantum recoil modifies wave-particle resonances.

For two counterstreaming degenerate beams with drift velocity $\pm v_0$, the dielectric function derived from the quantum linear response theory takes the form:

$$\epsilon(\omega, k) = 1 - \frac{2\omega_p^2}{k^2 v_F^2}\left[1 - \frac{\omega}{2kv_F}\ln\left(\frac{\omega + kv_F + kv_0}{\omega - kv_F + kv_0}\right)\right] + \frac{\hbar^2 k^4}{4m^2\omega^2},$$

where v_F is the Fermi velocity. The term $\hbar^2 k^4/(4m^2\omega^2)$ acts as a quantum regularizer, suppressing the instability at small scales. For sufficiently large v_0, the logarithmic term admits an imaginary component, resulting in a complex frequency $\omega = \omega_r + i\gamma$, where $\gamma > 0$ indicates instability. Quantum streaming instabilities are vital in ultracold plasma experiments, semiconductor superlattices, and early-universe pair production fronts [22].

6.5.5 Beam–Plasma Instabilities: Quantum Langmuir–Cherenkov Dynamics

Beam–plasma instabilities occur when a population of high-velocity particles (beam) penetrates a background plasma. The classical instability is driven by resonance between beam particles and Langmuir waves. In quantum plasmas, the dielectric function is modified by both quantum recoil and Fermi–Dirac statistics. The general dispersion relation is given by:

$$\varepsilon(\omega, k) = 1 - \frac{\omega_{pb}^2}{(\omega - kv_b)^2} - \frac{\omega_{pe}^2}{\omega^2 - 3k^2 v_{Fe}^2 - \hbar^2 k^4/4m^2},$$

where: ω_{pb} and v_b are the beam plasma frequency and velocity, ω_{pe}, v_{Fe} are the background plasma frequency and Fermi velocity. The growth rate peaks when $kv_b \approx \omega_{pe}$, satisfying the quantum-modified Cherenkov resonance condition. The presence of $\hbar^2$ in the denominator introduces a cutoff wavevector above which no unstable modes exist, defining a quantum stabilization scale $k_c \sim \left(\frac{4m^2 v_b^2}{\hbar^2}\right)^{1/4}$.

These instabilities are crucial in laser–plasma interactions, fast ignition schemes in inertial fusion, and astrophysical jets encountering cold magnetized backgrounds.

6.6 Quantum Turbulence in Magnetized Regimes

Quantum turbulence in magnetized plasma represents one of the most intricate frontiers in modern plasma physics, arising from the synthesis of quantum fluid dynamics and electromagnetic field theory under nontrivial boundary and initial conditions. Unlike classical turbulence, where vorticity and energy cascades are continuous and stochastic, quantum turbulence is deeply influenced by discrete, coherent structures such as quantized vortices, and by the fundamental quantum properties of matter waves. When an external magnetic field threads a quantum plasma whether composed of degenerate electrons, electron–positron pairs, or ultracold ionized gases the interplay of Lorentz forces, magnetic helicity, and quantum pressure leads to a regime that is neither classically turbulent nor strictly linear, but governed by a complex nonlinear quantum field theory framework.

At the heart of quantum turbulence in magnetized regimes lies the generalization of the hydrodynamic equations to include quantum effects. The governing framework typically extends the classical magnetohydrodynamic (MHD) system into the Quantum Magnetohydrodynamic (QMHD) regime, which incorporates quantum-statistical corrections through terms like the Bohm potential, spin contributions, and in high-density regimes, exchange–correlation effects. The foundational equations include the continuity equation for particle number conservation, the quantum-modified Navier–Stokes, or Euler equation with Lorentz force and quantum potential terms, and Maxwell's equations for electromagnetic fields. The quantum-modified momentum equation takes the form:

$$m\left(\frac{\partial \mathbf{v}}{\partial t} + (\mathbf{v} \cdot \nabla)\mathbf{v}\right) = -\nabla P + q(\mathbf{E} + \mathbf{v} \times \mathbf{B}) - \nabla V_Q,$$

where $V_Q = -\frac{\hbar^2}{2m}\frac{\nabla^2\sqrt{n}}{\sqrt{n}}$ represents the quantum Bohm potential arising from wave-function dispersion. This potential plays a fundamental role in dictating how quantum pressure acts to delocalize particle density and smooth out turbulent gradients that would otherwise steepen in classical systems.

A crucial diagnostic in turbulence studies is the vorticity field, defined as $\boldsymbol{\omega} = \nabla \times \mathbf{v}$. In the presence of magnetic fields, vorticity evolves according to an equation derived from the curl of the quantum-modified momentum equation. The resulting vorticity transport equation includes not only the standard vortex stretching term and baroclinic torque, but also contributions from magnetic curvature and quantum mechanical stress tensors:

$$\frac{\partial \boldsymbol{\omega}}{\partial t} + (\mathbf{v} \cdot \nabla)\boldsymbol{\omega} = (\boldsymbol{\omega} \cdot \nabla)\mathbf{v} - \boldsymbol{\omega}(\nabla \cdot \mathbf{v}) + \frac{q}{m}\nabla \times (\mathbf{v} \times \mathbf{B}) + \frac{1}{m}\nabla \times \left(-\nabla V_Q\right).$$

This equation underscores the fact that quantum effects can seed or dampen vorticity depending on the local curvature of density fields, while the magnetic field couples into vorticity dynamics through the frozen-in flux constraint and magnetic tension.

The structure of quantum turbulence in magnetized settings is highly anisotropic due to the directional nature of the external field. The energy cascade that underlies turbulent behaviour no longer remains isotropic in Fourier space; instead, energy preferentially cascades across field lines, with reduced transfer along the field. This is captured in the modified spectral energy density, which often reveals a dual cascade: a Kolmogorov-like inertial range transverse to **B** and a suppressed or altered spectrum along **B**. For instance, the spectral density $E(k)$ in quantum turbulence may follow a modified scaling law such as:

$$E(k_\perp) \propto k_\perp^{-\alpha}, \quad \alpha \in \{5/3, 3, 7/3\},$$

depending on the role of quantum diffraction, compressibility, and scale-locality of energy transfers. The nature of quantized vortices in quantum turbulence, wherein circulation is discretized in units of h/m, imposes topological constraints that modify the cascade process compared to classical fluids. These quantized vortices can reconnect, form vortex bundles, and self-organize into lattices or turbulent tangles, and their evolution is deeply influenced by magnetic field geometry.

From a kinetic theory perspective, the modeling of quantum turbulence employs generalizations of the Wigner–Moyal formalism or the quantum Vlasov equation. The Wigner function $f_W(x, p, t)$, interpreted as a quasi-distribution in phase space, evolves according to:

$$\frac{\partial f_W}{\partial t} + \frac{p}{m} \cdot \nabla_x f_W + \mathcal{Q}[f_W] = 0,$$

where $\mathcal{Q}[f_W]$ encapsulates nonlocal quantum corrections arising from the potential, including spin–orbit coupling, or exchange effects if incorporated. This approach captures phase-space vorticity, quantum interference fringes, and the entropic structure of quantum turbulent fields. In strong magnetic fields, the Landau quantization of energy levels further modifies the kinetic landscape, leading to discrete transitions and magnetized energy redistribution across quantum states.

Vorticity generation in such plasmas can arise through several nonclassical mechanisms. Baroclinic torque, resulting from misalignment between density and pressure gradients, is a traditional source of vorticity in classical flows. In quantum magnetized plasmas, however, vorticity can also be generated through magnetovorticity coupling, where gradients in the magnetic field act on spin-aligned particles or modulate the wavefunction phase to create rotational structures. Additionally, the quantum pressure tensor often neglected in semiclassical treatments can act as a seed for turbulent eddies in the absence of collisional mechanisms.

The study of quantum turbulence in magnetized regimes is not merely a theoretical pursuitit has practical implications across several domains. In astrophysical plasmas, turbulent reconnection processes, pulsar magnetospheres, and neutron star crusts may exhibit turbulence that is inherently quantum and strongly magnetized. In fusion plasmas, understanding and controlling quantum fluctuations and their nonlinear interactions with magnetic confinement fields is essential for the stability and efficiency of next-generation reactors. Moreover, recent experiments on ultracold quantum plasmas and Bose–Einstein condensates under synthetic gauge fields have opened up laboratory avenues to simulate magnetized quantum turbulence under controllable conditions.

Quantum turbulence in magnetized plasmas represents a rich interplay of nonlinearity, coherence, topology, and field-mediated interactions. A coherent mathematical description demands the merging of hydrodynamic, kinetic, and quantum mechanical formalisms, each of which reveals different aspects of the turbulent cascade. The inclusion of magnetic fields breaks the symmetry of energy transfer, introduces new instability channels, and reshapes vortex dynamics, leading to an anisotropic, quantized, and profoundly rich turbulence landscape unique to quantum plasma systems.

References

1. Medvedev, M. V. (2023). Plasma modes in QED super-strong magnetic fields of magnetars and laser plasmas. *Physics of Plasmas, 30*(9).
2. Smolyakov, A. I., Chapurin, O., Frias, W., Koshkarov, O., Romadanov, I., Tang, T., et al. (2016). Fluid theory and simulations of instabilities, turbulent transport and coherent structures in partially-magnetized plasmas of $\mathbf{e} \times \mathbf{b}$ discharges. *Plasma Physics and Controlled Fusion, 59*(1), 014041.
3. Kim, J. Y., Jang, J. Y., Choi, J., Wang, J. I., Jeong, W. I., Elgarhy, M. A. I., et al. (2021). Magnetic confinement and instability in partially magnetized plasma. *Plasma Sources Science and Technology, 30*(2), 025011.
4. Brown, L. S., & Yaffe, L. G. (2001). Effective field theory for highly ionized plasmas. *Physics Reports, 340*(1–2), 1–164.
5. Rylyuk, V. M., & Tkachenko, I. M. (2023). Electromagnetism of one-component plasmas of massless fermions. *Journal of Plasma Physics, 89*(4), 905890415.
6. Vazza, F., Locatelli, N., Rajpurohit, K., Banfi, S., Domínguez-Fernández, P., Wittor, D., et al. (2021). Magnetogenesis and the cosmic web: A joint challenge for radio observations and numerical simulations. *Galaxies, 9*(4), 109.
7. Zhang, C., Huang, C. K., & Joshi, C. (2023). Self-organization of photoionized plasmas via kinetic instabilities. *Reviews of Modern Plasma Physics, 7*(1), 34.
8. Cheremisin, M. V. (2014). On the quantum hall effect in mono (bi)-layer graphene. *Physica E: Low-Dimensional Systems and Nanostructures, 64*, 15–22.
9. Arutyunov, K. Y., Lehtinen, J. S., Radkevich, A. A., Semenov, A. G., & Zaikin, A. D. (2018). Smearing of electron density of states in quasi-one-dimensional superconducting channels due to quantum phase fluctuations. *Journal of Magnetism and Magnetic Materials, 459*, 356–358.
10. Ipp, A., Rebhan, A., & Strickland, M. (2011). Non-abelian plasma instabilities: Su (3) versus su (2). *Physical Review D—Particles, Fields, Gravitation, and Cosmology, 84*(5), 056003.
11. Rajawat, R. S., Shvets, G., & Khudik, V. (2025). Continuum damping of topologically protected edge modes at the boundary of cold magnetized plasmas. *Physical Review Letters, 134*(5), Article 055301.
12. Das Sarma, S., & Hwang, E. H. (2009). Collective modes of the massless Dirac plasma. *Physical Review Letters, 102*(20), Article 206412.
13. Tsujimura, T. I., Yanagihara, K., Goto, Y., & Kubo, S. (2021). Trajectory shift in propagation of electron cyclotron waves due to berry curvature in magnetized plasma. *Plasma and Fusion Research, 16*, 2401009–2401009.
14. Rat, V., André, P., Aubreton, J., Elchinger, M. F., Fauchais, P., & Lefort, A. (2001). Transport properties in a two-temperature plasma: Theory and application. *Physical Review E, 64*(2), Article 026409.
15. Naulin, V., Garcia, O. E., Nielsen, A. H., & Rasmussen, J. J. (2004). Statistical properties of transport in plasma turbulence. *Physics Letters A, 321*(5–6), 355–365.
16. Pogodaeva, M. K., Levchenko, S. V., & Drachev, V. P. (2023). Spin-dependent plasma frequency from all-electron ab initio calculations including spin-orbit coupling. *Physical Review B, 107*(4), Article 045113.
17. Danel, J. F., Kazandjian, L., & Zérah, G. (2008). Equation of state of a dense boron plasma by Thomas–Fermi–Dirac–Weizsäcker molecular dynamics. *Physics of Plasmas, 15*(7).
18. Hager, Y. A., Khaled, M. A., & Shukri, M. A. (2023). Magnetosonic waves propagation in a magnetorotating quantum plasma. *Physical Review E, 107*(5), Article 055202.
19. Kulsrud, R. M., Valeo, E. J., & Cowley, S. C. (1986). Physics of spin-polarized plasmas. *Nuclear Fusion, 26*(11), 1443.
20. Landau, L. (1950). Concerning WP Allis' criticism of my paper on coulomb interactions in a plasma. *Physical Review, 77*(4), 567.
21. Ternov, A. I. (2016). Matter-induced magnetic moment and neutrino helicity rotation in external fields. *Physical Review D, 94*(9), Article 093008.

22. Bass, F. G., & Tetervov, A. P. (1986). High-frequency phenomena in semiconductor superlattices. *Physics Reports, 140*(5), 237–322.
23. Dellweg, M. M., & Müller, C. (2017). Controlling electron spin dynamics in bichromatic Kapitza-Dirac scattering by the laser field polarization. *Physical Review, 95*(4), 042124.
24. Shuryak, E. (2017). Strongly coupled quark-gluon plasma in heavy ion collisions. Reviews of modern physics. *89*(3), 035001.

Chapter 7
Spin Polarization in Quantum Plasmas

The emergence of spin-polarized quantum plasma physics represents a natural yet nontrivial extension of the established theoretical edifice of quantum plasma dynamics. While preceding formulations have extensively addressed collective phenomena governed by charge, density, and energy transport in the quantum regime, the incorporation of intrinsic spin degrees of freedom demands an enriched kinematical and dynamical framework. Spin polarization becomes physically significant in plasma systems that are either strongly magnetized, extremely dense, or operate at ultra-low temperatures, where quantum degeneracy effects prevail and the spin degree of freedom ceases to be a passive label, instead actively participating in the collective evolution of the plasma state.

In astrophysical environments, particularly in the crusts of magnetars and white dwarfs, the magnetic field strength can reach or exceed the critical Landau scale $B_c = \frac{m_e^2 c^3}{e\hbar} \approx 4.4 \times 10^{13}$G, at which the magnetic energy of an electron becomes comparable to its rest mass. Under such conditions, the electron population is significantly spin-polarized due to the Zeeman splitting of Landau levels. Consequently, the plasma exhibits anisotropic transport properties, modified equation-of-state behaviour, and emergent spin-dependent collective excitations. Further, the degeneracy pressure of electrons in such regimes must be corrected for spin imbalance, leading to altered stability criteria and modified dispersion characteristics of plasma modes.

In terrestrial laboratory systems, such as laser-compressed plasmas and solid-density plasmas generated in ultrashort-pulse laser interactions, the temperature may remain sub-eV while densities reach $10^{22} - 10^{24}\mathrm{cm}^{-3}$. At such high densities, even in the absence of external magnetic fields, exchange and correlation effects between fermions render the spin dynamics non-negligible. Moreover, when spin-polarized electron beams are injected into plasmas or self-consistent magnetic fields develop via Weibel-like instabilities, spin currents and spin precession enter the dynamical equations, contributing to angular momentum transport and anisotropic heating.

P. Kumar, *Quantum Plasma*, https://doi.org/10.1007/978-981-95-0481-7_7

A third class of systems emerges from the burgeoning interface between quantum plasmas and spintronics. In these systems, spin-polarized carriers typically electrons are injected, manipulated, and detected in confined geometries where collective plasma behaviour (e.g., screening, wake formation, Langmuir oscillations) coexists with spin coherence and entanglement. This confluence opens new regimes of spin–charge coupled dynamics in quasi-one- and two-dimensional electron gases, quantum wells, and nano-plasma systems, where conventional drift–diffusion models become insufficient, and a hydrodynamic model enriched by spinor structure becomes essential.

From the perspective of theoretical plasma physics, the imperative to account for spin stems not merely from its presence in microscopic dynamics, but from its profound influence on macroscopic observables momentum transport, energy partitioning, magnetic structure formation, and stability thresholds all of which demand systematic inclusion of spin dynamics beyond perturbative or ad hoc treatments.

7.1 What is Spin Polarization?

Spin polarization, in its most fundamental sense, is a measure of asymmetry in the spin population distribution of fermionic plasma constituents. For an ensemble of electrons described by a spinor field $\psi(\mathbf{r}, t) \in \mathbf{C}^2$, the Spin polarization vector $\mathbf{S}(\mathbf{r}, t)$ is defined via the expectation value of the Pauli spin matrices:

$$\mathbf{S}(\mathbf{r}, t) = \psi^{\dagger}(\mathbf{r}, t)\widehat{\hat{\sigma}}\, \psi(\mathbf{r}, t),$$

where $\hat{\sigma} = \left(\hat{\sigma}_x, \hat{\sigma}_y, \hat{\sigma}_z\right)$ are the standard Pauli matrices.

In macroscopic quantum plasma models, especially those adopting a hydrodynamic or kinetic description, one transitions from the spinor field formalism to an ensemble-averaged representation. Here, the local Spin polarization vector is defined as

$$\langle \mathbf{S}(\mathbf{r}, t) \rangle = \sum_{s=\uparrow,\downarrow} f_s(\mathbf{r}, \mathbf{p}, t)\mathbf{s}_s,$$

where f_s is the spin-resolved distribution function and $\mathbf{s}_s \in \{\pm\hbar/2\}$ are the spin eigenvalues associated with the spin projection along a specified quantization axis. The total local polarization is then quantified by the dimensionless Spin polarization ratio:

$$\mathcal{P}(\mathbf{r}, t) = \frac{n_{\uparrow}(\mathbf{r}, t) - n_{\downarrow}(\mathbf{r}, t)}{n_{\uparrow}(\mathbf{r}, t) + n_{\downarrow}(\mathbf{r}, t)},$$

where $n_{\uparrow,\downarrow}$ are the spin-up and spin-down electron number densities. This scalar polarization index ranges from -1 (fully down-polarized) to $+1$ (fully up-polarized), with $\mathcal{P} = 0$ denoting an unpolarized plasma.

The physical origin of spin polarization in plasma systems arises from spin–magnetic coupling via the Zeeman interaction:

$$\mathcal{H}_Z = -\mu_B \hat{\sigma} \cdot \mathbf{B}(\mathbf{r}, t),$$

where $\mu_B = \frac{e\hbar}{2m_e}$ is the Landau levels. This interaction energetically favors alignment (or anti-alignment) of the spin vector with the local magnetic field, leading to a Boltzmann-weighted population imbalance in equilibrium:

$$\mathcal{P}_{\text{eq}} = \tanh\left(\frac{\mu_B B}{k_B T}\right),$$

which becomes significant at low temperatures and high magnetic fields, as typically encountered in degenerate quantum plasmas.

However, in non-equilibrium and dynamically evolving systems, the Spin polarization becomes a time- and space-dependent field obeying its own continuity and transport equations, influenced by spin precession, spin–orbit interaction, collective forces, and collision dynamics. Thus, unlike classical polarization, spin polarization in quantum plasmas possesses both intrinsic microscopic character and collective dynamical significance.

In practice, this necessitates the augmentation of the conventional QHD fluid variables density n, velocity $\mathbf{u}$, and pressure p with the spin vector field $\mathbf{S}(\mathbf{r}, t)$, and in some cases, its flux tensor or higher-order moments. These spin variables obey evolution equations derivable from the Pauli equation, Wigner–Pauli formalism, or moments of spin-resolved kinetic theories.

Critically, Spin polarization not only responds to electromagnetic fields but also sources magnetic fields through magnetization currents of the form:

$$\mathbf{J}_{\text{mag}} = \nabla \times \mathbf{M} = \nabla \times (\gamma \mathbf{S}),$$

where γ is the gyromagnetic ratio. This self-consistent feedback loop where Spin polarization modifies fields which in turn alter spin orientation introduces a nonlinear, nonlocal, and anisotropic structure into the governing dynamics of quantum plasmas [1].

7.1.1 Plasma Environments Where Spin is Relevant

The manifestation of spin effects in plasma environments is highly sensitive to the interplay between external constraints such as magnetic field strength, thermal energy, and plasma density and intrinsic quantum scales such as the thermal de Broglie wavelength, Fermi temperature, and cyclotron quantization energy. A systematic classification of regimes where Spin polarization acquires dynamical significance may be constructed by comparing relevant energy scales.

The Zeeman energy associated with a magnetic field **B** is given by

$$E_Z = \mu_B |\mathbf{B}| = \frac{e\hbar}{2m_e} |\mathbf{B}|,$$

while the characteristic thermal energy is $E_T = k_B T$, and the Fermi energy for a degenerate electron gas is

$$E_F = \frac{\hbar^2}{2m_e} \left(3\pi^2 n\right)^{2/3},$$

where n is the electron number density. Spin effects become non-negligible when either $E_Z \gtrsim E_T$ in low-temperature plasmas, or $E_Z \gtrsim E_F$ in degenerate systems. These criteria delineate the domains in which spin-resolved occupation, magnetic susceptibility, and spin transport phenomena must be included in a physically consistent plasma description.

(i) *Strongly Magnetized Astrophysical Plasmas*. In neutron stars and magnetars, magnetic fields often exceed the Landau critical field $B_c \approx 4.4 \times 10^{13}$G, where Landau quantization dominates particle motion transverse to the field. In such quantizing fields, electrons populate discrete Landau levels, and the ground state is spin-polarized due to the Zeeman shift. Further, the phase space available for opposite spin states becomes asymmetric, leading to substantial macroscopic spin polarization. The resulting effects include anisotropic pressure, modified thermodynamic stability, and magnetization-driven collective modes. Such effects are essential in modeling radiative processes, cooling dynamics, and wave–particle interactions in these extreme environments.

(ii) *Ultra-Cold Solid-Density Plasmas.* In modern high-intensity laser-solid interactions, plasmas with solid-like densities ($n \sim 10^{23}\text{cm}^{-3}$) and electron temperatures in the sub-eV to few-eV range are generated. These systems naturally approach a degenerate regime where $T << T_F$, , and the electrons occupy a narrow shell around the Fermi surface. The Spin polarization in such conditions can be enhanced either externally (through applied fields), or internally (through spin exchange, or magnetic ordering). Spin alignment modifies transport coefficients (e.g., spin-dependent conductivity and viscosity) and plays a pivotal role in nonlocal transport, self-magnetization, and current filamentation.

(iii) *Spintronic and Low-Dimensional quantum plasmas.* Semiconductor heterostructures, quantum wells, and nanowires give rise to effectively one-, or two-dimensional plasmas, where electron motion is quantized in confined directions. These systems exhibit spin-polarized carriers injected via ferromagnetic contacts or electric-field–driven spin separation (spin Hall effect). Even in the absence of large-scale magnetic fields, spin–orbit coupling generates effective spin-dependent fields, and the interplay between spin dynamics and collective plasma excitations becomes nontrivial. Phenomena such as spin-polarized plasma waves, spin-modified screening, and spin–charge coupled instabilities emerge in these systems, often with no classical analog.

These environments exemplify diverse settings where Spin polarization ceases to be an ancillary feature and instead constitutes a primary dynamical degree of freedom, necessitating theoretical treatments that fully incorporate spin into the collective evolution of the plasma.

7.1.2 Challenges: Incorporating Spin Consistently into Collective Plasma Models

While the necessity of including spin effects in quantum plasma dynamics is now well appreciated, the formal incorporation of spin into self-consistent plasma models presents deep conceptual and technical challenges. These difficulties arise due to the inherently quantum and non-commutative nature of spin, the lack of a direct classical analog, and the subtle coupling between spin and macroscopic fields via magnetization and current.

(i) *Spin as a Nonclassical Internal Degree of Freedom.* Unlike position and momentum, the spin operator does not admit a simultaneous eigenbasis with the Hamiltonian in general electromagnetic environments. Its dynamics are governed by operator-valued precession equations, derived from the Pauli equation:

$$i\hbar \frac{\partial \psi}{\partial t} = \left[\frac{1}{2m} (-i\hbar \nabla - e\mathbf{A})^2 - \mu_B \hat{\sigma} \cdot \mathbf{B} \right] \psi,$$

which includes both minimal coupling and Zeeman interaction. The resulting velocity and spin-current operators do not commute, leading to subtleties in defining conserved densities and fluxes for spin.

(ii) *Coupling Between Spin and Collective Fields.* Spin polarization modifies the magnetization $\mathbf{M}$, which in turn enters Maxwell's equations via the magnetization current $\nabla \times \mathbf{M}$, producing feedback effects on the plasma dynamics. At the same time, spatial and temporal inhomogeneities in the electromagnetic fields induce spin precession and spin torque effects, leading to nontrivial evolution of the spin vector field. The resulting spin–field coupling is inherently nonlinear and often nonlocal.

(iii) *Closure Problems in Moment Hierarchies.* In both kinetic and hydrodynamic approaches, incorporating spin entails extending the hierarchy of moments to include spin density, spin flux, and higher-order spin–velocity correlations. These higher-order moments do not naturally close and require either phenomenological closure schemes, or the development of tractable approximations, such as the spin relaxation time model, or spin-diffusion approximations. Ensuring consistency with known quantum limits (e.g., the Pauli's exclusion principle, Heisenberg uncertainty) further constrains these approximations.

(iv) *Lack of Universal Gauge-Invariant Spin Current.* The definition of a spin current density $\mathbf{J}_S$ is ambiguous in quantum field theory due to the non-conservation of spin in the presence of gauge interactions and relativistic effects. In low-energy approximations, pseudo-spin conservation may be assumed, but in general, spin torque and spin–orbit coupling act as sources or sinks. The implications for global conservation laws and local spin transport are not fully resolved in many-body quantum plasma contexts.

Together, these challenges necessitate a foundational reformulation of collective plasma models, one that treats spin on equal footing with charge and mass while preserving physical consistency, gauge invariance, and compatibility with quantum statistics.

7.2 Methods: From Quantum Kinetic Theory to Quantum Hydrodynamics

The incorporation of spin into plasma modeling can proceed from several levels of description, each offering distinct advantages and limitations. The two principal theoretical frameworks quantum kinetic theory (QKT) and quantum hydrodynamics (QHD) provide the foundation for most developments in spin-resolved quantum plasma theory. Each can be derived systematically from the spinor quantum many-body problem, subject to appropriate approximations.

(i) *Spin-Resolved Quantum Kinetic Theory.* At the most fundamental level, the evolution of the one-particle spin-resolved Wigner function

$$f(\mathbf{r}, \mathbf{p}, \mathbf{s}, t) = \frac{1}{(2\pi\hbar)^3} \int d^3\boldsymbol{\xi}\, e^{-i\mathbf{p}\cdot\boldsymbol{\xi}/\hbar} \left\langle \hat{\psi}^\dagger\left(\mathbf{r} + \frac{\boldsymbol{\xi}}{2}, t\right) \hat{\psi}\left(\mathbf{r} - \frac{\boldsymbol{\xi}}{2}, t\right) \right\rangle_{\mathbf{s}}$$

encodes the full quantum dynamics of spinor particles in phase space. Here, **s** denotes the Spin polarization degree of freedom, which may be either continuous (in semi-classical treatments) or discrete (in spinor bases). The evolution of f obeys a modified Vlasov equation, incorporating spin–field coupling, spin precession terms, and quantum diffraction corrections. In weakly coupled systems, the equation takes the form:

$$\frac{\partial f}{\partial t} + \frac{\mathbf{p}}{m} \cdot \nabla_{\mathbf{r}} f - (\nabla_{\mathbf{r}} U + \mu_B \nabla_{\mathbf{r}} \mathbf{B} \cdot \nabla_{\mathbf{s}}) \cdot \nabla_{\mathbf{p}} f + (\mu_B \mathbf{B} \times \nabla_{\mathbf{s}}) \cdot \nabla_{\mathbf{s}} f = 0.$$

This formulation enables the study of spin-resolved Landau damping, spin-polarized instabilities, and kinetic scale phenomena.

(ii) *Quantum Hydrodynamics with Spin.* By taking suitable velocity and spin-moment integrals of the quantum kinetic equation, one obtains a spin-extended quantum hydrodynamic (QHD) framework. This yields coupled equations for particle density n, fluid velocity $\mathbf{u}$, and spin density $\mathbf{S}$, augmented by quantum corrections such as the Bohm potential and spin–pressure tensors. The basic structure consists of:

$$\frac{\partial n}{\partial t} + \nabla \cdot (n\mathbf{u}) = 0,$$

$$m\left(\frac{\partial \mathbf{u}}{\partial t} + \mathbf{u} \cdot \nabla \mathbf{u}\right) = -\nabla P + ne(\mathbf{E} + \mathbf{u} \times \mathbf{B}) + \mathbf{F}_{\text{quantum}} + \mathbf{F}_{\text{spin}},$$

$$\frac{\partial \mathbf{S}}{\partial t} + \nabla \cdot (\mathbf{S}\mathbf{u}) = \gamma \mathbf{S} \times \mathbf{B} + \mathcal{T}_{\text{int}},$$

where $\mathcal{T}_{\text{int}}$ denotes spin torque terms due to exchange, spin–orbit, or interaction effects.

This hydrodynamic description is particularly powerful in analyzing collective phenomena such as spin–plasma waves, spin-modified shocks, magnetization instabilities, and turbulence in spin-polarized environments.

7.3 Quantum Hydrodynamic Framework with Spin Degrees of Freedom

The quantum hydrodynamic (QHD) formalism with spin degrees of freedom represents a mesoscopic description of collective plasma phenomena that emerge from the underlying quantum mechanics of individual particles, such as electrons and positrons. It systematically bridges microscopic quantum dynamics to macroscopic fluid behaviour, incorporating the essential effects of spin polarization, spin–orbital interactions, and spin-mediated force contributions within a continuum framework. The formulation proceeds from ensemble-averaged single-particle quantum mechanics and leverages quasi-probabilistic representations and statistical closures to extract a set of fluid equations suited to collective plasma behaviour [2].

7.3.1 Wigner–Pauli Approach and Madelung Decomposition

The Wigner function provides a phase-space representation of quantum states, offering a valuable intermediary between the Schrödinger (or Pauli) wave mechanics and a semiclassical kinetic, or hydrodynamic formalism. For a spin-½ particle species in a magnetized plasma, the spinor wavefunction $\Psi(\mathbf{r}, t)$ satisfies the Pauli equation in the presence of external electromagnetic fields, including the magnetic vector potential $\mathbf{A}(\mathbf{r}, t)$ and scalar potential $\phi(\mathbf{r}, t)$:

$$i\hbar\frac{\partial \Psi}{\partial t} = \left[\frac{1}{2m}(-i\hbar\nabla - q\mathbf{A})^2 + q\phi - \mu_B \boldsymbol{\sigma} \cdot \mathbf{B}\right]\Psi,$$

where $\boldsymbol{\sigma}$ is the vector of Pauli matrices, $\mu_B = q\hbar/2m$ is the Landau levels, and $\mathbf{B} = \nabla \times \mathbf{A}$ is the magnetic field. To convert this into a fluid description, one constructs the Wigner–Pauli function, 2×2 matrix-valued distribution $W_{ij}(\mathbf{r}, \mathbf{p}, t)$, defined via the Wigner transformation of the spinor density matrix. This function retains both positional and momentum dependencies, enabling the capture of quantum interference and spin coherence effects at the kinetic level.

To access hydrodynamic variables, one performs velocity moment integrations of the Wigner–Pauli function. However, for practical tractability and to yield equations amenable to closure, a Madelung-like decomposition of the spinor is often employed:

$$\Psi(\mathbf{r}, t) = \sqrt{n(\mathbf{r}, t)}e^{iS(\mathbf{r},t)/\hbar}\chi(\mathbf{r}, t),$$

where $n(\mathbf{r}, t)$ is the number density, $S(\mathbf{r}, t)$ is a scalar phase field (contributing to canonical momentum), and χ is a normalized spinor $\left(\chi^{\dagger}\chi = 1\right)$ that encodes the local spin orientation. This decomposition allows the construction of fluid variables by projecting the conserved Noether currents derived from global gauge and spin symmetries.

The resultant velocity field is defined as:

$$\mathbf{v} = \frac{1}{m}(\nabla S - q\mathbf{A}) + \frac{\hbar}{2m}\chi^{\dagger}\nabla\chi - \frac{\hbar}{2m}\nabla\chi^{\dagger}\chi,$$

where the latter two terms represent spinor-induced quantum vorticity and spin–orbit coupling corrections to the momentum flux.

7.3.2 *Ensemble-Averaged Dirac (Pauli) Equation Leading to Spin QHD*

Although the Dirac equation forms the fully relativistic foundation for spinor dynamics, its hydrodynamic reduction often leads to complications due to Zitterbewegung and higher-order time derivatives. In non-relativistic regimes typical of many quantum plasma scenarios, the Pauli's equation suffices, particularly when augmented with external fields and magnetic interactions. The quantum hydrodynamic equations are derived by taking statistical ensemble averages over single-particle Pauli wavefunctions, assuming a mixed state representation via the density operator $\hat{\rho}$.

The density operator evolution, governed by the Liouville–von Neumann equation [3],

$$i\hbar\frac{d\hat{\rho}}{dt} = [\hat{H}, \hat{\rho}],$$

is projected onto the space of hydrodynamic observables using spinor projectors and expectation values. The central quantities of interest number density, fluid velocity, and spin density are extracted by evaluating:

$$n(\mathbf{r}, t) = \mathrm{Tr}\big[\hat{\rho}\delta(\hat{\mathbf{r}} - \mathbf{r})\big], \quad \mathbf{v}(\mathbf{r}, t) = \frac{1}{n}\mathrm{Re}\left\{\mathrm{Tr}\left[\hat{\rho}\frac{\hat{\mathbf{p}}}{m}\delta(\hat{\mathbf{r}} - \mathbf{r})\right]\right\},$$
$$\mathbf{S}(\mathbf{r}, t) = \mathrm{Tr}\big[\hat{\rho}\boldsymbol{\sigma}\delta(\hat{\mathbf{r}} - \mathbf{r})\big].$$

Applying this formalism yields a closed set of coupled fluid equations, comprising the continuity equation, momentum balance equation (with spin forces), and spin evolution equation (resembling the Bloch equation in magnetized media). The spin–velocity coupling manifests in the form of a spin stress tensor, as well as via the effective magnetic torque $\boldsymbol{\tau} \sim \mathbf{S} \times \mathbf{B}$ that drives local spin precession. Importantly, spin–gradient couplings and quantum Bohm potential terms arise from non-commutative effects in phase-space representations, ensuring that the inherently quantum structure of the theory is preserved.

7.4 Governing Variables in Spin Quantum Hydrodynamics

The QHD framework enriched with spin degrees of freedom evolves a set of physically meaningful macroscopic observables, each representing a coarse-grained quantity derived from the underlying spinor field configuration. These variables not only encapsulate the usual plasma dynamics, but also encode spin-aligned microstructure and spin-driven transport phenomena, particularly relevant in degenerate and magnetized plasmas.

Number Density $n(\mathbf{r}, t)$): This scalar field denotes the probability density of finding a plasma particle (electron, positron, or ion) at a spatial point. It arises as the zeroth moment of the Wigner distribution and satisfies a continuity equation of the form:

$$\frac{\partial n}{\partial t} + \nabla \cdot (n\mathbf{v}) = 0,$$

reflecting local conservation of charge or particle number in the absence of pair production, or recombination processes.

Fluid Velocity ($\mathbf{v}(\mathbf{r}, t)$): The mean velocity field incorporates contributions from the gradient of the quantum phase, electromagnetic gauge potentials, and spin-coupled terms. In spin-polarized systems, the velocity field becomes subtly intertwined with the spin density gradient and local spin orientation, affecting vorticity generation and Hall-like transport.

Pressure Tensor ($P_{ij}(\mathbf{r}, t)$): Beyond isotropic thermal pressure, the momentum flux tensor must account for spin-dependent stresses and quantum fluctuations. The anisotropy introduced by spin alignment or magnetization gives rise to a gyrotropic pressure tensor [4] of the form:

$$P_{ij} = p\delta_{ij} + \pi_{ij}^{\text{spin}} + \pi_{ij}^{\text{quantum}},$$

where π_{ij}^{spin} encapsulates spin–strain and spin–magnetization contributions, while $\pi_{ij}^{\text{quantum}}$ arises from nonlocal quantum correlations, effectively capturing the influence of the quantum potential.

Spin Density Vector ($\mathbf{S}(\mathbf{r}, t)$): The spin vector field represents the expectation value of the Pauli matrices at each point in space, thus characterizing the local degree and direction of spin polarization. It satisfies an evolution equation of the generalized Bloch form:

$$\frac{d\mathbf{S}}{dt} + (\mathbf{v} \cdot \nabla)\mathbf{S} = \frac{2\mu_B}{\hbar}(\mathbf{S} \times \mathbf{B}_{\text{eff}}) + \mathcal{D}[\mathbf{S}],$$

where $\mathbf{B}_{\text{eff}}$ includes contributions from both the external magnetic field and the self-consistent magnetization field due to spin alignment, while $\mathcal{D}[\mathbf{S}]$ represents dissipative and decoherence effects including spin–spin and spin–orbit collisions [5].

Altogether, these variables furnish a complete dynamical description of a spin-1/2 quantum plasma fluid, suitable for analyzing magnetization dynamics, spin-induced instability mechanisms, and coherent spin–wave propagation. In contrast to classical magnetohydrodynamics, the inclusion of spin variables introduces additional conservation laws and couplings, including spin angular momentum conservation and magnetization vorticity transport, which play a pivotal role in dense astrophysical plasmas, spintronic devices, and strongly magnetized laboratory plasmas.

This formulation sets the foundation for advanced topics such as spin–magnetization waves, spin Hall effects in degenerate plasma, and spin–current generation in nonuniform magnetic textures. In subsequent sections, the implications of these governing equations will be explored in contexts where quantum and spin phenomena dominate the plasma behaviour, including dynamo effects, helicity generation, and spin–orbit-induced turbulence.

7.5 Spin Transport Equations in Quantum Hydrodynamics

The evolution of Spin polarization in a quantum plasma is fundamentally governed by the local conservation of spin angular momentum. This evolution is captured within the spin continuity equation, which, like the mass and charge continuity equations, describes the local balance between the temporal change in spin density and the spatial flux of spin. However, due to the intrinsic nature of spin as an internal angular momentum degree of freedom, its dynamics are inherently nontrivial, spin does not simply convect with the fluid, but also undergoes precession and torque due to electromagnetic and spin-interaction fields.

7.5.1 Conservation of Spin Density in the Absence of Spin Relaxation

In the absence of dissipation, collisions, or spin relaxation processes, the total spin angular momentum in a closed quantum system composed of spin-½ particles remains conserved. Locally, this conservation is expressed via a continuity-like equation for the spin density vector field $\mathbf{S}(\mathbf{r}, t)$, defined in terms of the expectation value of the Pauli matrices with respect to the quantum spinor wavefunction $\psi(\mathbf{r}, t) \in \mathbf{C}^2$:

$$\mathbf{S}(\mathbf{r}, t) = \frac{\hbar}{2}\psi^\dagger(\mathbf{r}, t)\boldsymbol{\sigma}\psi(\mathbf{r}, t).$$

The time evolution of $\mathbf{S}$ is then governed by a spin continuity equation of the form:

$$\frac{\partial \mathbf{S}}{\partial t} + \nabla \cdot \mathbf{J}_S = \mathbf{T}_S,$$

where $\mathbf{J}_S$ is the spin flux tensor and $\mathbf{T}_S$ represents the spin torque density. In the limit where the torque term vanishes, the equation assumes the form of a conservation law for spin angular momentum density. This idealized situation applies to regions of homogeneous magnetic field and vanishing spin–spin interactions, where no external or internal mechanisms exist to alter the local orientation of the spin vector field.

In realistic quantum plasmas, however, the torque term is nonzero due to the presence of magnetic fields, spin gradients, and potentially spin–orbit coupling, or exchange interactions. Nevertheless, this conservative form forms the backbone for spin hydrodynamic models and serves as a starting point for the derivation of fluid spin transport dynamics.

7.5.2 Derivation from the Pauli Equation

To derive the spin continuity equation, one begins with the non-relativistic Pauli equation for a spin-½ charged particle in an external electromagnetic field:

$$i\hbar\frac{\partial\psi}{\partial t} = \left[\frac{1}{2m}(-i\hbar\nabla - q\mathbf{A})^2 + q\phi - \mu_B\boldsymbol{\sigma}\cdot\mathbf{B}\right]\psi.$$

The evolution of the spin density is obtained by differentiating $\mathbf{S} = (\hbar/2)\psi^\dagger\boldsymbol{\sigma}\psi$ with respect to time and using the Pauli equation and its Hermitian conjugate:

$$\frac{\partial\mathbf{S}}{\partial t} = \frac{\hbar}{2}\left(\frac{\partial\psi^\dagger}{\partial t}\boldsymbol{\sigma}\psi + \psi^\dagger\boldsymbol{\sigma}\frac{\partial\psi}{\partial t}\right).$$

Substituting from the Pauli equation yields:

$$\frac{\partial\mathbf{S}}{\partial t} = \frac{i\hbar}{2}\left[\psi^\dagger\boldsymbol{\sigma}\hat{H}\psi - (\hat{H}\psi)^\dagger\boldsymbol{\sigma}\psi\right],$$

where $\hat{H}$ is the Pauli Hamiltonian operator. After some algebra and application of the standard commutation relations of the Pauli matrices, one arrives at the following expression:

$$\frac{\partial\mathbf{S}}{\partial t} + \nabla\cdot\mathbf{J}_S = \frac{2\mu_B}{\hbar}\mathbf{S}\times\mathbf{B},$$

where the spin flux tensor $\mathbf{J}_S$ is given by:

$$\mathbf{J}_S = \frac{\hbar}{4mi}\left[\psi^\dagger\boldsymbol{\sigma}\nabla\psi - (\nabla\psi^\dagger)\boldsymbol{\sigma}\psi\right] - \frac{q}{2mc}\mathbf{A}\psi^\dagger\boldsymbol{\sigma}\psi.$$

This flux tensor embodies both quantum convection and the electromagnetic coupling via minimal substitution. The right-hand side corresponds to the spin torque term, representing the precessional dynamics induced by the magnetic field. Importantly, this term is purely quantum mechanical and arises from the non-commutativity of the spin operators and the Hamiltonian.

7.5.3 *Role of Spin Torque and Precession*

The term $\mathbf{T}_S = (2\mu_B/\hbar)\mathbf{S} \times \mathbf{B}$ in the spin continuity equation plays a critical dynamical role. It describes the torque exerted on the magnetic moment associated with the spin vector by the local magnetic field. Physically, this leads to the Larmor precession of the spin vector about the direction of $\mathbf{B}$, with angular frequency:

$$\boldsymbol{\omega}_L = \frac{2\mu_B}{\hbar}\mathbf{B}.$$

As a result, even in the absence of any net flow or density variation, the spin density vector at a fixed point in space evolves due to the magnetic torque. This distinguishes spin from scalar quantities like mass, or charge density, which are convected or diffused but do not possess an intrinsic angular momentum that responds to local field gradients.

In hydrodynamic terms, the spin precession manifests as an internal mode of excitation in the plasma, distinct from charge oscillations or acoustic modes. These spin waves or magnon-like excitations are characterized by the interplay between the torque-induced precession and the convective, or diffusive transport encoded in the spin flux tensor. Their dispersion relations and damping properties depend on the magnetic field strength, gradient scales, and the coupling to other plasma degrees of freedom, such as velocity shear and density fluctuations [6].

In inhomogeneous, or turbulent magnetic environments, the spatial variation of $\mathbf{B}$ induces differential torques that can lead to spin decoherence, nonlocal spin transport, and the generation of spin vorticity. These effects are particularly pronounced in strongly magnetized Quantum Plasmas, such as those found in neutron star crusts or spin-polarized ultra-cold electron gases.

Moreover, in spin–orbit coupled systems, or under intense magnetic field gradients, higher-order corrections to the spin torque arise, including those involving the Berry curvature in momentum space or anomalous spin–current interactions. While not considered in this basic derivation, such terms become essential in relativistic, or topologically nontrivial regimes and will be revisited in later sections of this chapter.

The final form of the spin continuity equation, valid in the non-relativistic quantum hydrodynamic regime for spin-½ charged fermions under the Pauli Hamiltonian, is:

$$\frac{\partial \mathbf{S}}{\partial t} + \nabla \cdot \mathbf{J}_S = \frac{2\mu_B}{\hbar}\mathbf{S} \times \mathbf{B},$$

where: $\mathbf{S}(\mathbf{r}, t)$ is the spin density vector field. $\mathbf{J}_S$ is the spin flux tensor (including convective and magnetic contributions). The right-hand side captures the quantum torque responsible for spin precession.

This equation forms the cornerstone of spin quantum hydrodynamics, enabling the exploration of spin transport, spin–magnetization coupling, and collective spin phenomena in magnetized quantum plasmas. Its coupling with the momentum and continuity equations will be addressed in the subsequent sections as part of the complete spin-QHD system.

7.6 Spin Momentum

The momentum dynamics of spin degrees of freedom in quantum plasma are governed by a spin momentum equation that balances convective transport, magnetic torque, quantum force contributions, and inter-spin interactions. Unlike classical fluid momentum equations, the spin momentum equation encapsulates the transport of the intrinsic angular momentum density, reflecting both quantum-statistical features and the internal dynamics of spin precession and alignment under magnetic and interparticle forces.

7.6.1 Spin Force Contributions: Magnetic Torque and Gradient of Spin Pressure

We begin by noting that the spin momentum equation describes the evolution of the spin flux tensor $\mathbf{J}_S$ introduced in the spin continuity equation. Formally, its dynamics are governed by a time evolution equation of the form:

$$\frac{\partial \mathbf{J}_S}{\partial t} + \nabla \cdot \boldsymbol{\Pi}_S = \mathbf{F}_S,$$

where: $\mathbf{J}_S$ is the spin flux tensor, $\boldsymbol{\Pi}_S$ is the spin stress tensor, encoding spin–spin and spin–fluid correlations, $\mathbf{F}_S$ is the total spin force density, including magnetic, pressure, and quantum contributions.

A principal contribution to $\mathbf{F}_S$ originates from the magnetic torque term discussed earlier:

$$\mathbf{F}_{\text{mag}} = \frac{2\mu_B}{\hbar} \nabla(\mathbf{S} \times \mathbf{B}),$$

which arises from spatial inhomogeneities in the torque field and reflects the spatial redistribution of spin angular momentum due to magnetic field gradients.

A second essential component is the gradient of spin pressure, analogous to thermal pressure in scalar fluid dynamics but here associated with the anisotropy and magnitude of the Spin polarization vector:

$$\mathbf{F}_{\text{sp}} = -\nabla \cdot \mathbf{P}_S,$$

where $\mathbf{P}_S$ is the spin pressure tensor, whose elements encode the spin–spin correlation and polarization fluctuation effects. In magnetized quantum plasma, this term governs how local inhomogeneities in spin alignment generate restoring forces, akin to pressure gradients in classical plasmas. In isotropic spin-polarized systems, the trace of $\mathbf{P}_S$ corresponds to a scalar spin pressure $p_S \propto \nabla n_S$, where $n_S = |\mathbf{S}|$ is the scalar spin density.

These forces together give rise to a complex spin dynamics that incorporates both conservative redistribution (via the pressure gradient) and rotational realignment (via magnetic torque).

7.6.1.1 Bohm-Type Quantum Spin Force Terms

In addition to classical-like pressure and magnetic torque forces, quantum plasmas exhibit uniquely quantum corrections to the spin momentum equation through gradient terms in the quantum stress tensor. These are typically encapsulated in a quantum Bohm-like spin force:

$$\mathbf{F}_{\text{Bohm}} = \frac{\hbar^2}{4m^2}\nabla\left[\nabla^2\mathbf{S} - \frac{1}{n}(\nabla n \cdot \nabla\mathbf{S}) - \frac{1}{n}(\nabla \cdot \mathbf{S})\nabla n\right].$$

This term arises from second-order gradient corrections in the Madelung representation of the Pauli's equation and encodes the nonlocal, non-classical interaction between spin and density inhomogeneities. Physically, this term reflects the resistance of the spin texture to rapid spatial variations, analogous to quantum pressure in the scalar quantum hydrodynamic equations, but with directional and tensorial character due to the vectorial nature of spin.

These contributions become particularly important in systems with strong spatial modulation, such as spin domain walls, polarized quantum shocks, or spin-selective confinement, and they play a pivotal role in shaping coherent spin excitations and dispersive spin wave phenomena in dense quantum plasmas.

7.6.2 Coupling to Magnetization Current and External Magnetic Field

The collective spin state of a quantum plasma couples to the electromagnetic sector through the magnetization current, which supplements the standard conduction current in Maxwell's equations. The magnetization vector $\mathbf{M}$ is defined via the spin density:

$$\mathbf{M}(\mathbf{r}, t) = -\gamma \mathbf{S}(\mathbf{r}, t),$$

with $\gamma = gq/(2m)$ denoting the gyromagnetic ratio for electrons (where $g \approx 2$ for Dirac particles). The magnetization current $\mathbf{j}_M$ thus enters Maxwell–Ampère's equation via:

$$\mathbf{j}_M = \nabla \times \mathbf{M} = -\gamma \nabla \times \mathbf{S}.$$

This term leads to feedback between the spin structure and the self-consistent magnetic field generated by the plasma itself. In particular, the interplay of $\mathbf{S}$ and $\mathbf{B}$ can induce nontrivial magneto-spin dynamics, such as self-focusing of spin waves or the generation of spin-induced magnetic structures in ultradense or degenerate environments.

Consequently, the spin momentum equation closes only when coupled self-consistently to the modified Maxwell system, with the magnetization current entering the field equations and, reciprocally, the magnetic field influencing the spin torque and force. This mutual coupling underpins a variety of collective excitations spin-magnetosonic waves, spin-polarized Bernstein modes, and nonlinear magneto-spin solitons which form the subject of later sections [7, 8].

7.7 Spin Diffusion Equation

In parallel with the transport of charge and mass, Spin polarization undergoes spatial spreading due to gradient-driven redistribution processes. In quantum plasmas, the evolution of spin density includes both quantum-coherent and dissipative components. The spin diffusion equation models the relaxation of nonuniform spin distributions over space and time, incorporating quantum interference, spin–spin correlations, and spin-dependent transport channels.

Starting from the spin continuity equation and augmenting it with spin flux dynamics derived from kinetic, or hydrodynamic closure schemes, one arrives at a general spin diffusion equation of the form:

$$\frac{\partial \mathbf{S}}{\partial t} = D_S \nabla^2 \mathbf{S} + \mathbf{T}_S,$$

where: D_S is the quantum spin diffusivity, which may itself be tensorial or spatially dependent in anisotropic plasmas. $\mathbf{T}_S$ denotes the torque term introduced earlier. In quantum settings, D_S is not a simple phenomenological coefficient, but emerges from microscopic considerations, involving the Fermi velocity, spin mean free path, and interaction timescales. In degenerate systems, the spin diffusion is often anomalous due to Pauli blocking and spin-selective scattering rates.

Quantum diffusion of spin leads to coherent spreading of localized spin textures (e.g., skyrmions, spin packets) and defines the timescale over which initial spin imbalances equilibrate in absence of external driving or torques.

A distinctive feature of spin diffusion in quantum plasmas is the modification of the diffusive dynamics by spin–spin correlations and quantum interference effects. In dense or strongly interacting regimes, spin alignment of nearby particles leads to collective transport of polarization, which cannot be reduced to independent particle diffusion. This leads to non-Fickian diffusion, where higher-order gradients and memory effects alter the spreading rate and directionality [7].

Additionally, coherent quantum interference between spin states enhanced in low-temperature or weakly decohering environments can result in oscillatory spin transport, known in solid-state physics as spin beats or coherent spin precession. These effects require inclusion of phase-coherent off-diagonal spin density matrix elements in the transport description, as in spin Wigner or quantum Boltzmann formulations.

Spin transport in quantum plasmas can be decomposed into dissipative and non-dissipative contributions. The former corresponds to irreversible processes that drive the system toward local spin equilibrium and are often captured via relaxation-time approximations or collisional integral closures. These processes are parametrized by spin relaxation times τ_s, leading to exponential decay terms in the spin diffusion equation:

$$\left.\frac{\partial \mathbf{S}}{\partial t}\right|_{\text{diss}} = -\frac{\mathbf{S} - \mathbf{S}_{\text{eq}}}{\tau_s}.$$

Non-dissipative spin currents, on the other hand, arise from coherent precessional transport or topological spin flow and do not lead to entropy production. Such currents are divergence-free and include effects like spin Hall transport, magnetization precession, or Berry curvature–induced spin drift, though the latter emerge more prominently in relativistic or topological plasmas beyond the current non-relativistic framework.

Classically, spin diffusion in magnetized fluids is described by Fick's law generalized [8] for vector quantities:

$$\mathbf{j}_S = -D_s \nabla \mathbf{S},$$

with the dynamics governed by simple relaxation-diffusion equations without intrinsic precession, or quantum coherence. In such systems, spin transport is driven solely by collisional and magnetic alignment effects, and does not exhibit the nonlocal, oscillatory, or interference-driven phenomena intrinsic to quantum spin systems.

In contrast, the quantum spin diffusion equation presented here integrates several advanced quantum effects. It includes gradient terms analogous to quantum pressure, reflecting spatial variations in spin density, torque-induced dynamics that arise from the non-commutative nature of spin operators, nonlocal behaviour captured through gradient-corrected spin fluxes, and phase coherence effects represented by off-diagonal elements of the density matrix. Together, these components provide a nuanced and comprehensive framework for describing spin transport in quantum plasmas. These fundamental differences distinguish quantum spin plasma dynamics from classical magnetized spin fluids, leading to a rich variety of nontrivial equilibrium and nonequilibrium behaviour s absent in the classical limit.

$$\text{Spin momentum evolution}: \frac{\partial \mathbf{J}_S}{\partial t} + \nabla \cdot \mathbf{\Pi}_S = \mathbf{F}_{\text{mag}} + \mathbf{F}_{\text{sp}} + \mathbf{F}_{\text{Bohm}}$$

$$\text{Quantum spin diffusion equation}: \frac{\partial \mathbf{S}}{\partial t} = D_S \nabla^2 \mathbf{S} + \mathbf{T}_S - \frac{\mathbf{S} - \mathbf{S}_{\text{eq}}}{\tau_s}$$

These equations collectively determine the dynamical evolution of spin polarization in a quantum plasma, enabling the study of spin transport, wave propagation, magnetic structure formation, and dissipative relaxation within a unified hydrodynamic framework. Their extension to relativistic and topological regimes will be explored in the concluding sections of this chapter.

7.8 Outlook and Future Directions

The exploration of spin polarization phenomena within quantum plasma physics stands at the nexus of quantum many-body theory, relativistic electrodynamics, and emerging experimental technologies.

7.8.1 Extensions to Relativistic Spin Quantum Hydrodynamics (Spin-QHD)

While non-relativistic spin-QHD models have offered valuable insights into spin dynamics and collective excitations, many physically relevant regimes such as ultra-intense laser-plasma interactions and magnetospheres of compact astrophysical objects demand a fully relativistic treatment that consistently incorporates spin degrees of freedom.

Relativistic spin-QHD extends traditional quantum fluid models by embedding spinor field structures within a covariant hydrodynamic framework. The fundamental dynamical variables become the relativistic spin four-vector $S^\mu(x)$ alongside the fluid four-velocity $u^\mu(x)$ and particle density $n(x)$. The governing equations are derived from the Dirac equation, or the relativistic Wigner function formalism, yielding a set of coupled conservation laws and spin precession dynamics:

$$m\left(u^\nu\partial_\nu u^\mu + \frac{\hbar}{2m}S^\nu\partial^\mu S_\nu\right) = qF^{\mu\nu}u_\nu - \partial^\mu P + \cdots,$$

$$u^\nu\partial_\nu S^\mu = \frac{2\mu_B}{\hbar}F^{\mu\nu}S_\nu + \mathcal{O}(\hbar^2),$$

where $F^{\mu\nu}$ is the electromagnetic field tensor, P the pressure scalar, and higher-order terms include spin–orbit and quantum corrections.

Physically, relativistic spin-QHD captures phenomena such as spin precession under strong electromagnetic fields, relativistic spin currents, and spin-polarized radiation reaction effects. This framework enables consistent modeling of spin-polarized quantum Plasmas in extreme environments, such as neutron star magnetospheres and next-generation laser facilities.

The mathematical challenge lies in closing the moment hierarchy generated by the relativistic Wigner equation while preserving gauge invariance and Lorentz covariance. Ongoing efforts focus on developing suitable truncation schemes and identifying effective potentials that encapsulate spin correlations.

7.8.2 Inclusion of Spin–Spin and Exchange Correlations

The mean-field treatment of spin polarization captures leading-order effects, but neglects many-body correlations critical for accurate descriptions of dense quantum plasmas where interparticle spacing approaches the thermal de Broglie wavelength.

Spin–spin interactions, arising from magnetic dipole–dipole couplings, and exchange correlations, originating from antisymmetrization of fermionic wavefunctions, modify the effective spin Hamiltonian. The total spin-dependent interaction energy includes terms of the form:

$$\hat{H}_{\text{spin}} = -\sum_{i\neq j} J_{ij}\mathbf{S}_i\cdot\mathbf{S}_j + \sum_{i\neq j}\frac{\mu_0}{4\pi r_{ij}^3}\left[\mathbf{S}_i\cdot\mathbf{S}_j - 3(\mathbf{S}_i\cdot\hat{\mathbf{r}}_{ij})(\mathbf{S}_j\cdot\hat{\mathbf{r}}_{ij})\right],$$

where J_{IJ} denotes the exchange integral and the second term describes the magnetic dipole interaction between spins $\mathbf{S}_i$, $\mathbf{S}_j$.

Incorporating these effects requires advanced many-body techniques such as spin-density functional theory (SDFT) extended to plasma conditions, and diagrammatic perturbation methods within spinor quantum field theory. Physically, these correlations influence spin-wave spectra, enhance or suppress spin coherence, and contribute to novel collective excitations such as spin plasmons and spin-charge coupled modes.

Mathematically, this entails solving the Bethe–Salpeter equations [9] for two-particle spin correlation functions or employing 2-particle irreducible (2PI) effective action techniques adapted to spinor fields, facilitating a self-consistent treatment of spin correlations beyond Hartree–Fock approximations.

7.8.3 Coupling to Quantum Kinetic and 2PI Spinor Field Theory Approaches

Quantum kinetic theories provide a microscopic description of spin-polarized plasma dynamics including non-equilibrium effects, collisions, and decoherence. The spin-dependent Wigner function formalism encapsulates the phase-space distribution of spin-1/2 particles with the spin density matrix $\rho_s(\mathbf{r}, \mathbf{p}, t)$. The Wigner equation generalized for spin reads:

$$\frac{\partial \rho_s}{\partial t} + \frac{\mathbf{p}}{m} \cdot \nabla_{\mathbf{r}} \rho_s + \frac{q}{\hbar}\left(\mathbf{E} + \frac{\mathbf{p}}{m} \times \mathbf{B}\right) \cdot \nabla_{\mathbf{p}} \rho_s + \frac{2\mu_B}{\hbar} \mathbf{B}_{\text{eff}} \times \rho_s = \mathcal{C}[\rho_s],$$

where $\mathbf{B}_{\text{eff}}$ incorporates external and self-consistent magnetic fields, and $\mathcal{C}$ denotes the collision integral including spin-flip processes.

The 2PI (two-particle irreducible) effective action framework extends this kinetic theory by allowing non-perturbative resummations of correlation functions, crucial for strongly coupled or out-of-equilibrium spinor plasmas. Here, the spinor Green's functions $G_{\alpha\beta}(x, y)$ satisfy Dyson–Schwinger equations derived from the 2PI generating functional $\Gamma[G]$:

$$\frac{\delta \Gamma[G]}{\delta G_{\alpha\beta}(x, y)} = 0,$$

which self-consistently incorporates self-energy corrections and vertex functions.

Physically, this approach captures transient spin coherence, dynamical spin relaxation, and quantum memory effects beyond Boltzmann or mean-field descriptions. It is particularly pertinent to ultrafast laser–plasma experiments and strongly correlated astrophysical plasmas.

Mathematically, the coupling of spin-QHD and 2PI spinor field theories requires hybrid multi-scale modeling techniques, marrying hydrodynamic variables with quantum Green's functions to accurately describe the interplay of collective and microscopic spin dynamics [9].

7.8.4 *Opportunities in Quantum Simulation Platforms and Spintronic Plasmas*

Quantum simulation technologies such as cold atomic gases with engineered spin interactions, trapped ion arrays, and superconducting qubits offer controlled environments to emulate spin-polarized quantum plasma phenomena. These platforms permit tunable interactions, precise initialization of spin states, and real-time measurement of spin coherence and transport, inaccessible in conventional plasma experiments.

From a theoretical standpoint, mapping spin-polarized plasma Hamiltonians onto quantum simulators enables exploration of non-perturbative regimes, topological spin textures, and quantum phase transitions induced by spin-dependent interactions.

Simultaneously, the nascent field of spintronic plasmas plasmas with intrinsic spin current generation and manipulation capabilities opens transformative possibilities for information processing and energy transport. Spintronic effects in plasmas, including spin Hall currents and spin transfer torques, extend classical plasma wave and instability theory to incorporate spin-charge coupled modes, necessitating novel kinetic and fluid theories.

Mathematically, these phenomena motivate development of spin-dependent magneto-hydrodynamic models augmented with spin current continuity equations and quantum torque terms:

$$\frac{\partial \mathbf{J}_s}{\partial t} + \nabla \cdot \mathbf{\Pi}_s = \mathbf{T}_s,$$

where $\mathbf{J}_s$ is the spin current density, $\mathbf{\Pi}_s$ the spin stress tensor, and $\mathbf{T}_s$ spin torque density.

The convergence of quantum plasma physics and spintronics, facilitated by quantum simulation, promises breakthroughs in coherent spin control, low-dissipation spin transport, and hybrid quantum–classical devices.

References

1. Brizard, A. (1992). Nonlinear gyrofluid description of turbulent magnetized plasmas. *Physics of Fluids B: Plasma Physics, 4*(5), 1213–1228.
2. Ichimaru, S. (2018). *Basic principles of plasma physics: A statistical approach.* CRC Press.
3. Lewis, H. R., Barnes, D. C., & Melendez, K. J. (1987). The Liouville theorem and accurate plasma simulation. *Journal of Computational Physics, 69*(2), 267–282.
4. Che, H., Schiff, C., Le, G., Dorelli, J. C., Giles, B. L., & Moore, T. E. (2018). Quantifying the effect of non-Larmor motion of electrons on the pressure tensor. *Physics of Plasmas, 25*(3).
5. Escande, D. F., & Elskens, Y. (2003). Microscopic dynamics of plasmas and chaos: The wave–particle interaction paradigm. *Plasma Physics and Controlled Fusion, 45*(12A), A115.
6. Shinohara, S., Matsuoka, N., & Matsuyama, S. (2001). Establishment of strong velocity shear and plasma density profile modification with associated low frequency fluctuations. *Physics of Plasmas, 8*(4), 1154–1158.

7. Vázquez, F., & Márkus, F. (2010). Non-Fickian particle diffusion in confined plasmas and the transition from diffusive transport to density waves propagation. *Physics of Plasmas, 17*(4).
8. Ménard, M., & Paynter, R. W. (2005). Fick's law of diffusion depth profiles applied to the degradation of oxidized polystyrene during ARXPS analysis. *Surface and Interface Analysis: An International Journal Devoted to the Development and Application of Techniques for the Analysis of Surfaces, Interfaces and Thin Films, 37*(5), 466–471.
9. Schroedter, E., & Bonitz, M. (2024). Two-time quantum fluctuations approach and its relation to the Bethe-Salpeter equation. *Physica Status Solidi (B), 261*(9), 2300564.

Chapter 8
Quantum Plasma in Astrophysics

The present chapter aims to provide a comprehensive and detailed exposition of the behavior, role, and consequences of quantum plasma in astrophysical environments, especially in high-density and low-temperature regimes where quantum mechanical principles become indispensable. While prior chapters have introduced general plasma physics, classical and quantum electrodynamics, and the foundational constructs of quantum plasma, this chapter transitions from abstract theory to applied astrophysical contexts. The intention is to rigorously analyze how quantum plasma principles operate within compact astrophysical objects and extreme cosmic environments, where conditions significantly deviate from terrestrial or laboratory plasmas and instead reflect regimes governed by strong gravitational fields, relativistic speeds, and intense magnetic flux densities.

8.1 White Dwarfs

In astrophysical systems characterized by extremely high densities but comparatively low temperatures, such as white dwarfs, the governing physical framework transcends classical thermodynamics and enters the domain of degenerate quantum plasma. The principal component of this plasma is a highly compressed, relativistically degenerate electron gas that counterbalances gravitational collapse via degeneracy pressure an inherently quantum mechanical phenomenon dictated by the Pauli's exclusion principle. The theoretical structure describing white dwarfs is rooted in the physics of Fermi–Dirac statistics, quantum electrodynamics in dense media, and the equilibrium solutions to general relativistic stellar structure equations.

The white dwarf interior is composed primarily of fully ionized nuclei (typically carbon and oxygen in the case of intermediate-mass progenitors) immersed in a sea of degenerate electrons. These electrons constitute a quantum plasma, but unlike in conventional high-temperature plasmas, thermal energy is negligible compared to

P. Kumar, *Quantum Plasma*, https://doi.org/10.1007/978-981-95-0481-7_8

the Fermi energy, $k_B T > E_F$ where $E_F = \sqrt{(p_F c)^2 + (m_e c^2)^2}$. The electron number density n_e is extremely high of the order $10^{29} - 10^{33}\text{cm}^{-3}$, which elevates the Fermi momentum p_F to relativistic magnitudes. The Fermi momentum is related to the density via:

$$p_F = \hbar\left(3\pi^2 n_e\right)^{1/3}.$$

In this regime, the kinetic energy of the electrons is relativistic, and one must transition from the non-relativistic expression $E = p^2/2m_e$ to the fully relativistic dispersion:

$$\varepsilon(p) = \sqrt{p^2c^2 + m_e^2c^4}.$$

The total pressure of the degenerate electron gas at zero temperature (the Chandrasekhar approximation) is obtained from integrating the statistical momentum distribution function:

$$P_e = \frac{8\pi}{3h^3}\int_0^{p_F} \frac{p^4c^2}{\sqrt{p^2c^2 + m_e^2c^4}}dp.$$

Evaluating this integral yields an expression that interpolates between the non-relativistic and ultra-relativistic limits:

Non-relativistic limit ($p_F \ll m_e c$): $P_e \propto n_e^{5/3}$,

Ultra-relativistic limit ($p_F \gg m_e c$): $P_e \propto n_e^{4/3}$.

This result reflects the stiffening of the equation of state due to quantum degeneracy and underlies the gravitational stability of white dwarfs. Physically, this quantum pressure is a manifestation of the Heisenberg's uncertainty principle, preventing the electrons from occupying the same quantum state and thus resisting further compression.

A key outcome of relativistic quantum plasma theory in the context of white dwarfs is the derivation of the Chandrasekhar's equation of state, which enables the computation of equilibrium configurations by solving the equations of hydrostatic balance:

$$\frac{dP}{dr} = -\frac{GM(r)\rho(r)}{r^2}, \quad \frac{dM}{dr} = 4\pi r^2\rho(r).$$

Incorporating the relativistic degeneracy pressure $P_e(\rho)$ into these equations and solving numerically yields the mass-radius relation:

$$R \propto \frac{1}{M^{1/3}}, \quad M < M_{\text{Ch}},$$

and defines the Chandrasckhar mass limit:

$$M_{\text{Ch}} = \frac{5.83}{\mu_e^2} M_{solar},$$

where μ_e is the mean molecular weight per electron. This mass limit ($\sim 1.4 M_{solar}$) corresponds to the maximum mass that can be supported by electron degeneracy pressure alone. Beyond this point, the electrons are forced into energies exceeding the neutronization threshold, triggering inverse beta decay and collapse into a neutron star.

The relativistic nature of the degenerate electron plasma necessitates proper treatment of the energy–momentum tensor $T^{\mu\nu}$ within the Einstein field equations. For a perfect fluid approximation, appropriate at macroscopic scales, the energy–momentum tensor takes the form:

$$T^{\mu\nu} = (\varepsilon + P)u^{\mu}u^{\nu} - Pg^{\mu\nu},$$

where ε is the energy density and u^{μ} the four-velocity field. For degenerate matter, ε includes the rest mass of ions and the relativistic electron energy density:

$$\varepsilon = \rho c^2 + \varepsilon_e(n_e).$$

The pressure is entirely due to electrons, since the nuclei contribute minimally to pressure due to their large mass and classical statistical behavior. The inclusion of $T^{\mu\nu}$ in the Tolman–Oppenheimer–Volkoff (TOV) equation:

$$\frac{dP}{dr} = -\frac{G(\varepsilon + P)\left(M + 4\pi r^3 P/c^2\right)}{r^2\left(1 - 2GM/rc^2\right)},$$

provides a fully general relativistic description of the white dwarf structure and is crucial for configurations approaching the Chandrasekhar limit.

Although thermal motion is suppressed, the quantum plasma in white dwarfs still supports collective effects governed by quantum electrodynamic interactions. The Thomas–Fermi screening length λ_{TF}, which describes the scale over which charge perturbations are screened, is significantly shorter than in classical plasma and is given by:

$$\lambda_{\text{TF}}^{-2} = \frac{4\pi e^2}{\partial \mu_e / \partial n_e}.$$

In the relativistic regime, this derivative involves the relativistic density of states and the modified Fermi velocity, indicating the intimate dependence of screening on the equation of state.

Further, high-frequency Langmuir oscillations persist in the degenerate plasma, albeit with modifications in their dispersion relation:

$$\omega^2 = \omega_p^2 + \frac{3}{5} v_F^2 k^2,$$

where $\omega_p = \sqrt{4\pi e^2 n_e / m_e}$ is the classical plasma frequency, and $v_F = p_F / \sqrt{p_F^2 + m_e^2 c^2}$ is the relativistic Fermi velocity. These oscillations reflect the compressibility of the electron sea and can contribute to observable phenomena such as pulsation modes and thermal transport suppression.

Gravitational stability is modulated by the interplay between quantum pressure and relativistic gravity. As the mass increases, the compressibility increases due to the softening of the equation of state (from $n^{5/3} \rightarrow n^{4/3}$), ultimately leading to dynamical instability. The onset of secular, or dynamical instability can be assessed via variational principles or by examining the sign of the second derivative of the total energy with respect to radius [1, 2].

In many white dwarfs, significant magnetic fields ($10^5 - 10^9$G) and rotational effects ($\Omega \sim 10^{-3} - 1$ Hz) are observed. These contribute additional quantum plasma phenomena. Under strong magnetic fields, Landau Quantization of electron motion perpendicular to the field alters the density of states:

$$n_e = \frac{eB}{2\pi^2 \hbar c} \sum_{n=0}^{n_{\max}} g_n \sqrt{p_F^2 - 2ne\hbar B/c},$$

where g_n is the degeneracy factor of the nth Landau level. This quantization modifies the pressure and energy density, impacting the stellar structure and cooling processes. In extreme cases, the plasma enters the quantum Hall regime, where anisotropies arise in conductivity and pressure tensors, demanding tensorial generalizations of the TOV equation.

8.2 Neutron Stars

Neutron stars embody one of the most extreme realizations of quantum plasma in the universe, comprising a highly dense, relativistic, and degenerate environment. Unlike white dwarfs, where electron degeneracy alone halts gravitational collapse, neutron stars sustain their structure through the combined effects of neutron degeneracy, strong nuclear interactions, and relativistic gravitational equilibrium. The dense interior consists of a multi-component plasma made up of neutrons (n), protons (p), and electrons (e), forming a composite relativistic Fermi system in β-equilibrium, with the possible inclusion of muons, hyperons, and quark matter at higher densities.

At densities beyond the neutron drip threshold ($\rho \gtrsim 4 \times 10^{11}$ g/cm^3), matter is compressed into a configuration where neutrons begin to dominate the mass content. The electrons and protons are highly degenerate and relativistic, and the entire matter is stabilized under β-equilibrium, a condition arising from weak interactions that interconvert neutrons, protons, and electrons via:

$$n \rightleftharpoons p + e^- + \overline{\nu}_e.$$

In equilibrium, this reaction balances such that the chemical potentials satisfy:

$$\mu_n = \mu_p + \mu_e,$$

where μ_j denotes the relativistic chemical potential of species j. This relation, combined with charge neutrality ($n_p = n_e$), determines the proton fraction and governs the microphysical structure of the plasma.

All three species, neutrons, protons, and electrons are modeled as relativistic degenerate Fermi gases. Their respective Fermi momenta are given by:

$$p_{Fj} = \hbar\left(3\pi^2 n_j\right)^{1/3},$$

with associated chemical potentials (including rest mass contributions):

$$\mu_j = \sqrt{p_{Fj}^2 c^2 + m_j^2 c^4}.$$

As the density increases, the electrons become ultrarelativistic ($p_{Fe} \gg m_e c$), the protons moderately relativistic, and the neutrons can transition from non-relativistic to relativistic depending on depth within the star. The resulting plasma is therefore a multi-component, relativistically degenerate quantum plasma, where both degeneracy pressure and quantum interactions are essential.

The total pressure in the n-p-e plasma arises from the contributions of all constituent fermions:

$$P = P_n + P_p + P_e.$$

Each term is derived from the Fermi–Dirac distribution at zero temperature, using:

$$P_j = \frac{1}{3\pi^2\hbar^3}\int_0^{p_{F_j}} \frac{p^4}{\sqrt{p^2c^2 + m_j^2c^4}} dp.$$

In the limit $p_{Fj} \gg m_j c$, the pressure scales relativistically as $P_j \sim p_{Fj}^4$, while in the non-relativistic regime it behaves as $P_j \sim p_{Fj}^5/m_j$. The strong variation in the degeneracy pressure across the radial depth of the neutron star reflects the spatial stratification of relativistic effects within the star's structure.

This composite equation of state (EOS) must also incorporate nuclear interaction potentials, especially for neutrons and protons, which interact via the strong force. A realistic EOS requires inclusion of the symmetry energy, nucleon-nucleon correlations, and three-body forces, typically derived from relativistic mean-field

(RMF) theory, or chiral effective field theory. The pressure due to nuclear interactions modifies the energy density and enters into the general relativistic structure equations [3, 4].

The conditions of β-equilibrium and charge neutrality serve as algebraic constraints interlinking the partial densities of each species. Together, they form the minimal thermodynamic closure:

$$\text{Chemical equilibrium:}\mu_n - \mu_p = \mu_e.$$

$$\text{Charge neutrality:}n_p = n_e.$$

Solving these simultaneously for a given baryon number density $n_b = n_n + n_p$ determines the individual number densities n_n, n_p, and n_e. These densities then serve to construct the EOS, closing the system for further analysis in gravitational equilibrium.

At densities exceeding $2 \times 10^{14}\ \text{g/cm}^3$, the electron chemical potential exceeds the muon rest mass, and muons are energetically favorable:

$$e^- \to \mu^- + \overline{\nu}_\mu + \nu_e, \quad \mu^- \to e^- + \overline{\nu}_e + \nu_\mu.$$

Inclusion of muons leads to an extended chemical equilibrium condition:

$$\mu_e = \mu_\mu,$$

which modifies the lepton content and softens the EOS, having direct implications on the mass–radius relation and cooling behavior.

The highly compact nature of neutron stars mandates a general relativistic treatment. The equilibrium structure is governed by the Tolman–Oppenheimer–Volkoff (TOV) equation:

$$\frac{dP(r)}{dr} = -\frac{G}{r^2}\left[\varepsilon(r) + \frac{P(r)}{c^2}\right]\left[M(r) + 4\pi r^3 \frac{P(r)}{c^2}\right]\left(1 - \frac{2GM(r)}{rc^2}\right)^{-1},$$

supplemented by the mass continuity equation:

$$\frac{dM(r)}{dr} = 4\pi r^2 \varepsilon(r).$$

Here, $\varepsilon(r)$ is the total energy density, including rest mass, kinetic energy, and interaction energies. The integration begins from the core with central pressure P_c and proceeds outward until the pressure vanishes, defining the stellar radius. The maximum stable mass analogous to the Chandrasekhar limit is sensitively dependent on the EOS and is observationally constrained to $M_{\max} \gtrsim 2M_{solar}$.

Even though thermal motion is negligible, quantum collective effects persist in the n-p-e plasma. The screening of electric fields is characterized by the relativistic Thomas–Fermi screening length:

$$\lambda_{\text{TF}}^{-2} = \frac{4\pi e^2}{\hbar^2} \frac{\partial n_e}{\partial \mu_e},$$

which quantifies the spatial response of the degenerate electron sea to perturbations in charge distribution. In the relativistic limit, this screening becomes highly efficient, confining electric field penetration to sub-femtometer scales.

Quantum oscillations, such as plasmon and proton–electron sound modes, may manifest under certain conditions. However, the extremely high degeneracy and rapid relaxation due to nuclear collisions suppress most hydrodynamic-like modes. The relevant dynamical degrees of freedom often involve superfluid neutron components and superconducting protons, requiring microscopic quantum field-theoretic treatment of pairing phenomena.

At core temperatures $T \lesssim 10^9$ K, which are significantly below the Fermi temperatures ($T_F \sim 10^{12}$ K), neutrons and protons undergo Cooper pairing, leading to superfluid and superconducting phases respectively. This introduces a BCS-type energy gap $\Delta(k)$ in the excitation spectrum [5]:

$$E_k = \sqrt{(\varepsilon_k - \mu)^2 + \Delta(k)^2},$$

where ε_k is the single-particle energy. The resulting superfluidity alters transport coefficients, suppresses neutrino emissivity, and supports phenomena such as glitches sudden changes in neutron star rotation due to angular momentum exchange between crust and superfluid core. The presence of superconducting protons leads to the formation of quantized magnetic flux tubes, modifying the magnetic field distribution and dynamics. These flux tubes can couple to neutron superfluid vortices, giving rise to vortex pinning, a key mechanism in explaining rotational irregularities.

8.3 Quark-Gluon Plasma

In the extreme inner cores of neutron stars, where baryon densities exceed several times the nuclear saturation density ($\rho_0 \approx 2.8 \times 10^{14}$ g/cm^3), the traditional hadronic degrees of freedom may dissolve into their constituent quarks, forming a deconfined state of matter known as the quark-gluon plasma (QGP). Unlike the QGP generated in high-temperature environments such as heavy-ion collisions, the QGP in neutron stars exists under conditions of ultra-high baryon chemical potential and low temperature a regime dominated by degenerate quantum statistics, quantum chromodynamics (QCD) asymptotic freedom, and strong coupling effects at intermediate scales. The nature of this dense quark matter admits the possibility of colour superconductivity, a phase in which quarks form Cooper-like pairs via attractive channels mediated by gluon exchange, breaking the color gauge symmetry dynamically [6].

The transition to QGP in neutron star cores is expected when the baryonic chemical potential (μ_B) crosses a critical threshold. This is quantified by considering the Gibbs conditions for phase equilibrium, i.e., equal pressure and chemical potentials across the hadronic and quark phases. The system can be described by the QCD Lagrangian:

$$\mathcal{L}_{\text{QCD}} = \overline{\psi}\left(i\gamma^{\mu} D_{\mu} - m\right)\psi - \frac{1}{4} G^{a}_{\mu\nu} G^{a\mu\nu},$$

where $D_{\mu} = \partial_{\mu} - ig_s T^a A^a_{\mu}$ is the covariant derivative with A^a_{μ} being the gluon fields and $G^a_{\mu\nu}$ their field strength tensor. At high μ_B, the non-perturbative QCD vacuum structure allows chiral symmetry restoration and deconfinement.

In analogy to BCS theory of superconductivity, quarks at the Fermi surface can pair due to the residual attractive force via gluon exchange. The Fermi sea of degenerate quarks supports the formation of color Cooper pairs, leading to color superconductivity. The simplest pairing structure is the 2-flavor color superconducting (2SC) phase, where up and down quarks pair in a color-antitriplet channel, leading to a condensate of the form:

$$\left\langle \psi^{i}_{\alpha} C\gamma_5 \psi^{j}_{\beta} \right\rangle \propto \Delta\ \epsilon^{ij} \epsilon_{\alpha\beta 3},$$

where i, j denote flavor indices and α, β denote color indices. For higher densities, all three light quark flavors may participate in the pairing, forming the Color-Flavor Locked (CFL) phase. This phase exhibits spontaneous breaking of baryon number, chiral, and color symmetries, leading to massive gluons and modified Goldstone modes.

The equation of state of quark matter with color superconductivity significantly affects the mass-radius relationship of neutron stars. The pressure and energy density in such a phase can be derived from the thermodynamic potential, for instance:

$$\Omega = \Omega_{\text{free}} - \frac{3}{\pi^2} \Delta^2 \mu^2 + \mathcal{O}(\Delta^4),$$

where Δ is the superconducting gap and μ is the quark chemical potential. The gap can reach values on the order of 10–100 MeV depending on the density, implying significant corrections to the stiffness of the EoS. These modifications influence the maximum mass of the neutron star and can potentially explain the observed $2M_{solar}$ pulsars.

While direct detection of quark-gluon plasma (QGP), or color superconductivity within neutron stars remains beyond current observational capabilities, several astrophysical signatures may offer indirect evidence of their presence. One such indicator is the thermal evolution of neutron stars, where cooling curves can be altered due to suppressed neutrino emission in color superconducting phases, particularly through modifications to URCA processes. Additionally, gravitational wave emissions may arise from phase transitions between hadronic and quark matter or from structural rearrangements within crystalline color superconducting phases, producing distinct signals. Further, variations in a neutron star's internal composition can influence its rotational behavior, potentially affecting glitch events and the subsequent recovery of pulsar spin rates.

The presence of a quark-gluon plasma core and the onset of color superconductivity within neutron stars represent one of the most fascinating intersections between quantum field theory and astrophysics. These phenomena lie at the confluence of non-Abelian gauge theories, condensed matter physics, and general relativity. Understanding them demands a multiscale treatment, from microscopic pairing dynamics governed by QCD to macroscopic observables constrained by general relativistic hydrostatic equilibrium. Continued developments in quantum many-body techniques, lattice QCD at finite density, and multi-messenger astrophysics promise to refine our understanding of this exotic matter phase and its role in compact astrophysical environments [7, 8].

8.4 Brown and Red Dwarfs

Brown and red dwarfs, which occupy the mass range between the largest planets and the smallest main-sequence stars, present a distinct regime for quantum plasma physics characterized by low temperature and intermediate degeneracy. In these astrophysical bodies, the plasma neither behaves as a fully classical ideal gas, nor as an ultra-degenerate Fermi system, but instead exhibits transitional features where both quantum statistical and collective effects are relevant. This intermediate degeneracy regime necessitates a nuanced theoretical framework that synthesizes principles from quantum statistics, non-relativistic many-body theory, and low-temperature plasma electrodynamics.

The interior of brown dwarfs, with masses below the hydrogen-burning limit ($\sim 0.075\, M_{solar}$), and red dwarfs just above that threshold, predominantly comprises molecular hydrogen, helium, and traces of heavier elements. Ionization occurs primarily due to pressure ionization rather than thermal processes. The temperature in their interiors typically ranges from 10^5 to 10^6 K, and central densities lie in the range 10^2 to 10^3 g/cm^3. Under such conditions, the electrons form a weakly, or moderately degenerate Fermi gas, while the ions remain non-degenerate and possibly strongly coupled [9].

The degeneracy parameter $\chi = \frac{k_B T}{E_F}$, where E_F is the electron Fermi energy, quantifies the degree of degeneracy. For brown and red dwarfs, $\chi \sim \mathcal{O}(1)$, implying that the system resides in the transitional regime where neither classical, nor fully degenerate approximations suffice. This necessitates quantum kinetic and thermodynamic treatments that interpolate between the Maxwell–Boltzmann and Fermi–Dirac limits.

The thermodynamic properties of the electron gas in such objects are modified by quantum exchange and correlation effects. The leading order expression for the pressure of a moderately degenerate electron gas is given by:

$$P_e = \frac{8\pi}{3h^3}\int_0^\infty \frac{p^4}{\sqrt{p^2+m_e^2}}\left[\frac{1}{e^{(\epsilon-\mu)/k_BT}+1}\right]dp,$$

where $\epsilon = \sqrt{p^2+m_e^2}$ is the single-particle energy and μ the chemical potential. In the low-temperature expansion, this integral can be treated perturbatively in k_BT/μ to extract corrections due to partial degeneracy. Exchange contributions due to antisymmetrization of the many-body wavefunction also modify the internal energy and compressibility.

The resulting equation of state (EoS) softens as compared to a fully degenerate gas, influencing the hydrostatic equilibrium and radius-mass relation of the star. The degenerate electron pressure $P_e \sim \rho^{5/3}$ in the non-relativistic limit gradually transitions to a $P \sim \rho$ dependence in the classical regime. This hybrid scaling dictates the stellar structure equations, yielding nearly constant radii over a wide range of stellar masses in the sub-stellar domain.

The plasma in brown and red dwarfs also exhibits pronounced quantum screening effects due to partial degeneracy. The dielectric response of such a plasma is modified by the finite-temperature Lindhard function $\Pi(\omega, \mathbf{k})$ [10], replacing the classical Debye screening length with a quantum Debye-Hückel length [11], which accounts for Fermi velocity and thermal smearing:

$$\lambda_D^{-2} = \frac{4\pi e^2}{k_BT}\left[n_e - \int \frac{d^3p}{(2\pi)^3}\frac{\partial f(p)}{\partial \epsilon}\right],$$

where *f(p)* is the Fermi–Dirac distribution. The reduction in screening length compared to the classical case enhances collective plasma modes, such as ion-acoustic and Langmuir oscillations, and leads to partially coherent quasi-particle excitations.

The ions, though typically classical, may approach the strong-coupling regime depending on the coupling parameter $\Gamma = \frac{Z^2e^2}{a_ik_BT}$, where a_i is the interionic spacing. When $\Gamma \gtrsim 1$, the ionic component of the plasma behaves as a correlated Coulomb liquid, potentially exhibiting plasma crystallization near the outer layers or in planetary-mass brown dwarfs.

Brown and red dwarfs often exhibit strong surface magnetic fields (up to several kilogauss), which influence the transport properties and plasma modes within the degenerate regions. The quantum magnetohydrodynamic behaviour of the partially degenerate plasma requires incorporation of Landau quantization of electron orbits. When the magnetic field strength satisfies $\hbar\omega_c > k_BT$, with $\omega_c = \frac{eB}{m_ec}$ as the cyclotron frequency, electrons populate discrete Landau levels, and the density of states acquires an oscillatory structure:

$$g(\epsilon) = \frac{eB}{hc}\sum_{n=0}^{n_{\max}} \frac{1}{\sqrt{\epsilon^2-\left(m_e^2+2n\hbar\omega_c\right)}},$$

leading to quantum oscillations in thermodynamic and conductivity profiles (e.g., de Haas–van Alphen effect). This directly affects thermal conduction and magnetic field diffusion, modifying the long-term cooling behaviour and magnetic braking mechanisms of the dwarf.

The Quantum plasma properties in brown and red dwarfs play a pivotal role in shaping their observable features. First, luminosity and cooling in brown dwarfs are governed by gravitational contraction and residual thermal energy rather than sustained nuclear fusion. Quantum degeneracy pressure prevents further collapse, leading to a gradual cooling process in which Quantum plasma effects strongly influence thermal conductivity and radiative opacity. Second, magnetic activity is evident despite the lack of a conventional convective dynamo, with observed flares and radio emissions likely arising from plasma-wave interactions in partially ionized quantum environments, potentially involving mechanisms like electron cyclotron maser instabilities. Lastly, these plasma characteristics are crucial for defining the planet–star transition, helping to distinguish massive exoplanets from the smallest stellar objects and informing models of low-mass stellar evolution.

8.5 Quantum Plasma in Accretion Disks and Magnetospheres

The presence of magnetized quantum plasma in accretion disks encircling compact astrophysical objects such as black holes and neutron stars introduces a regime of plasma dynamics that merges relativistic gravitation, intense magnetic fields, and quantum collective effects. In these extreme environments, where densities approach, or exceed nuclear density, temperatures reach up to $10^7 - 10^9$K, and magnetic field strengths surpass $10^{10} - 10^{15}$G, the plasma constituents (primarily electrons, ions, positrons, and possibly muons or strange quark components) no longer obey classical kinetic theory. Instead, quantum statistics, Landau quantization, spin polarization, and modified dispersion relations fundamentally alter transport, emissivity, and stability profiles of the disk.

The appropriate theoretical framework for magnetized quantum plasma in such relativistic and degenerate environments is quantum magnetohydrodynamics (QMHD). Unlike classical MHD, QMHD incorporates quantum statistical pressure, spin magnetization, and tunneling effects, all of which are essential near the inner regions of accretion disks where gravitational potentials are steep and particle phase space is compressed.

Let ρ, $\mathbf{v}$, P, and $\mathbf{B}$ denote the mass density, fluid velocity, quantum pressure, and magnetic field, respectively. The generalized momentum conservation equation in QMHD reads:

$$\rho\left(\frac{\partial \mathbf{v}}{\partial t} + (\mathbf{v} \cdot \nabla)\mathbf{v}\right) = -\nabla P + \frac{1}{c}\mathbf{J} \times \mathbf{B} + \nabla \cdot \mathbf{T}_Q - \rho\nabla\Phi,$$

where Φ is the gravitational potential, $\mathbf{J}$ is the current density, and $\mathbf{T}_Q$ is the quantum stress tensor. The term $\mathbf{T}_Q$, stemming from the Madelung-Bohm quantum potential, encapsulates quantum diffraction effects and is given by:

$$\mathbf{T}_Q = -\frac{\hbar^2}{2m_e}\nabla\left(\frac{\nabla^2\sqrt{n_e}}{\sqrt{n_e}}\right),$$

with n_e denoting the electron number density. This quantum stress becomes dominant in regions where density gradients are steep, such as at the innermost stable circular orbit (ISCO) or near boundary layers between the disk and magnetosphere.

In the presence of ultra-strong magnetic fields, the motion of charged particles perpendicular to the field lines is quantized into discrete Landau levels. For electrons, the energy eigenvalues under such quantization become:

$$E_n = \sqrt{m_e^2c^4 + p_z^2c^2 + 2ne\hbar Bc}, \quad n = 0, 1, 2, \ldots,$$

where p_z is the momentum along the magnetic field direction. At $B \gg B_c = \frac{m_e^2c^3}{e\hbar} \approx 4.4 \times 10^{13}$G, only the ground Landau level is populated (n = 0), leading to one-dimensional motion and highly anisotropic pressure and heat conduction. The magnetization pressure becomes direction-dependent:

$$P_\parallel \neq P_\perp,$$

which significantly affects disk stability, magneto-rotational instabilities (MRI), and the dynamics of jet launching from the disk surface. The modified equation of state incorporating Landau quantization for degenerate electrons in a strong magnetic field is nontrivial and must be evaluated via the density of states in a magnetic field, yielding:

$$n_e = \frac{eB}{2\pi^2\hbar c}\sum_{n=0}^{n_{\max}} g_n \int_0^\infty \frac{dp_z}{1 + \exp\left(\frac{E_n-\mu}{k_BT}\right)},$$

where $g_n = 2 - \delta_{n0}$ accounts for spin degeneracy, and μ is the chemical potential. This integral formulation governs both thermodynamic quantities and transport coefficients in the disk plasma [12].

In quantum magnetized accretion plasmas, classical Spitzer-Härm transport theory fails, and the kinetic coefficients must be derived from the quantum Boltzmann equation, or Wigner function formulations. Specifically, the longitudinal ($\kappa_\parallel$) and transverse ($\kappa_\perp$) thermal conductivities exhibit strong suppression due to Landau quantization and magnetic confinement. For degenerate plasmas, the quantum thermal conductivity parallel to the field is approximated as:

$$\kappa_{\parallel} \sim \frac{\pi^2 k_B^2 T n_e \tau}{3m_e}, \quad \kappa_{\perp} \ll \kappa_{\parallel},$$

where τ is the effective relaxation time influenced by electron–electron and electron–ion quantum scattering rates. Viscosity coefficients, critical for angular momentum transport and MRI regulation, are similarly affected, with magneto-quantum modifications altering the standard Shakura-Sunyaev α-viscosity prescriptions.

In addition to charge-based quantum effects, the electron spin also plays a non-negligible role under astrophysical magnetic fields. The magnetization density **M** arising from spin alignment introduces an additional force in the momentum equation:

$$\mathbf{F}_{\text{spin}} = \mu_B n_e \nabla B,$$

where μ_B is the Bohr magneton. This spin-induced force can lead to spin-separation instabilities and affect the equilibrium profile of density and pressure in the disk. Moreover, the spin susceptibility, and thereby the magnetic response of the plasma, becomes non-linear and strongly temperature- and field-dependent:

$$\chi_{\text{spin}} \sim \frac{\mu_B^2 n_e}{k_B T}\left[1 - \frac{1}{3}\left(\frac{\mu_B B}{k_B T}\right)^2 + \cdots\right].$$

In ultra-cold, dense disk segments (e.g., neutron star crustal boundaries), the spin alignment may even trigger ferromagnetic-like behaviour or modulate Alfvén wave propagation.

The presence of quantum degeneracy and Landau quantization modifies the dispersion relation of collective modes, including electrostatic Langmuir waves and electromagnetic Alfvén-type modes. For instance, in the presence of a strong magnetic field and degenerate electrons, the Langmuir dispersion relation generalizes to:

$$\omega^2 = \omega_{pe}^2 + \frac{3k^2 v_{Fe}^2}{5} + \frac{\hbar^2 k^4}{4m_e^2},$$

where $v_{Fe} = \sqrt{2E_F/m_e}$ is the Fermi velocity. The third term represents the Bohm potential correction arising from quantum tunneling. These corrections are significant in short-wavelength regimes and near the inner edge of the disk. Further, the MRI critical wavelength is modified by quantum effects, thereby influencing the onset of turbulence and angular momentum redistribution.

The inclusion of general relativistic effects becomes indispensable near the ISCO around black holes and in rapidly rotating neutron stars. The spacetime curvature modifies the conservation equations, and the GR-covariant formulation of quantum MHD must be invoked [13]. The 4-momentum conservation reads:

$$\nabla_\mu T^{\mu\nu}_{\text{total}} = 0,$$

where the total stress-energy tensor includes contributions from matter, magnetic fields, and quantum corrections. Frame-dragging effects and GR redshift influence both the thermodynamic properties and radiative transfer processes, particularly in the context of quantum -modified emissivity and spectral features. Moreover, the Quantum plasma's interaction with the strong-field metric (e.g., Kerr, or Schwarzschild background) leads to non-trivial modifications of transport equations, which require covariant Wigner function techniques for accurate modeling.

8.6 Quantum Plasma in Nebulae and Stellar Birth Regions

In the context of astrophysical environments where gravitational contraction and radiative cooling mechanisms act simultaneously such as dense molecular clouds and pre-stellar cores the emergence of quantum plasma behaviour at low temperatures becomes critically significant. These regions represent the earliest phases of star formation, where matter is cold, dense, and weakly ionized, often forming a complex milieu of dust, neutral species, and partially degenerate charged particles. The quantum statistical mechanics and collective behaviour of such systems diverge markedly from classical plasma treatments, and necessitate a rigorous incorporation of quantum degeneracy, zero-point pressure, and quantum tunneling effects into the plasma description.

The typical conditions in dense molecular clouds temperatures $T \sim 10 - 100$ K and particle number densities $n \sim 10^4 - 10^8\,\text{cm}^{-3}$ imply that electrons, and in some cases protons or dust-trapped ions, may approach degeneracy. The degeneracy parameter, defined as the ratio of the thermal energy $k_B T$ to the Fermi energy E_F, quantifies the onset of quantum effects:

$$\Theta = \frac{k_B T}{E_F}, \quad \text{with} \quad E_F = \frac{\hbar^2}{2m_e}\left(3\pi^2 n_e\right)^{2/3}.$$

For $\Theta \lesssim 1$, the electron component becomes quantum degenerate, and the pressure becomes dominated by Pauli exclusion rather than thermal agitation. Importantly, even for relatively low densities $n_e \sim 10^6\,\text{cm}^{-3}$, the corresponding Fermi temperature is:

$$T_F = \frac{E_F}{k_B} \approx 3.3 \times 10^{-3}\left(\frac{n_e}{10^6 \text{cm}^{-3}}\right)^{2/3} \text{K},$$

which, while small, implies that quantum degeneracy may arise under specific compression or cooling conditions near the gravitational collapse threshold.

8.7 Quantum Plasma During Star Formation and Protostellar Collapse

The star formation process, particularly during the gravitational collapse of dense molecular cloud cores into protostars, provides an astrophysical environment of extreme complexity in which quantum plasma physics becomes fundamentally relevant. In this stage, the interplay between gravity, degeneracy pressure, quantum diffusion, and radiative energy loss must be carefully modeled within a rigorous framework that accounts for quantum mechanical modifications to both the equation of state and the dynamical evolution of the collapsing plasma. The characterization of the plasma during protostellar collapse must thus include quantum statistical degeneracy effects, quantum hydrodynamic corrections, modified screening lengths, and quantum modified instabilities all of which influence the onset of fragmentation, angular momentum transport, and the development of magnetized substructures.

At the earliest stages of collapse, as density increases and temperature remains low due to efficient radiative cooling, quantum degeneracy becomes significant particularly for electrons. The equation of state (EOS) for a degenerate plasma departs from the classical ideal gas law and assumes a Fermi–Dirac statistical form. For a completely degenerate, non-relativistic electron gas, the pressure is governed by the degeneracy pressure:

$$P_e = \frac{\left(3\pi^2\right)^{2/3}}{5}\frac{\hbar^2}{m_e}n_e^{5/3},$$

where n_e is the electron number density. This form is derived by integrating the Fermi distribution at zero temperature and physically represents the resistance of the system to compression due to the Pauli exclusion principle. As the core contracts, this pressure becomes comparable to or exceeds the thermal pressure and acts as a non-thermal, non-dissipative counterbalance to gravity.

As the collapse proceeds and the central density reaches $n_e \sim 10^{21}\,\mathrm{cm}^{-3}$, Relativistic effects begin to modify the EOS. The pressure for a relativistically degenerate electron gas is then given by:

$$P_e^{\mathrm{rel}} = \frac{\pi m_e^4 c^5}{3h^3}\left[x\left(2x^2-3\right)\sqrt{1+x^2}+3\sinh^{-1}(x)\right],$$

where $x = p_F/m_e c$, and p_F is the Fermi momentum. This transition from non-relativistic to relativistic degeneracy defines the critical threshold at which quantum pressure alone becomes insufficient to halt collapse, signaling the onset of further stages leading to brown dwarf formation, white dwarfs, or further collapse into neutron stars.

8.8 Quantum Effects in Dusty Astrophysical Plasma

In the astrophysical context, dusty plasmas are ubiquitous in environments ranging from planetary rings and cometary tails to star-forming regions and supernova remnants. These plasmas are composed of electrons, ions, neutral atoms, and charged dust grains, typically micrometer to submicrometer in size, often carrying multiple electronic charges. While the classical plasma physics of such environments has been extensively studied, the incorporation of quantum mechanical effects, particularly tunneling phenomena and collective quantum behaviors, becomes critical in high-density, low-temperature, or strongly magnetized regimes prevalent in the inner regions of molecular clouds, protoplanetary disks, and compact object accretion zones [14].

Quantum tunneling in dusty plasmas primarily manifests in the context of particle transport across potential barriers generated by electrostatic fields, or spatial confinement in potential wells formed by collective interactions. In dense astrophysical plasmas, quantum degeneracy of electrons, or even ions implies the presence of significant zero-point energy, allowing for particle tunneling even when classical trajectories are forbidden.

The quantum mechanical tunneling probability for a particle of energy E incident upon a one-dimensional potential barrier V(x), in the semi-classical (WKB) approximation, is given by:

$$T \approx \exp\left(-\frac{2}{\hbar}\int_{x_1}^{x_2} \sqrt{2m[V(x) - E]dx}\right)$$

where x_1, x_2 are the classical turning points. For dusty plasmas, V(x) may represent the screened Coulomb potential due to charged dust grains, modeled via the Yukawa form:

$$V(r) = \frac{Q_d e}{4\pi\varepsilon_0 r}\exp\left(-\frac{r}{\lambda_D}\right)$$

where Q_d is the dust grain charge and λ_D is the effective Debye length of the plasma. In dense, cold regions, λ_D may be comparable to the interparticle separation, and quantum corrections to the potential become essential.

When the electron de Broglie wavelength $\lambda_{dB} = h/\sqrt{2\pi m_e k_B T}$ becomes comparable to the intergrain spacing d, quantum diffraction and tunneling effects dominate the transport and localization properties of the electrons. In particular, this condition introduces nonlocality in the electron response and invalidates purely local dielectric models.

Collective quantum effects become pronounced when the characteristic length scales associated with inter-particle interactions, such as the Thomas–Fermi screening length λ_{TF} or the Fermi wavelength λ_F, are of the same order or smaller than the inter-grain or inter-electron spacing. In such regimes, linear and nonlinear wave propagation phenomena exhibit quantum corrections.

For example, the dispersion relation for quantum dust-ion acoustic waves (QDIAWs), which describe low-frequency modes in dusty plasmas with inertial ions and massive, negatively charged dust grains, incorporates quantum pressure and tunneling effects:

$$\omega^2 = \frac{k^2 C_s^2}{1 + k^2 \lambda_q^2}$$

where $C_s = \sqrt{\frac{Z_i k_B T_e}{m_i}}$ is the ion acoustic speed, $\lambda_q = \sqrt{\frac{\hbar^2}{4m_e^2 C_s^2}}$ is the quantum correction length scale, and k is the wave number. This equation shows how quantum pressure (via the Bohm potential) regularizes short-wavelength fluctuations and leads to modified group velocities and stability thresholds.

In non-linear regimes, these effects lead to the formation of quantum modified solitons, governed by the quantum Korteweg-de Vries (q-KdV) equation, or its generalizations, where dispersive and quantum pressure terms counterbalance nonlinearity. The typical soliton width is modified by λ_q, highlighting the non-negligible role of tunneling in determining coherent structure formation in astrophysical dusty plasmas.

Quantum tunneling and collective effects in dusty plasmas have profound implications in the early stages of star formation, especially within dense molecular cores where thermal energy is insufficient to overcome Coulomb barriers for grain coagulation or ion–electron recombination. Tunneling allows charged particles to surmount potential barriers, facilitating processes such as ambipolar diffusion, recombination cooling, and dust coagulation key mechanisms in the gravitational collapse and fragmentation of clouds.

Moreover, tunneling-induced enhancements in charge mobility influence the conductivity and magnetic diffusion properties of protostellar plasmas, thereby affecting the magneto-hydrodynamic (MHD) evolution, including the onset of magnetic reconnection and the efficiency of magnetic braking.

Collective quantum effects also modify wave damping and dispersion, influencing energy Transport and radiative emission signatures in dusty protostellar environments. For instance, the shift in wave frequencies due to quantum corrections can be linked to anomalous broadening or displacement of infrared or radio lines, which are often used to infer plasma conditions in dense interstellar regions.

8.9 Instabilities in Quantum Astrophysical Plasmas

8.9.1 Quantum-Modified Weibel, Jeans, and Rayleigh–Taylor Instabilities in Astrophysical Plasmas

In the extreme environments characteristic of compact astrophysical systems, quantum mechanical corrections to classical plasma instabilities become critically important. Such corrections arise from quantum statistical pressure, quantum diffraction (modeled via the Bohm potential), and spin-related effects. In this context, one must revisit the classical Weibel, Jeans, and Rayleigh–Taylor instabilities each originally formulated within the kinetic, or hydrodynamic regimes now within the framework of quantum plasma theory. These instabilities, which govern the onset of anisotropic, gravitational, or buoyancy-driven structure formation, acquire qualitatively new features in high-density and strongly magnetized astrophysical settings such as neutron stars, white dwarfs, and primordial cosmological plasmas. Below, each instability is analyzed rigorously with quantum modifications explicitly delineated and interpreted.

Quantum-Modified Weibel Instability

The Weibel instability, originally formulated as a kinetic anisotropy-driven instability in unmagnetized plasmas, arises when there exists a temperature anisotropy $T_\perp > T_\parallel$, where $T_\perp$ and $T_\parallel$ denote the perpendicular and parallel temperatures relative to the perturbation wave vector $\mathbf{k}$. In quantum plasmas, the presence of Fermi pressure and quantum recoil modifies the dielectric response of the medium.

To formalize the analysis, consider a quantum Vlasov-Maxwell system in the presence of anisotropic Fermi–Dirac distribution. The dielectric tensor becomes dependent on the quantum corrected distribution function $f(\mathbf{v}, \mathbf{r}, t)$, which, under linear perturbation and Fourier analysis, yields the dispersion relation for transverse modes:

$$1 - \frac{\omega_p^2}{k^2 v_{\rm th}^2}(1 + \xi Z(\xi)) + \frac{\hbar^2 k^2}{4m^2} = 0$$

Here, ω_p is the plasma frequency, v_{th} is the thermal velocity, $\xi = \omega/(k v_{\rm th})$, and $Z(\xi)$ is the plasma dispersion function. The last term, proportional to $\hbar^2$, is the quantum diffraction correction, representing the quantum pressure (Bohm term). It stabilizes short-wavelength modes and modifies the growth rate γ of the instability. In dense degenerate environments, the growth rate becomes bounded and is given by:

$$\gamma^2 \approx \omega_p^2\left(\frac{T_\perp}{T_\parallel} - 1\right) - \frac{\hbar^2 k^4}{4m^2}$$

This result reveals a critical wavenumber k_c beyond which quantum dispersion suppresses the instability entirely, a feature absent in the classical regime. In the astrophysical context, such quantum regulated anisotropic instabilities can be significant during the thermalization of electron beams, or shock-heated regions in protoneutron star crusts or pulsar magnetospheres, where anisotropic distribution functions naturally arise.

Quantum-Modified Jeans Instability

The classical Jeans instability governs the gravitational collapse of self-gravitating media and arises when the internal pressure is insufficient to counteract the gravitational self-attraction. In quantum plasmas, especially those governed by degenerate electrons, the pressure term is augmented by both Fermi degeneracy pressure and quantum diffraction. For a quantum plasma composed of degenerate electrons and massive ions, the modified hydrodynamic equations yield the dispersion relation:

$$\omega^2 = c_q^2 k^2 - 4\pi G\rho_0 + \frac{\hbar^2 k^4}{4m_e m_i}$$

Here, c_q is the effective sound speed incorporating Fermi pressure:

$$c_q^2 = \frac{\partial P}{\partial \rho} = \frac{1}{3}\left(\frac{3\pi^2}{2}\right)^{2/3} \frac{\hbar^2}{m_e m_i} n_0^{2/3}$$

and ρ_0 is the equilibrium mass density. The term $\hbar^2 k^4/(4m_e m_i)$ is the quantum Bohm potential correction. The critical Jeans wavenumber becomes:

$$k_J^2 = \frac{4\pi G\rho_0}{c_q^2 + \frac{\hbar^2 k^2}{4m_e m_i}}$$

This relation demonstrates that quantum pressure, arising from degeneracy and diffraction, enhances stability against gravitational collapse for small-scale perturbations, thereby altering the mass spectrum of collapsing regions. In white dwarf and neutron star interiors, this leads to a suppression of small-scale fragmentation, impacting the morphology and thermal evolution of dense stellar matter. Further, during cosmological recombination, quantum modified Jeans criteria could affect baryonic structure formation in the early universe, particularly in regions where quantum statistical pressure dominates over classical thermal effects.

Quantum-Modified Rayleigh–Taylor Instability

The Rayleigh–Taylor (RT) instability classically arises at an interface between fluids of differing densities when the lighter fluid pushes against the heavier one in a gravitational or inertial field. The quantum modification arises when such an interface is embedded within a dense, low-temperature quantum plasma, where quantum tunneling and degeneracy effects significantly modify the hydrodynamics.

In the presence of a uniform gravitational field **g**, and for a planar interface subject to perturbation $\phi = \phi_0 e^{i(kx-\omega t)}$, the dispersion relation in quantum hydrodynamic approximation becomes:

$$\omega^2 = -gk\frac{\rho_2 - \rho_1}{\rho_2 + \rho_1} + \frac{\hbar^2 k^4}{4m_e m_i}$$

Here, ρ_1 and ρ_2 are the mass densities on either side of the interface. The second term in the expression arises from the quantum Bohm potential and stabilizes the system at short wavelengths. The critical wavelength λ_c below which the RT instability is suppressed is given by:

$$\lambda_c = 2\pi\left(\frac{\hbar^2}{4m_e m_i g(\rho_2 - \rho_1)}\right)^{1/3}$$

This suppression has astrophysical significance in stratified quantum plasmas, such as proto-white dwarfs, supernova interiors, or collapsing star layers, where density inversions and magnetic field gradients may develop. The presence of quantum pressure smoothens interfaces and suppresses fine-scale RT-driven turbulence, thereby influencing the mixing and Transport processes within stellar environments.

The quantum corrections to these canonical plasma instabilities underscore a common theme: the suppression or modification of small-scale structure growth due to intrinsic quantum effects. These corrections are not merely theoretical refinements but dictate the fundamental evolution of structure and energy dissipation in extreme astrophysical settings. From the anisotropic electron beams in pulsar winds (Weibel), to the gravitational fragmentation of compact object crusts (Jeans), and to density gradient interfaces in supernova remnants (RT), quantum plasma theory introduces non-classical thresholds and cutoff scales that profoundly affect dynamics. The synthesis of quantum mechanics with fluid and kinetic instabilities offers a new lens through which to examine astrophysical evolution, especially in regions of ultra-high density, low temperature, and intense magnetic or gravitational fields [15, 16].

8.9.2 Quantum Exchange-Driven and Spin-Induced Instabilities in Astrophysical Quantum Plasmas

In high-density astrophysical plasmas, where Fermi degeneracy, quantum coherence, and strong magnetic fields prevail, instabilities can arise not merely from macroscopic gradients in density, pressure, or velocity, but from the intrinsic quantum mechanical properties of the plasma constituents. In this regard, quantum exchange interactions arising from the anti-symmetrization of many-body wavefunctions and spin-dependent forces rooted in the magnetic moment coupling of electrons to external and self-generated fields can give rise to a novel class of plasma instabilities. These instabilities are absent in classical theory and become crucial in environments such as white dwarf interiors, magnetars, and early Universe baryon-electron plasmas, where spin and exchange effects are not negligible.

Quantum Exchange-Driven Instabilities

Exchange interactions originate from the Pauli's exclusion principle and are fundamentally nonlocal in nature, introducing an effective potential that reflects the quantum statistical correlation among identical fermions. In the context of a dense degenerate plasma, particularly with polarized electrons, this exchange interaction modifies the total internal energy density and, consequently, the pressure tensor. For a spin-polarized degenerate electron gas, the exchange correction to the energy density $\mathcal{E}_{\text{ex}}$ is approximately given by:

$$\mathcal{E}_{\text{ex}} \approx -\frac{3}{4}\left(\frac{3}{\pi}\right)^{1/3} \frac{e^2}{r_s} n_e^{4/3}$$

where n_e is the electron number density and r_s is the Wigner–Seitz radius. This correction induces an effective negative pressure contribution, which, when included in the momentum balance of the QHD equations, modifies the plasma's susceptibility to perturbations. The dispersion relation for longitudinal electrostatic modes, incorporating exchange corrections, is expressed as:

$$\omega^2 = \omega_p^2 + \frac{3}{5} v_F^2 k^2 - \alpha_{\text{ex}} k^2 + \frac{\hbar^2 k^4}{4m^2}$$

where ω_p is the plasma frequency, v_F is the Fermi velocity, $\alpha_{\text{ex}} \sim n_e^{1/3} e^2/m$ is the exchange coefficient, and the final term represents the quantum Bohm correction. A sufficiently large negative exchange term $\alpha_{\text{ex}} k^2$ can reduce the restoring force below the destabilizing threshold for specific wavenumbers, leading to the emergence of growing modes. This gives rise to exchange-driven electrostatic instabilities that dominate over thermal or fluid instabilities at short wavelengths, particularly in the context of strongly coupled, spin-polarized fermionic plasmas in compact astrophysical cores. Physically, these instabilities reflect the competition between quantum pressure (stabilizing), exchange repulsion (destabilizing), and quantum dispersion (scale-selective stabilization). Their astrophysical relevance is pronounced in environments where spin polarization becomes energetically favorable, such as ultra-strong magnetic fields near magnetars, where Landau quantization results in spin-split Fermi surfaces [17].

Spin-Induced Instabilities

Spin-induced instabilities arise from the interaction of the intrinsic magnetic moment of charged particles primarily electrons with external magnetic fields and internal spin-generated magnetization. In a magnetized quantum plasma, the spin dynamics must be described via a generalized Bloch equation coupled with the plasma fluid dynamics. The spin magnetization current density contributes an additional term in

Maxwell's equations through the magnetization **M**, given by:

$$\mathbf{M} = -\mu_B n_e \langle \boldsymbol{\sigma} \rangle$$

where μ_B is the Bohr magneton and $\langle \boldsymbol{\sigma} \rangle$ denotes the average spin orientation. The total current in Maxwell's equations becomes:

$$\mathbf{J}_{\text{tot}} = \mathbf{J}_{\text{free}} + \nabla \times \mathbf{M}$$

This magnetization current can feed back into the plasma dynamics, modifying the dispersion of both electromagnetic and electrostatic modes. When these spin-aligned currents couple to density or field fluctuations, spin-induced modulational, or drift instabilities can develop. The dispersion relation for such spin-modified electromagnetic waves in the presence of a uniform external field $\mathbf{B}_0$ takes the form:

$$\omega^2 = \omega_p^2 + k^2 c^2 + \frac{\mu_B^2 B_0^2 k^2}{m^2}$$

In this relation, the last term represents the magnetization energy stored in spin alignment. For systems close to spin-saturation, this term becomes large and may lead to an effective negative mass behaviour in certain transverse modes, triggering spin-modulational instability. Further, in situations where spin-precession resonances align with plasma oscillations, spin-cyclotron resonances can amplify wave-particle interactions, leading to anomalous Transport, or turbulence.

In gravitationally bound, spin-polarized layers such as those found in the crust of magnetars or the degenerate outer layers of white dwarfs, such spin-induced instabilities may lead to the spontaneous formation of magnetized filaments, domain-like spin structures, or even magnetic reconnection via localized current collapse. The impact on energy Transport and magnetic field evolution in these regions is potentially profound, especially when such spin instabilities nonlinearly couple to large-scale modes or trigger magnetic phase transitions in quantum degenerate matter.

Astrophysical Implications and Unified Perspective

From a broader perspective, both exchange-driven and spin-induced instabilities are manifestations of the deeper quantum coherence that governs the behaviour of high-density matter in astrophysics. Unlike classical instabilities that emerge from macroscopic field and flow gradients, these instabilities arise from microscopic quantum interactions exchange correlation and spin alignment underscoring the importance of many-body quantum field theoretic effects in astrophysical plasma dynamics.

Their incorporation into astrophysical modeling is essential, particularly for understanding anomalous field generation, small-scale turbulence, and nonlocal transport in magnetized neutron stars, white dwarfs, and the early universe. For instance, the evolution of magnetic topology in young neutron stars may be significantly influenced by quantum spin-induced instabilities, while in dense accretion disks, exchange instabilities could regulate the fragmentation threshold and suppress turbulence at small scales.

In conclusion, quantum exchange and spin interactions not only alter the plasma's linear stability spectrum, but also modify the nonlinear evolution of instabilities, with potential implications ranging from wave dissipation, heat conduction, and magnetogenesis to the structural evolution of compact astrophysical objects. Their treatment thus forms a fundamental pillar of quantum plasma astrophysics and is indispensable for any consistent theory of plasma instabilities in the quantum astrophysical regime.

8.9.3 Quantum Hall Drift Instabilities and Turbulence Under Strong Confinement

In regimes of extreme magnetic confinement and low dimensionality, as often encountered in the outer crusts of magnetars, dense neutron star envelopes, or localized filamentary structures in astrophysical accretion systems, the interplay between quantum degeneracy, strong Lorentz forces, and discrete energy-level structure introduces novel instability mechanisms not captured by classical magnetofluid theories. One of the most prominent among these is the class of quantum Hall drift instabilities, which emerge due to the breakdown of equilibrium in systems where the quantization of cyclotron orbits, Hall drift, and shear flow act synergistically in a reduced-dimensional plasma geometry. Under such conditions, turbulence develops through the destabilization of collective excitations that couple quantum mechanical quantization to macroscopic current flows, resulting in unique signatures in Transport, dissipation, and magnetic field evolution.

Quantum Hall Framework in Magnetized Quantum Plasmas

In the presence of an ultra-strong magnetic field $\mathbf{B} = B_0\hat{z}$, electron motion in the plane perpendicular to the field becomes quantized into discrete Landau levels, while motion along the field remains continuous. The Hamiltonian for a non-relativistic spinless electron in such a field is:

$$\hat{H} = \frac{1}{2m}(\mathbf{p} + e\mathbf{A})^2$$

In the symmetric gauge $\mathbf{A} = \frac{1}{2}\mathbf{B} \times \mathbf{r}$, the energy eigenstates are the familiar Landau levels:

$$E_n = \hbar\omega_c\left(n + \frac{1}{2}\right), \quad n = 0, 1, 2, \ldots$$

where $\omega_c = \frac{eB_0}{m}$ is the cyclotron frequency. The degeneracy of each Landau level per unit area is given by $g_B = \frac{eB_0}{2\pi\hbar}$. At high magnetic fields and low temperatures, only the lowest few Landau levels are populated, resulting in an effectively two-dimensional electron gas (2DEG) in astrophysical environments.

The Hall current in such a setting arises from the drift velocity:

$$\mathbf{v}_D = \frac{\mathbf{E} \times \mathbf{B}}{B^2}$$

leading to a transverse Hall conductivity. In the quantum regime, this conductivity becomes quantized in integer, or fractional units depending on the filling factor $\nu = n_e/g_B$, producing the quantum Hall effect (QHE). When gradients, or shear in the drift velocity occur, or when inhomogeneities in the magnetic field, or density profile exist, the system becomes prone to drift instabilities, especially when particle, or wave-packet dynamics become resonant with the drift frequency.

Instability Mechanism and Dispersion Characteristics

Quantum Hall drift instabilities can be modeled by extending the quantum Vlasov, or Wigner-Maxwell formalism in the strongly magnetized limit. In the linearized electrostatic regime, considering perturbations $\sim \exp\left[i\left(k_x x + k_y y - \omega t\right)\right]$ and a background Hall drift $v_{D(y)}$, one obtains a modified dispersion relation:

$$\omega - k_y v_D(y) = \Omega(k) + i\gamma(k)$$

Here, Ω(k) encodes the real frequency of the mode, and γ(k) is the growth rate. The instability arises when the Hall shear $\partial v_D/\partial y$ exceeds a threshold set by the balance of quantum pressure, magnetic tension, and Bohm dispersion. More rigorously, in a fully quantized treatment, the effective dielectric function $\varepsilon(k, \omega)$ incorporates the quantized orbital motion of electrons through Landau Quantization, yielding a susceptibility of the form:

$$\chi(\omega, k) = \sum_{n=0}^{\infty} \frac{f_n - f_{n+1}}{\hbar\omega - (E_{n+1} - E_n) + i\eta} \mathcal{F}_{n,n+1}(k)$$

where f_n is the Fermi–Dirac distribution in the nth Landau level, and $\mathcal{F}_{n,n+1}(k)$ is the form factor governing inter-level transitions. Instability manifests when $\mathrm{Im}[\varepsilon(k, \omega)] < 0$, enabling growing modes via resonant absorption or emission of energy from drift-sheared Hall flows.

Turbulence and Nonlinear Saturation

The development of quantum Hall drift instabilities initiates the onset of mesoscale turbulence in magnetized quantum plasmas. Unlike classical turbulence, which cascades from large to small scales via eddy formation, the turbulence here operates through the dephasing of quantized orbits and the nonlinear interaction of discrete eigenmodes. In reduced 2D magnetized systems, such as crustal layers of neutron stars or confined filaments in magnetized accretion tori, the turbulent state may not reach a Kolmogorov-type cascade but rather saturate through phase mixing, or quantum dephasing. The characteristic saturation amplitude $\delta n/n_0$ of the density fluctuations can be estimated from energy balance considerations between drift shear energy and quantum pressure:

$$\frac{1}{2}m(\delta v)^2 \sim \frac{\hbar^2}{2m}\left(\frac{\delta n}{n_0\lambda}\right)^2$$

implying a quantum limited turbulent amplitude scaling as $\delta n/n_0 \sim \lambda/\lambda_q$, where $\lambda_q = \hbar/\sqrt{2mE}$ is the local de Broglie wavelength. The resulting turbulence spectrum is anisotropic, and characterized by spectral flattening at scales approaching the magnetic length $l_B = \sqrt{\hbar/eB}$.

Quantum Hall drift instabilities and their nonlinear consequences are highly relevant in astrophysical contexts where strong magnetic fields, low dimensionality, and shear flows coincide. In magnetar crusts, where magnetic fields may exceed 10^{14}G, the electronic structure becomes essentially two-dimensional, and thermal or tectonic stresses can drive Hall drift. In these settings, the emergence of quantum drift instabilities may trigger magnetic micro-reconnection, localized heating, or crustal failures potentially linked to magnetar bursts or quasi-periodic oscillations in giant flares.

Similarly, in stratified layers of white dwarfs, or neutron star envelopes, the combination of strong vertical confinement and lateral drift can initiate low-frequency drift modes that feed into localized field dissipation or rearrangement. Over cosmological scales, early Universe baryonic plasmas, subject to primordial magnetic fields and relativistic degeneracy, could also have experienced similar drift-induced turbulence, influencing the evolution of magnetic helicity and seed field amplification.

References

1. Livadiotis, G., & McComas, D. J. (2014). Large-scale quantization from local correlations in space plasmas. *Journal of Geophysical Research: Space Physics, 119*(5), 3247–3258.
2. Livadiotis, G., & McComas, D. J. (2013). Evidence of large-scale quantization in space plasmas. *Entropy, 15*(3), 1118–1134.
3. Buividovich, P. V., & Ulybyshev, M. V. (2016). Numerical study of chiral plasma instability within the classical statistical field theory approach. *Physical Review D, 94*(2), Article 025009.
4. Schober, J., Brandenburg, A., & Rogachevskii, I. (2020). Chiral fermion asymmetry in high-energy plasma simulations. *Geophysical and Astrophysical Fluid Dynamics, 114*(1–2), 106–129.
5. Mackie, M., Dannenberg, O., Piilo, J., Suominen, K. A., & Javanainen, J. (2004). New directions in degenerate dipolar molecules via collective association. *The European Physical Journal D-Atomic, Molecular, Optical and Plasma Physics, 31*, 273–282.
6. Bartlett, R. (2023). Alternating current and superconductivity quantum entanglement of waves replaces nuclear fusion as the power source in stars. Available at SSRN 4556339
7. Manuel, C., & Torres-Rincon, J. M. (2015). Dynamical evolution of the chiral magnetic effect: Applications to the quark-gluon plasma. *Physical Review D, 92*(7), Article 074018.
8. Shuryak, E. (2017). Strongly coupled quark-gluon plasma in heavy ion collisions. *Reviews of Modern Physics, 89*(3), Article 035001.

9. Reid, N., & Hawley, S. (2006). New light on dark stars: Red dwarfs, low-mass stars, brown stars.
10. Archubi, C. D., & Arista, N. R. (2020). Unified description of interactions and energy loss of particles in dense matter and plasmas. *Physical Review A, 102*(5), Article 052811.
11. Janev, R. K., Zhang, S., & Wang, J. (2016). Review of quantum collision dynamics in Debye plasmas. *Matter and Radiation at Extremes, 1*(5), 237–248.
12. Scudder, J. D. (2019). The long-standing closure crisis in coronal plasmas. *The Astrophysical Journal, 885*(2), 148.
13. Shende, M. (2022). *Relativistic outflows from active galactic nuclei (AGN) and their connection to accretion disks* (Doctoral dissertation, Department of Physics).
14. Garcia de Andrade, L. C., & Ferrández, A. (2011). Decay of magnetic fields in de Sitter and FRW universes. *The European Physical Journal Plus, 126*, 1–6.
15. Uzdensky, D. A., & Rightley, S. (2014). Plasma physics of extreme astrophysical environments. *Reports on Progress in Physics, 77*(3), Article 036902.
16. Lebedev, S. V., Frank, A., & Ryutov, D. D. (2019). Exploring astrophysics-relevant magnetohydrodynamics with pulsed-power laboratory facilities. *Reviews of Modern Physics, 91*(2), Article 025002.
17. Chou, C. T., Ghosh, S., McGoldrick, B. C., Nguyen, T., Gurung, G., Tsymbal, E. Y., Li, M., Mkhoyan, K. A., & Liu, L. (2024). Large Spin polarization from symmetry-breaking antiferromagnets in antiferromagnetic tunnel junctions. *Nature Communications, 15*(1), 7840.

Chapter 9
Quantum Plasma in Fusion Research

The pursuit of controlled nuclear fusion stands as one of the paramount technological ambitions of modern physics. Its potential to yield a virtually inexhaustible, clean, and efficient energy source has motivated extensive research across plasma physics, high-energy density physics, and nuclear engineering. As fusion research has matured, the limitations of classical plasma theories adequate at moderate densities and temperatures have become increasingly evident in regimes characterized by extreme compression, intense magnetic fields, and high particle degeneracy. It is within these frontiers that quantum plasma theory emerges as an indispensable framework, bridging quantum mechanics, statistical physics, and collective plasma phenomena under conditions pertinent to fusion experiments and reactor designs.

Fusion reactions, such as the Deuterium–Tritium (D–T) fusion reaction,

$$\mathrm{D} + \mathrm{T} \rightarrow^{4} \mathrm{He} + n + 17.6\ \mathrm{MeV},$$

require plasma temperatures on the order of 10^8 K to overcome the Coulomb barrier via thermal motion. However, achieving and sustaining such conditions entails managing instabilities, energy transport, and confinement processes that are substantially modified when quantum mechanical considerations are included.

In ICF schemes, quantum effects impact the electron degeneracy pressure during the compression phase, modify the conductivity and opacity of the plasma, and influence the laser-plasma coupling efficiency.

In MCF devices such as tokamaks and stellarators, under certain operating scenarios (e.g., advanced fuels like $\mathrm{D} -^{3} \mathrm{He}$, extremely strong magnetic fields for compact devices), spin alignment effects, quantum corrections to drift kinetics, and quantum magnetohydrodynamics (QMHD) become nontrivial.

In gravitationally confined fusion analogs (e.g., laboratory analogs inspired by stellar interiors), the interplay between quantum statistical pressure and gravitational forces mirrors that seen in white dwarf and neutron star matter, requiring a fully quantum mechanical treatment of plasma equilibrium.

P. Kumar, *Quantum Plasma*, https://doi.org/10.1007/978-981-95-0481-7_9

The study of quantum plasma in fusion research constitutes a specialized and advanced domain, synthesizing theoretical constructs developed for isolated quantum systems and applying them to highly coupled, high-energy–density laboratory plasmas. It marks a convergence of quantum theory, plasma dynamics, and practical engineering, pushing the boundaries of our understanding toward realizing controlled fusion as a sustainable energy source.

9.1 Overview of Confinement Strategies

The confinement of a high-temperature plasma constitutes the central challenge in achieving sustained nuclear fusion. The primary objective of confinement strategies is to maintain the plasma density, temperature, and energy density over a time period sufficient for nuclear fusion reactions to proceed at a net energy gain. The mathematical and physical frameworks governing confinement involve complex interplay between electromagnetic fields, hydrodynamic pressure, inertial forces, and, in extreme regimes, gravitational binding. In the context of quantum plasma research, confinement strategies must also accommodate quantum mechanical corrections to classical dynamics, particularly in high-density and strong-field conditions.

9.1.1 Magnetic Confinement

Magnetic confinement exploits the Lorentz force to restrict the motion of charged particles:

$$\mathbf{F} = q(\mathbf{E} + \mathbf{v} \times \mathbf{B}),$$

where, q is the particle charge, $\mathbf{v}$ is the velocity, and $\mathbf{E}, \mathbf{B}$ are the electric and magnetic fields, respectively. In high-temperature fusion plasmas, the $\mathbf{E}$ field is often screened, rendering the confinement primarily dependent on the magnetic component.

Charged particles spiral along magnetic field lines with a Larmor radius:

$$r_L = \frac{mv_\perp}{|q|B},$$

where, $v_\perp$ is the velocity component perpendicular to the field. The collective behaviour of particles leads to macroscopic confinement within magnetic configurations such as tokamaks, stellarators, and mirror machines, governed by magnetohydrodynamic (MHD) equilibrium conditions, notably the Grad-Shafranov equation.

The stability of magnetically confined plasmas is traditionally analyzed via energy principles, such as the ideal MHD energy integral:

$$\delta W = \frac{1}{2} \int \left(\gamma p (\nabla \cdot \xi)^2 + \mathbf{B}_0^2 (\nabla \times \xi)^2 + \cdots \right) d^3x,$$

where, ξ is the plasma displacement vector, p is the pressure, γ is the adiabatic index, and $\mathbf{B}_0$ is the equilibrium magnetic field.

In quantum plasma regimes, classical MHD must be extended to quantum magnetohydrodynamics (QMHD) by including additional terms arising from quantum effects:

i. Quantum Pressure from Fermi–Dirac statistics:

$$p_q = \frac{\left(3\pi^2\right)^{2/3} \hbar^2}{5m} n^{5/3},$$

modifying the momentum conservation equation.

ii. Quantum Bohm Potential contributions, giving rise to dispersive corrections in the magnetofluid equations:

$$\mathbf{F}_{\text{Bohm}} = -\nabla V_B = \frac{\hbar^2}{2m} \nabla \left(\frac{\nabla^2 \sqrt{n}}{\sqrt{n}} \right).$$

iii. Spin Quantum Magnetization terms, accounting for collective spin alignments in strong magnetic fields:

$$\mathbf{M}_q = \chi_q \mathbf{B},$$

where χ_q is the quantum magnetic susceptibility.

Magnetic confinement in quantum plasmas involves a richer structure, necessitating modified equilibrium, transport, and stability analyses that incorporate dispersive, nonlocal, and spin effects, beyond classical gyrokinetic and MHD approximations [1].

9.1.2 Inertial Confinement

Inertial confinement fusion (ICF) aims to achieve fusion conditions by rapidly compressing a small pellet of fuel (typically D-T) to extremely high densities and temperatures before significant hydrodynamic disassembly occurs. The compression is driven by lasers, ion beams, or magnetic fields, generating inwardly propagating shock waves.

The fundamental hydrodynamic equations governing classical ICF are the compressible Euler equations, supplemented by radiative energy transport and thermal conduction equations. The key confinement condition is that the inertia of the plasma must resist disassembly long enough for significant fusion burn, quantified by the Lawson criterion:

$$n\tau_E \geq \frac{12k_BT}{\langle\sigma v\rangle Q},$$

where, n is the particle density, τ_E is the energy confinement time, $\langle\sigma v\rangle$ is the reactivity, and Q is the energy release per reaction.

At the extreme densities ($n \sim 10^{26}\text{cm}^{-3}$) and moderate temperatures achieved during compression, quantum mechanical effects alter the dynamics in profound ways:

(i) Electron degeneracy pressure becomes significant and must be included alongside thermal pressure:

$$p_{\text{deg}} = \frac{\left(3\pi^2\right)^{2/3}}{5}\frac{\hbar^2}{m_e}n_e^{5/3}.$$

(ii) Quantum corrected heat transport: The classical Spitzer-Härm conductivity is modified, and heat conduction becomes suppressed due to Fermi degeneracy and quantum collisional effects.

(iii) Quantum hydrodynamics with radiation transport: A consistent quantum-radiative hydrodynamic model is necessary, incorporating photon pressure, quantum diffraction, and radiation field quantization effects.

(iv) Nonlocal electromagnetic corrections: The electromagnetic field coupling with plasma during laser compression is modified, influencing wave absorption, reflection, and laser-plasma instabilities (e.g., filamentation, two-plasmon decay).

The resulting implosion symmetry, shock convergence, and hotspot formation are critically dependent on these quantum-modified dynamics, rendering purely classical treatments inadequate for predictive modeling in modern ICF scenarios [2].

9.1.3 Gravitational Confinement

Gravitational confinement naturally occurs in astrophysical bodies, where the self-gravity of massive objects counterbalances internal pressure gradients. The classical balance condition is encapsulated in the hydrostatic equilibrium equation:

$$\frac{dP}{dr} = -\frac{GM(r)\rho(r)}{r^2},$$

where, P is the pressure, G is the gravitational constant, M(r) is the enclosed mass, and $\rho(r)$ is the mass density.

In the context of laboratory fusion, gravitational confinement is not feasible due to the insufficient mass of fusion plasmas. However, gravitational confinement analogs using extremely strong magnetic fields, or inertial forces to mimic gravitational binding are of conceptual and theoretical interest.

In dense stellar environments such as white dwarfs and neutron stars, quantum degeneracy pressure provides the principal counterforce to gravitational collapse:

(i) Electron degeneracy stabilizes white dwarfs (Chandrasekhar limit).
(ii) Neutron degeneracy stabilizes neutron stars (Tolman–Oppenheimer–Volkoff limit).

The equilibrium structure of such objects is governed by quantum-statistical equations of state and, in relativistic cases, general relativity. Analogously, certain advanced laboratory experiments investigate high-density quantum plasmas under strong external confinement fields, seeking to replicate quantum-gravitational equilibrium conditions.

Moreover, under extreme conditions where quantum electrodynamic (QED) corrections become significant (e.g., vacuum polarization, photon splitting), generalized theories of plasma dynamics must account for nonlinear field interactions. Although gravitational confinement is presently a theoretical model for laboratory plasma physics, its study enriches the understanding of how quantum plasma behaviour modifies classical gravitational equilibria, informing both astrophysical theory and advanced confinement concepts.

Each confinement strategy magnetic, inertial, and gravitational possesses a rich structure that, when extended to the quantum plasma domain, demands substantial revision of classical paradigms. The inclusion of quantum statistics, nonlocal field effects, spin dynamics, and generalized electromagnetic theories introduces novel transport behaviours, stability considerations, and confinement limits. Consequently, the development of a unified, quantum-corrected description of confinement is not merely an academic exercise but a practical necessity for advancing the design of next-generation fusion devices [3].

9.2 Magnetic Confinement Fusion in the Quantum Regime

Magnetic confinement fusion (MCF) devices such as tokamaks and stellarators have traditionally been analyzed within classical magnetohydrodynamic (MHD) frameworks where macroscopic plasma parameters are treated under continuum approximations. However, with the advent of ultra-dense fuel pellets, advanced field shaping, and strongly magnetized operational regimes, particularly in next-generation compact fusion devices and high-field tokamak concepts, the role of quantum mechanical effects becomes non-negligible.

In the magnetized environment of a tokamak or stellarator, the plasma can be described using quantum magnetohydrodynamics (QMHD), an extension of classical MHD incorporating quantum corrections via modified momentum and energy equations. The general momentum equation for a one-fluid quasineutral QMHD plasma reads:

$$\rho\left(\frac{\partial \mathbf{v}}{\partial t} + \mathbf{v}\cdot\nabla\mathbf{v}\right) = -\nabla P + \mathbf{J}\times\mathbf{B} + \mathbf{F}_Q + \mathbf{F}_S,$$

where:

ρ is the mass density,
$\mathbf{v}$ is the fluid velocity,
P is the total pressure (including degeneracy pressure),
$\mathbf{J} = \nabla \times \mathbf{B}/\mu_0$ is the current density,
$\mathbf{F}_Q$ denotes the quantum Bohm force,
$\mathbf{F}_S$ represents the force due to spin-magnetic coupling.

The quantum Bohm force term is given by:

$$\mathbf{F}_{\underline{Q}} = -\frac{\hbar^2}{2m_e}\nabla\left(\frac{\nabla^2\sqrt{n}}{\sqrt{n}}\right),$$

which introduces a dispersive pressure-like correction stemming from the nonlocality of the quantum wavefunction.

In classical MHD theory, the stability of tokamak plasmas is governed by criteria such as the Kruskal–Shafranov limit for kink instabilities and Mercier criteria for interchange modes [4, 5]. In the quantum regime, these criteria are modified due to the quantum pressure gradient ∇P_Q, where:

$$P_Q = \frac{\hbar^2}{5m_e}\left(3\pi^2\right)^{2/3} n^{5/3}.$$

This Fermi–Dirac pressure resists compressive perturbations and stabilizes short-wavelength modes. The incorporation of P_Q into linear stability analysis introduces a scale-dependent dispersion term in the perturbed MHD equations. The modified dispersion relation for interchange modes in slab geometry becomes:

$$\omega^2 = -gk_y\left(\frac{d\ln P}{dx}\right) + \frac{k^4\hbar^2}{4m_e^2},$$

where the second term on the right-hand side is stabilizing and increases with wavenumber k. This implies that in degenerate, or semi-degenerate regimes, the small-scale interchange instabilities are strongly suppressed, altering the growth rate spectrum of edge-localized modes (ELMs) and resistive ballooning instabilities in the edge pedestal [6].

The intrinsic spin of electrons and ions in a magnetized quantum plasma introduces additional magnetization contributions to the total current. The spin force density is defined by:

$$\mathbf{F}_S = \mu_B n \nabla B,$$

where $\mu_B = e\hbar/2m_e$ is the Bohr magneton. The total magnetization current is then:

$$\mathbf{J}_M = \nabla \times \mathbf{M}, \quad \text{with} \quad \mathbf{M} = -\mu_B \tanh\left(\frac{\mu_B B}{k_B T}\right)\hat{B}.$$

This spin-induced current modifies the equilibrium force balance and can contribute to diamagnetic stabilization, or enhancement of magnetic shear depending on the spatial gradient of the magnetic field strength. Particularly in stellarators, where magnetic field topology is inherently three-dimensional and field lines are twisted in complex helical patterns, the effect of spin-aligned magnetization currents introduces anisotropic corrections to neoclassical transport coefficients.

The classical collisional transport theory, exemplified by Braginskii's formulation, becomes insufficient in quantum regimes where the mean interparticle spacing approaches the de Broglie wavelength. In this limit, heat conduction and momentum transport must be modified to include quantum nonlocality. The quantum-corrected thermal conductivity κ_Q includes both Fermi-surface effects and quantum scattering corrections, given schematically as:

$$\kappa_{\underline{Q}} \approx \frac{\pi^2 k_B^2 T n_e \tau}{3m_e}\left(1 + \frac{\lambda_{db}^2}{\lambda_m^2}\right),$$

where $\lambda_{db} = h/\sqrt{2\pi m_e k_B T}$ is the thermal de Broglie wavelength and λ_m is the collisional mean free path. The presence of λ_{db} enhances transport suppression in highly degenerate regimes, thereby reducing electron heat flux across magnetic surfaces, improving confinement, and steepening edge pressure gradients.

In turbulent transport, quantum diffraction effects suppress high-wavenumber components of microturbulence by modifying the fluctuation spectra. This leads to decreased anomalous diffusivity in gyro-Bohm scaling, shifting the effective transport barrier formation threshold.

Tokamaks:

In high-field compact tokamaks, such as SPARC and future spherical designs, where strong compression leads to densities exceeding 10^{28}m^{-3}, quantum corrections impact both core confinement and pedestal dynamics. The incorporation of quantum modified stability criteria may enable operation closer to ideal limits with reduced reliance on external stabilization schemes. Quantum suppression of drift-wave turbulence could allow operation at lower magnetic shear, improving confinement at reduced field costs.

Stellarators:

For stellarators, whose advantage lies in steady-state operation and absence of large-scale current-driven instabilities, quantum magnetization effects may offer novel equilibrium shaping mechanisms. The intrinsic three-dimensionality allows tuning of quantum spin pressures and magnetic shear profiles through non-axisymmetric shaping, potentially enabling deeper access to transport barriers and stabilizing neoclassical tearing modes.

Quantum corrections to MHD stability and transport require hybrid modeling frameworks:

i. Quantum-MHD Codes: Extensions of standard MHD codes incorporating quantum stress tensors and nonlocal potentials (e.g., extended Grad-MHD frameworks).
ii. Wigner–Moyal Simulations: For kinetic-scale quantum effects, phase-space solvers employing Wigner functions can capture nonclassical evolution of distribution functions in toroidal geometry.
iii. Spinor Magnetofluid Models: Employing SU(2)-based spinor representations of magnetization to dynamically evolve spin densities in the presence of magnetic field gradients.

These approaches are under active development and are indispensable for predictive modeling in regimes where fusion plasmas operate near quantum degeneracy or under strong magnetization [7].

Quantum corrections to magnetically confined plasmas manifest as modifications to stability thresholds, transport barriers, and equilibrium configurations. While these effects are negligible in traditional low-density high-temperature tokamaks, they become critical in compact, high-density regimes, and in magneto-inertial configurations with intense fields. The fusion community must systematically incorporate quantum plasma formalism into the design, diagnostics, and control of next-generation confinement devices. Tokamaks and stellarators, as leading configurations, stand to benefit significantly from these insights, both in confinement optimization and operational stability.

9.2.1 Quantum-Modified Drift Kinetics and Gyrokinetics

The investigation of micro-scale instabilities and turbulent transport in magnetically confined plasmas necessitates a kinetic description beyond the fluid framework. In classical fusion plasma theory, this is accomplished through drift-kinetic and gyrokinetic formulations, which systematically average over the fast gyromotion of charged particles in strong magnetic fields, thereby enabling the study of slow drifts and fluctuations responsible for cross-field transport. In quantum plasmas, however, wave-particle duality, quantum statistics, spin dynamics, and non-locality significantly alter the underlying kinetic structures.

The classical drift-kinetic equation (DKE) governs the evolution of the reduced distribution function $f\left(\mathbf{R}, v_{||}, \mu, t\right)$, where $\mathbf{R}$ is the guiding-center position, $v_{||}$ is the parallel velocity, and $\mu = mv_{\perp}^2/2B$ is the magnetic moment. It reads:

$$\frac{\partial f}{\partial t} + \left(v_{\|}\mathbf{b} + \mathbf{v}_d\right) \cdot \nabla f + \frac{dv_{\|}}{dt}\frac{\partial f}{\partial v_{\|}} = C\left[f\right],$$

where $\mathbf{b} = \mathbf{B}/B$, $\mathbf{v}_d$ includes magnetic and electric drifts, and $C\left[f\right]$ is the collision operator.

Quantum modifications enter in two key ways:

1. Through the replacement of the classical distribution function with a Wigner quasi-probability distribution f_W,
2. Via additional quantum potentials and spin-coupled terms in the particle drifts and forces.

The quantum drift-kinetic equation (Q-DKE) therefore generalizes the classical DKE to:

$$\frac{\partial f_W}{\partial t} + \left(v_{\|}\mathbf{b} + \mathbf{v}_d^Q\right) \cdot \nabla f_W + \left(\frac{dv_{\|}}{dt}\right)^Q \frac{\partial f_W}{\partial v_{\|}} = C_Q\left[f_W\right],$$

where all terms are quantum-corrected.

The guiding center drifts are modified by the inclusion of quantum diffraction, spin–orbit coupling, and magnetization currents. Specifically, the magnetic drift velocity $\mathbf{v}_d^Q$ gains contributions from quantum stress tensors and spin polarization. To first order, the quantum-corrected drift velocity is:

$$\mathbf{v}_d^Q = \mathbf{v}_E + \frac{v_{\|}^2}{\Omega_c}\mathbf{b} \times (\mathbf{b} \cdot \nabla \mathbf{b}) + \frac{\mu}{qB}\mathbf{b} \times \nabla B + \mathbf{v}_Q + \mathbf{v}_S,$$

where:

$\mathbf{v}_E = \frac{\mathbf{E}\times\mathbf{B}}{B^2}$ is the classical $\mathbf{E} \times \mathbf{B}$ drift,
$\mathbf{v}_Q = -\frac{\hbar^2}{2m^2}\nabla\left(\frac{\nabla^2\sqrt{n}}{\sqrt{n}}\right) \times \mathbf{b}/\Omega_c$,
$\mathbf{v}_S = \frac{\mu_B}{m}\nabla B \times \mathbf{b}/\Omega_c$ is the spin-induced drift,
$\Omega_c = qB/m$ is the cyclotron frequency.

The presence of $\mathbf{v}_Q$ introduces a fundamentally nonlocal transport channel due to the dispersive nature of the quantum Bohm potential. These corrections are especially significant in regimes of high field strength, or steep density gradients, such as near the plasma edge or within internal transport barriers.

To move beyond semiclassical intuition, one adopts the Wigner–Moyal formalism, where the phase-space distribution is replaced by the Wigner function $f_W(\mathbf{r}, \mathbf{p}, t)$, satisfying the Moyal equation:

$$\frac{\partial f_W}{\partial t} + \frac{\mathbf{p}}{m} \cdot \nabla_{\mathbf{r}} f_W - (\nabla V \star f_W - f_W \star \nabla V) = 0,$$

with the Moyal star product defined as:

$$A{\star}\mathrm{B} = \mathrm{A}\exp\left[\frac{\mathrm{i}\hbar}{2}\left(\overleftarrow{\nabla}_{\mathrm{r}}\cdot\vec{\nabla}_{\mathrm{p}} - \overleftarrow{\nabla}_{\mathrm{p}}\cdot\vec{\nabla}_{\mathrm{r}}\right)\right]\mathrm{B}.$$

For electromagnetic potentials $V(\mathbf{r}, t)$, the above equation retains the full quantum mechanical structure and allows treatment of interference and tunneling phenomena. Within magnetic confinement, this yields kinetic equations preserving gauge invariance and quantum coherence.

The extension to gyrokinetics requires the averaging over the fast gyromotion while retaining quantum corrections to all slow dynamics. This is achieved by redefining the gyrocenter coordinates $(\mathbf{R}_g, v_{||}, \mu)$ in the presence of quantum fields and applying Lie-transform perturbation theory to the quantum Hamiltonian. The gyrocenter Hamiltonian is modified to:

$$H_g = \frac{1}{2}mv_{||}^2 + \mu B + q\langle\phi\rangle + Q(\mu, B, n),$$

where $\langle\phi\rangle$ is the gyroaveraged potential and Q includes quantum corrections such as the Bohm potential term, spin–orbit coupling, and Landau level quantization.

The resulting quantum gyrokinetic equation (Q-GKE) takes the form:

$$\frac{\partial f_g}{\partial t} + \dot{\mathbf{R}}_g \cdot \nabla f_g + \dot{v}_{||}\frac{\partial f_g}{\partial v_{||}} = \left(\frac{\partial f_g}{\partial t}\right)_{\text{coll}},$$

where: $\dot{\mathbf{R}}_g = \mathbf{v}_E + \mathbf{v}_{\nabla B}^Q + \mathbf{v}_{\text{curv}}^Q + \mathbf{v}_S$, $\dot{v}_{||} = -\mu\nabla B/m + F_{||}^Q/m$, and all terms incorporate quantum drift and potential forces. An essential quantum feature is the discretization of perpendicular motion into Landau levels, which modifies the phase-space structure and affects energy exchange between fluctuating fields and particles, particularly near magnetic nulls and in weak-collisional regimes.

In turbulent fusion plasmas, drift-wave instabilities (e.g., ion temperature gradient (ITG) and trapped electron modes (TEM)) govern cross-field transport [8]. The inclusion of quantum effects modifies their linear growth rates and nonlinear saturation dynamics. For instance, in the presence of quantum corrections, the dispersion relation for drift waves reads:

$$\omega = \omega_*\left[1 + \frac{k_\perp^2\lambda_q^2}{1 + k_\perp^2\rho_s^2}\right] - i\gamma_q,$$

where $\lambda_q \sim \hbar/\sqrt{2m_eT}$ is the quantum screening length and γ_q denotes quantum-induced damping. The stabilizing term in the real frequency reflects the dispersive nature of quantum fluctuations, while γ_q incorporates quantum Landau damping and tunneling.

In the nonlinear regime, the generation of zonal flows a key mechanism for turbulence regulation is also influenced. The Reynolds stress tensor gains quantum contributions due to modified particle orbits and coherence effects, altering the growth and damping of zonal modes and potentially leading to improved confinement scenarios under quantum regimes.

The practical application of quantum gyrokinetic equations (Q-GKE) requires significant enhancements to existing gyrokinetic simulation platforms. This involves integrating quantum stress tensors and nonlocal potentials into well-established codes like GENE, GS2, and GYRO. Additionally, spin dynamics are modeled through the implementation of spin kinetic equations using fluid closures based on Pauli algebra. Validation of these extended models is carried out by comparing simulation outcomes with experimental data obtained from high-magnetic-field compact tokamaks and laser-compressed plasma setups where $T < 10^3$eV and $n > 10^{28}\text{m}^{-3}$, particularly under conditions where quantum effects are pronounced. Quantum spectroscopic diagnostics, particularly in the soft X-ray and EUV range, provide experimental access to quantum-modified distribution functions via line-shape asymmetries and fine structure splitting under magnetization.

9.2.2 *Quantum Hall-MHD Effects in Strong-Field Environments*

The interplay of quantum coherence, Hall currents, and magnetohydrodynamic (MHD) flow in magnetically confined quantum plasmas gives rise to a distinct regime known as Quantum Hall-MHD. This regime becomes particularly significant in fusion plasmas where the characteristic scales approach, or fall below the ion skin depth and where strong external magnetic fields impose a hierarchy between electron and ion dynamics. The Hall term, classically negligible in large-scale MHD, becomes dominant at small spatial scales and early timescales, leading to the breakdown of classical frozen-in flux and the emergence of dispersive and nonlocal field dynamics. In the quantum domain, additional modifications arise from quantum pressure, tunneling, spin–orbit coupling, and charge density quantization, all of which deeply modify the Hall-MHD evolution.

Classical Hall-MHD extends ideal MHD by retaining the Hall current term in the generalized Ohm's law. The evolution equations for a two-fluid, quasi-neutral plasma are typically written as:

$$\frac{\partial \rho}{\partial t} + \nabla \cdot (\rho \mathbf{v}) = 0,$$

$$\rho\left(\frac{\partial \mathbf{v}}{\partial t} + \mathbf{v} \cdot \nabla \mathbf{v}\right) = -\nabla P + \mathbf{J} \times \mathbf{B} + \nabla \cdot \mathbf{\Pi},$$

$$\frac{\partial \mathbf{B}}{\partial t} = -\nabla \times \mathbf{E},$$

$$\mathbf{E} + \mathbf{v} \times \mathbf{B} = \frac{\mathbf{J} \times \mathbf{B}}{ne} - \frac{\nabla P_e}{ne} + \eta \mathbf{J}.$$

The Hall term $\mathbf{J} \times \mathbf{B}/ne$ becomes non-negligible at spatial scales approaching the ion skin depth $d_i = c/\omega_{pi}$ and timescales shorter than the ion cyclotron period. At these scales, the assumption of co-moving electrons and ions fails, and electron inertia and dispersive effects emerge, giving rise to fast reconnection, kinetic Alfvén waves, and whistler dynamics.

In the quantum regime where the de Broglie wavelength $\lambda_B \sim \hbar/\sqrt{2m_e T}$, interparticle distance $n^{-1/3}$, and magnetic length $l_B = \sqrt{\hbar/eB}$ become comparable the classical closure fails. The field variables must now incorporate quantum pressure, magnetization, and non-locality inherent to quantum kinematics.

To construct the quantum Hall-MHD system, one begins with the two-fluid quantum hydrodynamic equations and performs a mass-weighted averaging under the assumption of quasi-neutrality and scale separation. The electron momentum equation, incorporating quantum and Hall terms, reads:

$$m_e \left(\frac{\partial \mathbf{v}_e}{\partial t} + \mathbf{v}_e \cdot \nabla \mathbf{v}_e \right) = -e(\mathbf{E} + \mathbf{v}_e \times \mathbf{B}) - \frac{\nabla P_e}{n_e} + \mathbf{F}_Q^e + \mathbf{F}_S^e,$$

where $\mathbf{F}_Q^e = \frac{\hbar^2}{2m_e} \nabla \left(\frac{\nabla^2 \sqrt{n_e}}{\sqrt{n_e}} \right)$ is the Bohm potential force, and $\mathbf{F}_S^e \sim \mu_B \nabla(\mathbf{S} \cdot \mathbf{B})$ represents the spin force due to magnetic coupling.

Combining the ion and electron equations and invoking quasi-neutrality $n_i \approx n_e = n$, one obtains the momentum balance and magnetic induction equations for quantum Hall-MHD:

$$\rho \left(\frac{\partial \mathbf{v}}{\partial t} + \mathbf{v} \cdot \nabla \mathbf{v} \right) = -\nabla P + \mathbf{J} \times \mathbf{B} + \mathbf{F}_Q + \mathbf{F}_S,$$

$$\frac{\partial \mathbf{B}}{\partial t} = \nabla \times \left[\mathbf{v} \times \mathbf{B} - \frac{\mathbf{J} \times \mathbf{B}}{ne} + \frac{\nabla P_e}{ne} - \frac{\hbar^2}{2m_e e} \nabla \left(\frac{\nabla^2 \sqrt{n}}{\sqrt{n}} \right) + \eta \mathbf{J} \right].$$

The new terms in the induction equation reflect quantum corrections: the Bohm-like quantum electric field due to density fluctuations and a quantum spin-motive force. Notably, the quantum Hall electric field leads to enhanced magnetic field line breaking and supports dispersive whistler modes modified by quantum dispersion.

The quantum Hall-MHD regime supports a range of dispersive modes that are absent in classical MHD. Linearizing around a uniform background, one obtains a quantum-corrected dispersion relation for whistler-like modes:

$$\omega = \frac{k^2 B_0}{\mu_0 n e} \left[1 + \frac{\hbar^2 k^2}{4 m_e^2} + \frac{\mu_B B_0}{T} \right],$$

illustrating that quantum corrections raise the phase velocity of the wave and introduce spin polarization dependence. The second term reflects quantum diffraction, and the third accounts for spin alignment effects, relevant in ultra-strong magnetic fields such as those in compact fusion devices or laser-compressed plasmas.

The system also exhibits quantum analogues of topological effects, including quantized Hall currents when the plasma becomes strongly degenerate and partially spin-polarized. In this context, Landau quantization of transverse motion induces a Hall conductivity quantized in units of e^2/h, influencing current sheet structure and reconnection dynamics. The effective Hall parameter $\chi_H = J/(nev)$ now encodes quantum contributions via density of states modifications and spin-dependent transport.

Magnetic reconnection, essential to magnetic energy dissipation in fusion plasmas, is fundamentally altered in the Q-Hall-MHD regime. The reconnection electric field gains quantum contributions from both electron tunneling across magnetic barriers and spin-aligned acceleration. The generalized reconnection rate becomes:

$$E_{rec} = \eta J + \frac{1}{ne} JB - \frac{\hbar^2}{2m_e e} \nabla \left(\frac{\nabla^2 \sqrt{n}}{\sqrt{n}} \right),$$

where the last term is the quantum correction, enhancing reconnection even in the limit of vanishing resistivity $\eta \to 0$. This leads to fast, collisionless reconnection channels, relevant in the edge-localized modes (ELMs) and sawtooth oscillations in tokamaks.

The presence of quantum Hall terms modifies the magnetic topology by enabling fine-scale current filaments, flux ropes, and helical structures with topologically nontrivial winding numbers. These structures are stabilized, or destabilized by the interplay between spin-aligned currents and quantum pressure, potentially leading to self-organized states or confinement degradation depending on parameters [9, 10].

Quantum Hall-MHD turbulence differs fundamentally from classical MHD turbulence due to modified cascade properties and scale coupling. The inclusion of quantum pressure and Hall currents causes a spectral break in the inertial range and introduces an intermediate dispersive range between MHD and kinetic scales. The energy spectrum typically obeys:

$$E(k) \sim k^{-\alpha}, \quad \alpha \approx \begin{cases} 5/3 & \text{(classical MHD range)}, \\ 7/3 & \text{(Hall-MHD range)}, \\ 4 & \text{(quantum dispersive range)} \end{cases}$$

The quantum regime supports coherent structures, such as quantum vortices and spin-aligned current channels, which can persist across multiple scales. These structures can act as intermittent barriers or transport enhancers, modulating heat and particle confinement in magnetized plasmas.

The Hall term introduces asymmetry between parallel and perpendicular cascades, and quantum spin effects can introduce chirality into the turbulence, altering parity properties of the cascade and influencing dynamo processes. Quantum turbulence diagnostics in fusion plasmas may thus rely on spin-resolved magnetic fluctuation spectra and correlations in synchrotron emission polarization [8].

Quantum Hall-MHD effects become experimentally accessible in:

i. High-field spherical tokamaks with field strengths $B > 10\mathrm{T}$,
ii. Dense plasma focus (DPF) devices approaching $n \sim 10^{28}\mathrm{m}^{-3}$,
iii. Laser-compressed fusion targets with relativistically degenerate electrons.

Experimental diagnostics for Q-Hall-MHD include:

(a) Faraday rotation and Cotton-Mouton effects modified by spin-induced birefringence,
(b) Quantum spin noise spectroscopy via Thomson scattering,
(c) Zeeman splitting in electron cyclotron emission (ECE) lines revealing spin population distributions,
(d) Detection of quantized transverse resistivity via contactless conductivity probes.

9.2.3 *Spin Alignment and Quantum Vorticity in Magnetized Plasma Flows*

In strongly magnetized quantum plasmas, particularly those encountered in high-density inertial confinement fusion (ICF) targets and next-generation magnetic confinement systems, the collective behaviour of spin-polarized particles significantly alters the transport, stability, and dynamical evolution of the plasma. Unlike classical hydrodynamic vorticity, quantum vorticity in these systems emerges from the phase structure and spin dynamics of the constituent particles, leading to phenomena not captured by classical fluid, or even standard quantum fluid models.

The spin degree of freedom, though often neglected in classical magnetohydrodynamics (MHD), plays a fundamental role in quantum plasmas where the thermal energy $k_B T$ becomes comparable to or less than the magnetic spin coupling $\mu_B B$. For an electron gas, this corresponds to spin polarization becoming significant when:

$$\mu_B B \geq k_B T \quad \Rightarrow \quad B \geq \frac{k_B T}{\mu_B} \sim 1.5\mathrm{T}\left(\frac{T}{10^4\mathrm{K}}\right).$$

In fusion-grade plasmas with intense magnetic fields and partially degenerate electrons, this condition is readily met, especially near the plasma edge, or within magnetized microstructures.

The spinor fluid model begins with the Pauli equation for a non-relativistic spin-½ particle in an electromagnetic field:

$$i\hbar\frac{\partial \Psi}{\partial t} = \left[\frac{1}{2m}(\hat{\mathbf{p}} - q\mathbf{A})^2 + q\phi - \mu_B \boldsymbol{\sigma} \cdot \mathbf{B}\right]\Psi,$$

where $\Psi(\mathbf{r}, t)$ is a two-component spinor, $\boldsymbol{\sigma}$ are the Pauli matrices, and $\mu_B = e\hbar/2m$ is the Bohr magneton. The corresponding continuity and momentum equations for the spinor fluid can be derived by applying the Madelung transformation:

$$\Psi(\mathbf{r}, t) = \sqrt{n(\mathbf{r}, t)}\chi(\mathbf{r}, t)e^{iS(\mathbf{r},t)/\hbar}, \quad \text{with} \quad \chi^\dagger\chi = 1,$$

yielding a decomposition into scalar density, phase, and normalized spinor. This transformation yields a set of fluid-like equations for mass, momentum, and spin evolution, which incorporate quantum and spin effects explicitly.

The evolution of the spin density vector $\mathbf{s} = \hbar\chi^\dagger\boldsymbol{\sigma}\chi/2$ is governed by the spin precession equation:

$$\frac{d\mathbf{S}}{dt} = \frac{2\mu_B}{\hbar}\mathbf{S} \times \mathbf{B} + (\text{spin - torque and gradient terms}),$$

indicating Larmor precession of spins in the external magnetic field, modified by spin-torque interactions due to spin density gradients. The magnetization $\mathbf{M} = n\mathbf{S}$ enters Maxwell's equations through the modified Ampère law:

$$\nabla \times \mathbf{B} = \mu_0\left(\mathbf{J}_f + \nabla \times \mathbf{M} + \varepsilon_0\frac{\partial \mathbf{E}}{\partial t}\right),$$

where $\nabla \times \mathbf{M}$ contributes to the total current via the magnetization current density. This current plays a central role in the self-consistent evolution of fields and spin polarization in the plasma.

Importantly, the spin pressure tensor arising from the Pauli kinetic term contributes additional anisotropy to the momentum balance equation:

$$\rho\left(\frac{\partial \mathbf{v}}{\partial t} + \mathbf{v} \cdot \nabla\mathbf{v}\right) = -\nabla P + \mathbf{J} \times \mathbf{B} + \nabla \cdot \mathbf{P}_Q + \nabla \cdot \mathbf{P}_S,$$

with $\mathbf{P}_Q$ denoting the quantum Bohm pressure tensor, and $\mathbf{P}_S$ representing the spin-induced pressure anisotropy. In strongly magnetized plasmas, this anisotropy can stabilize certain instabilities or generate new wave modes.

Quantum vorticity arises not merely from velocity field circulation, but from the topology of the wavefunction phase and spin orientation [11]. The canonical quantum vorticity tensor is defined via the total phase and spin connection:

$$\Omega^{\mu V} = \partial^\mu P^V - \partial^V P^\mu + \varepsilon^{\mu\nu\alpha\beta}\partial_\alpha A_\beta,$$

where P^μ is the canonical momentum, and A_μ the gauge field. For a spinor fluid, the velocity circulation includes contributions from the Berry curvature $\mathcal{F}_{\mu\nu}$, derived from the spinor field configuration:

$$\mathcal{F}_{ij} = \partial_i\mathcal{A}_j - \partial_j\mathcal{A}_i + i\left[\mathcal{A}_i, \mathcal{A}_j\right], \quad \text{with} \quad \mathcal{A}_i = -i\chi^\dagger\partial_i\chi.$$

This structure results in non-trivial spin vorticity even in the absence of classical shear or rotation. The effective quantum vorticity vector $\boldsymbol{\omega}_Q$ thus becomes:

$$\boldsymbol{\omega}_Q = \nabla \times \mathbf{v} + \frac{1}{m}\nabla \times \left(\frac{\nabla\sqrt{n}}{\sqrt{n}}\right) + \nabla \times \mathbf{A}_S,$$

where $\mathbf{A}_S$ is the spin gauge potential. The resulting vorticity is intimately linked with the spin texture of the plasma, and supports quantum vortex filaments analogous to those observed in superfluid systems. These vortex structures carry quantized circulation:

$$\oint \mathbf{v} \cdot d\mathbf{l} = \frac{h}{m}(n+s), \quad s \in \mathbf{Z}/2,$$

where s encodes the spin contribution. Such quantized vorticity introduces topological stability, affects angular momentum transport, and modifies reconnection and rotation in confined plasma regions.

The coupling between spin alignment and vorticity is bidirectional. While magnetic fields align spins via the Zeeman interaction, vorticity gradients can in turn align spins through the spin-vorticity coupling term in the Hamiltonian:

$$H_{SV} = \gamma \mathbf{S} \cdot (\nabla \times \mathbf{v}),$$

where γ is a spin-vorticity coupling constant derived from relativistic corrections or effective Lagrangian approaches. This term implies that plasma rotation induces spin polarization even in the absence of external fields a phenomenon related to the Barnett effect in rotating conductors.

Consequently, in toroidally confined fusion plasmas with shear and rotation, spin alignment emerges naturally, generating additional magnetization currents and altering turbulence and transport coefficients. Moreover, nontrivial helicity structures combining velocity and spin fields contribute to generalized helicity conservation laws:

$$\frac{d}{dt}\int (\mathbf{A} + \mathbf{A}_s) \cdot (\mathbf{B} + \mathbf{B}_s) d^3x = 0,$$

where $\mathbf{B}_s = \nabla \times \mathbf{A}_s$ is the spin magnetic field. The conservation of total quantum helicity imposes constraints on relaxation processes and supports inverse cascades in turbulent plasmas.

Spin-aligned vorticity effects are expected to emerge prominently in high-field plasma regimes, particularly in systems like magnetized target fusion (MTF) where rotating plasmas play a key role. These effects may also be observed in dense Z-pinch, or field-reversed configurations that develop axial spin current channels, as well as in compact tokamaks, especially during edge-localized modes (ELMs), which generate sheared vorticity layers [12].

Experimental diagnostics for detecting spin-aligned vorticity effects in plasmas involve several advanced techniques. Spin-polarized neutron scattering can reveal net spin alignment in dense plasma conditions, while circular polarization observed in bremsstrahlung and synchrotron radiation may indicate the presence of net helicity. Additionally, nuclear magnetic resonance (NMR) like precession signals from spin-aligned ions in rotating plasma configurations can serve as a signature of spin dynamics. Laser-induced fluorescence (LIF), when combined with spin-resolved optical pumping, also offers a sensitive method for probing spin population distributions within the plasma.

Moreover, spin-vorticity interactions may influence anomalous transport barriers, act as seed fields for dynamo processes, and even regulate confinement transitions (e.g., L–H mode) through self-consistent magnetization feedback.

9.2.4 Quantum Landau Damping

In collisionless plasmas, damping of collective oscillations arises not from particle collisions, but from phase-space resonances between waves and particles. Landau damping first formulated in the classical context represents this archetypal wave–particle interaction. However, in quantum plasmas, particularly in dense magnetized systems relevant to fusion research, the classical paradigm is significantly modified. Quantum corrections to the kinetic description become necessary, and wave–particle interactions must be analyzed in the context of wavefunction coherence, quantum statistics, and discrete energy level structures induced by magnetic quantization.

The starting point is the Wigner–Moyal equation for the quantum one-particle distribution function $f(\mathbf{r}, \mathbf{p}, t)$, which serves as the quantum analog of the classical Vlasov equation. In the presence of an external magnetic field $\mathbf{B} = B\hat{z}$, the evolution of f is governed by:

$$\frac{\partial f}{\partial t} + \frac{\mathbf{p}}{m} \cdot \nabla_{\mathbf{r}} f + q\left[\mathbf{E} + \frac{\mathbf{p} \times \mathbf{B}}{m}\right] \cdot \nabla_{\mathbf{p}} f = Q[f],$$

where $Q[f]$ denotes the quantum corrections from the Moyal bracket expansion. For weak perturbations $\delta f \ll f_0$, and assuming quasi-stationarity, the linearized Wigner equation reduces to:

$$\frac{\partial \delta f}{\partial t} + \mathbf{v} \cdot \nabla \delta f + \frac{q}{m}(\delta \mathbf{E} + \mathbf{v} \times \mathbf{B}) \cdot \nabla_{\mathbf{v}} f_0 = \mathcal{O}\left(\hbar^2\right),$$

where f_0 is the equilibrium distribution function, which, for a degenerate plasma, is the Fermi–Dirac distribution. The term $\mathcal{O}\left(\hbar^2\right)$ represents the lowest-order quantum diffraction effects stemming from the non-local Wigner potential. In the long-wavelength limit, these corrections lead to an effective dispersive term known as the Bohm potential, while in the kinetic formulation, they manifest as finite-width resonances and energy level quantization.

In classical Landau theory, the damping arises from particles whose velocity v matches the phase velocity of the wave ω/k, leading to energy exchange through phase coherence. The quantum analog involves resonance with discrete momentum states and quantized cyclotron orbits. The resonance condition is modified by magnetic quantization:

$$\omega - k_{||}v_{||} = n\Omega_c + \delta_q,$$

where $n \in \mathbf{Z}$ labels the Landau level, $\Omega_c = eB/m$ is the cyclotron frequency, and $\delta_q \sim \hbar k^2/2\mathrm{m}$ captures quantum recoil, or tunneling corrections. In quantum plasmas, particles cannot change their velocity continuously due to the Pauli exclusion principle and discreteness of Landau levels. Therefore, the conventional continuous phase-space integration used in classical damping must be replaced by summation over Landau levels and integration over longitudinal momenta:

$$\gamma_q = \pi \sum_n \int \delta\left(\omega - n\Omega_c - kv_{||}\right)\left(\frac{q^2}{\varepsilon_0 m^2}\left|\hat{E}_k\right|^2 \frac{\partial f_0}{\partial v_{||}}\right)dv_{||}.$$

The quantum Landau damping rate γ_q thus depends discretely on n, and vanishes if the wave phase velocity falls between the discrete energy spacings. This gives rise to damping thresholds and selective mode dissipation, distinguishing quantum Landau damping from its classical counterpart.

The occupation of energy states under Fermi–Dirac statistics introduces an additional quantum mechanical constraint on the availability of resonant particles. At zero temperature, all states below the Fermi energy E_F are fully occupied and inaccessible to further energy exchange, implying suppression of Landau damping for phase velocities $v_\phi < v_F$. The effective damping window is thus given by:

$$\mathrm{v}_\phi \gtrsim \mathrm{v}_\mathrm{F} \Rightarrow \gamma_\mathrm{q} \sim 0, \text{ for } \mathrm{v}_\phi < \mathrm{v}_\mathrm{F}.$$

In partially degenerate plasmas, finite-temperature smearing allows partial occupation, and the damping rate becomes a function of both T/T_F and the wavevector k. The finite width of the Fermi surface defines the effective population of resonant particles. Quantum suppression of damping in degenerate regimes can explain the enhanced lifetime of electrostatic oscillations in ultra-dense fusion-relevant plasmas, and the relative inefficiency of collisionless energy transfer at small phase velocities [13].

Quantum Landau damping manifests differently for various wave types:

a. Electrostatic Langmuir Waves: The dispersion relation for quantum Langmuir waves becomes:

$$\omega^2 = \omega_p^2 + \frac{3}{5} v_F^2 k^2 + \frac{\hbar^2 k^4}{4m^2},$$

where ω_p is the plasma frequency. The imaginary part of the frequency, determining damping, is now sensitive to $\hbar$, k, and T/T_F. As $k \to 0$, the phase velocity $v_\phi \to \infty$, and damping vanishes. For large k, quantum recoil increases and can create new resonance bands where energy exchange is enhanced.

b. Electromagnetic Waves in Magnetized Traps: For magnetized configurations such as mirror traps or tokamak edge plasmas, transverse electromagnetic waves couple to both longitudinal velocity and spin degrees of freedom. In these settings, the damping is governed by the transverse susceptibility $\chi_\perp(\omega, \mathbf{k})$, modified by spin polarization and Landau-level occupation. The result is polarization-selective damping, with enhanced damping for right-handed modes (whistler waves) under electron spin alignment, and mode-selective transparency due to quantum resonances [14].

In magnetic confinement fusion systems, trapped particle populations play a critical role in driving instabilities such as Alfvén eigenmodes, ion temperature gradient (ITG) modes, and electron fishbone oscillations. The introduction of quantum Landau damping brings additional mechanisms that can either stabilize, or destabilize these modes, depending on factors like plasma degeneracy, coherence effects, and the specific magnetic configuration [15]. For instance, in mirror and cusp traps, the quantization of bounce motion modifies the wave–particle resonance structure, leading to discrete bounce-resonant modes and altering the typical energy dissipation channels. In toroidal confinement systems, especially for modes with high toroidal mode numbers (n), quantum Landau damping significantly impacts radial mode structures, with finite Larmor radius (FLR) effects becoming crucial. Further, in regimes with low phase velocities, classical damping may be suppressed, allowing coherent modes to persist, which can enhance transport barriers or initiate zonal flows. Additionally, asymmetries in quantum damping between co- and counter-propagating waves may result in net momentum transfer, providing novel methods for driving plasma rotation, an important control parameter in fusion performance.

Beyond the linear regime, wave–particle interactions enter nonlinear phases characterized by phase mixing, nonlinear Landau damping, and wave packet decoherence. In quantum plasmas, these effects are influenced by quantum dephasing due to the uncertainty in energy levels and the superposition of wavefunctions. The non-linear Wigner–Poisson system captures such dynamics:

$$\frac{\partial f}{\partial t} + \mathbf{v} \cdot \nabla_{\mathbf{r}} f = \frac{q}{\hbar} \int d^3 s e^{i\mathbf{p}-s/\hbar} \Big[\phi\Big(\mathbf{r} + \frac{\mathbf{s}}{2}\Big) - \phi\Big(\mathbf{r} - \frac{\mathbf{s}}{2}\Big)\Big] f(\mathbf{r}, \mathbf{p}, t),$$

whose solutions exhibit dynamical dephasing of wave packets, quantum echoes, and irreversible mixing. These processes limit the effective coherence time of driven modes and impact the development of non-linear saturation of instabilities. In fusion-grade plasmas with ultrafast heating (e.g., in laser-driven inertial schemes), such nonlinear quantum damping determines the thermalization timescale.

9.3 Inertial Confinement Fusion (ICF) and Quantum Effects

Inertial Confinement Fusion (ICF) is predicated on compressing a small mass of fuel typically a Deuterium–Tritium (D–T) mixture to extremely high densities and temperatures over timescales shorter than hydrodynamic disassembly. The core physical process in ICF involves the creation and manipulation of dense, transient plasmas in which both classical and quantum effects become intertwined. At fuel stagnation and during peak compression, the plasma reaches densities on the order of 10^{25}cm^{-3}, and temperatures in the keV regime, rendering it partially degenerate, strongly coupled, and quantum-mechanically screened.

Under ICF conditions, the free electron population becomes degenerate due to the high density and relatively moderate temperature. The degeneracy parameter $\Theta \equiv T/T_F$, where T_F is the Fermi temperature, provides a quantitative measure:

$$T_F = \frac{\hbar^2}{2m_e k_B}\left(3\pi^2 n_e\right)^{2/3},$$

where n_e is the electron number density. For densities in the range $n_e \sim 10^{26}\text{cm}^{-3}$, typical in ICF fuel cores during stagnation, one finds $T_F \sim 10^6\text{K}$, which implies $\Theta < 1$, and the plasma is therefore degenerate.

In such regimes, electrons fill quantum states up to the Fermi level, exerting a degeneracy pressure independent of thermal motion:

$$P_{\text{deg}} = \frac{\left(3\pi^2\right)^{2/3}}{5}\frac{\hbar^2}{m_e}n_e^{5/3},$$

in the zero-temperature limit. This pressure competes with the thermal ion pressure and provides an effective stabilizing mechanism against collapse. From a hydrodynamic viewpoint, the degeneracy pressure enters the momentum equation through the quantum-modified pressure tensor, yielding altered sound speeds, compressibility, and shock dynamics during implosion and stagnation phases.

Moreover, degeneracy inhibits the free compression of the electron gas, leading to stiffness in the equation of state (EOS) and altering compressibility metrics. This becomes especially relevant in hydrodynamic modeling, where the polytropic index transitions from $\gamma = 5/3$ to effective lower values under degeneracy constraints. Fusion ignition models must therefore be modified to include degeneracy-corrected equations of state for electrons, ensuring accurate evaluation of pressure balance and ignition thresholds [16].

Electron screening modifies the effective Coulomb potential experienced by two fusing nuclei. In dense ICF plasmas, this effect becomes non-negligible due to the high electron number density and limited thermal motion. Classically, Debye–Hückel theory describes screening via a statically shielded potential:

$$\phi(r) = \frac{Ze}{4\pi\varepsilon_0 r} e^{-r/\lambda_D},$$

where the Debye length λ_D is given by:

$$\lambda_D = \left(\frac{\varepsilon_0 k_B T_\varepsilon}{n_\varepsilon e^2}\right)^{1/2}.$$

In quantum plasmas, however, the Debye length must be replaced by the Thomas–Fermi screening length, arising from the linear response of a degenerate electron gas:

$$\lambda_{TF} = \left(\frac{\pi}{k_F e^2}\frac{\partial n_e}{\partial \mu}\right)^{-1/2} = \left(\frac{e^2}{\pi^2 \varepsilon_0 \hbar v_F}\right)^{-1/2},$$

with v_F and k_F being the Fermi velocity and wavenumber, respectively. Screening reduces the Coulomb barrier that ions must tunnel through, thereby enhancing the fusion cross section. This is captured via a screening enhancement factor f_{scr}, typically expressed as:

$$f_{\text{scr}} = \exp\left(\frac{U_e}{k_B T}\right),$$

where U_e is the screening potential energy, approximated in the quantum case as:

$$U_e \approx \frac{Z_1 Z_2 e^2}{\lambda_{TF}}.$$

The effective increase in the tunneling probability due to screening modifies the thermonuclear reaction rate per pair of nuclei R_{12} as:

$$R_{12} = R_0 f_{\text{scr}} = \left(\frac{n_1 n_2}{1+\delta_{12}}\right)\langle\sigma v\rangle_0 \exp\left(\frac{U_e}{k_B T}\right),$$

where $\langle\sigma v\rangle_0$ is the unenhanced rate, and δ_{12} accounts for identical particle statistics. This enhancement becomes particularly important in marginal ignition scenarios where small increases in reactivity can tip the plasma into an ignited state.

At extreme compression, the plasma coupling parameter $\Gamma = \frac{Z^2 e^2}{4\pi\epsilon_0 a k_B T}$, where a is the Wigner–Seitz radius, becomes larger than unity, indicating strong coupling. In such regimes, electrons exhibit collective behaviour akin to that of a degenerate electron liquid. The conventional mean-field models fail, and one must incorporate short-range correlation effects, often via quantum hypernetted-chain (QHNC) equations, or density functional theory (DFT) approaches [17].

Electron screening in the strongly coupled regime is non-linear and non-local. The effective pair potential between ions includes a many-body-induced attractive tail, which enhances clustering and modifies fusion reactivity. The screened potential in such cases often adopts a Yukawa–type form corrected by local field effects:

$$V_{\text{eff}}(r) = \frac{Z^2 e^2}{4\pi \epsilon_0 r} e^{-r/\lambda_{\text{aff}}} + \Delta V_{\text{corr}}(r),$$

where ΔV_{corr} includes local field corrections due to exchange and correlation among electrons. These modifications affect ion–ion structure factors, energy deposition profiles, and transport coefficients such as thermal conductivity and electrical resistivity. Moreover, the collective oscillation modes quantum ion acoustic waves, degenerate Langmuir oscillations are altered due to coupling-induced spectral shifts [18].

The incorporation of degeneracy pressure and quantum screening into the hydrodynamic models of ICF alters both the ignition threshold and burn propagation behaviour. For instance, the classical Lawson criterion must be reformulated to include quantum pressure contributions and enhanced reactivity:

$$n\tau_E \gtrsim \frac{3\mathrm{k_B T}}{\langle \sigma \mathrm{v} \rangle \mathrm{Q f_{scr}}}, \quad \text{with} \quad \mathrm{Q} = \text{fusion energy per reaction}.$$

Quantum effects increase $\langle \sigma v \rangle$ through screening, while also modifying the temperature dependence of the burn wave. In particular, the burn front speed is affected by the modified equation of state and energy transport coefficients:

$$v_b \sim \left(\frac{\kappa_{\text{eff}}}{\rho C_p} \right)^{1/2},$$

where κ_{eff} is the quantum-corrected thermal conductivity, which is typically reduced due to electron degeneracy and increased electron–ion scattering.

Degeneracy impedes electron heat conduction in the ablator and fuel interfaces, modifying preheat dynamics and thermal smoothing. This has implications for hydrodynamic stability, as ablative Rayleigh–Taylor growth is sensitive to preheat profiles. Accounting for degeneracy, screening, and strong coupling becomes essential in predictive ICF models, especially in regimes near the high-foot, or shock-ignition configurations, where plasma conditions are driven to maximal compressions.

9.3.1 Quantum Kinetic Models in Shock and Compression Dynamics

Shock and compression phenomena in dense plasmas are central to the physics of inertial confinement fusion (ICF), fast ignition schemes, and laboratory astrophysics. Classical fluid models based on the Navier–Stokes, or magnetohydrodynamic (MHD) equations capture macroscopic discontinuities, but fail to resolve kinetic-scale and quantum-scale features that arise under extreme compression. As ICF plasmas transition into regimes of high density, partial degeneracy, and short de Broglie wavelengths, a quantum kinetic description becomes indispensable.

In classical hydrodynamics, shock waves are governed by the Rankine–Hugoniot relations, which describe conservation of mass, momentum, and energy across discontinuities. These relations assume local thermodynamic equilibrium and neglect the granularity of phase space distributions. However, in quantum plasmas generated during ICF implosions, the spatial gradients become comparable to characteristic quantum lengths such as the Fermi wavelength λ_F, or thermal de Broglie wavelength λ_{dB}, leading to breakdowns in the assumptions underlying classical closure.

Key indicators of this breakdown include:

i. Knudsen number ($K_n = \lambda_{\text{mfp}}/L \sim 1$): Mean free path becomes comparable to the system size.
ii. Quantum degeneracy $\mathbf{T/T_F} < \mathbf{1}$: Fermi–Dirac statistics dominate electron behavior.
iii. Dispersive and non-local behavior: Classical flux-gradient laws (e.g., Fourier, Ohm) become invalid.

These breakdowns necessitate a quantum kinetic treatment based on phase-space distribution functions that incorporate non-locality, wave-particle coherence, and quantum statistics.

The cornerstone of quantum kinetic theory is the Wigner–Boltzmann equation, which governs the evolution of the Wigner quasi-distribution function $f_W(\mathbf{r}, \mathbf{p}, t)$, a quantum analogue of the classical distribution function. It is defined via the Wigner transform of the density matrix $\rho(\mathbf{r}, \mathbf{r}', t)$:

$$f_W(\mathbf{r}, \mathbf{p}, t) = \frac{1}{(2\pi\hbar)^3} \int d^3 s e^{-i\mathbf{p}\cdot s/\hbar} \rho\left(\mathbf{r} + \frac{\mathbf{s}}{2}, \mathbf{r} - \frac{\mathbf{s}}{2}, t\right).$$

The time evolution of f_W under an external, or self-consistent potential $V(\mathbf{r})$ follows the quantum Boltzmann–Wigner equation:

$$\frac{\partial f_W}{\partial t} + \frac{\mathbf{p}}{m} \cdot \nabla_{\mathbf{r}} f_W - \Theta\left[f_W, V\right] = \left(\frac{\partial f_W}{\partial t}\right)_{\text{coll}},$$

where $\Theta\left[f_W, V\right]$ is the non-local quantum potential term, given by:

$$\Theta\left[f_W, V\right] = \frac{1}{i\hbar}\left[V\left(\mathbf{r} + \frac{i\hbar}{2}\nabla_{\mathbf{p}}\right) - V\left(\mathbf{r} - \frac{i\hbar}{2}\nabla_{\mathbf{p}}\right)\right] f_W.$$

Expanding this term to third order yields the quantum corrections to the classical Vlasov equation:

$$\Theta\left[f_W, V\right] \approx \nabla_{\mathbf{r}} V \cdot \nabla_{\mathbf{p}} f_W - \frac{\hbar^2}{24} \nabla^3 V \cdot \nabla_{\mathbf{p}}^3 f_W + \cdots,$$

which manifest as dispersive terms in the macroscopic fluid closure. These dispersive terms account for quantum diffraction, zero-point energy spreading, and non-local tunneling phenomena near shock fronts.

The right-hand side $\left(\frac{\partial f_W}{\partial t}\right)_{\text{coll}}$ incorporates quantum versions of collisional interactions, typically modeled using quantum extensions of BGK (Bhatnagar–Gross–Krook) or Uehling–Uhlenbeck forms, which include Pauli blocking and detailed balance.

The presence of quantum pressure and non-locality alters the very structure of shock fronts. In classical theory, dissipation leads to monotonic shock profiles. In quantum plasmas, dispersive shock waves (DSWs) may emerge nonlinear waves stabilized by dispersion rather than viscosity.

The fluid moment hierarchy derived from the Wigner–Boltzmann system includes a quantum pressure tensor and Bohm potential contributions. The one-dimensional quantum hydrodynamic closure leads to the following momentum conservation law for a degenerate electron fluid:

$$\frac{\partial u}{\partial t} + u\frac{\partial u}{\partial x} + \frac{1}{mn}\frac{\partial P_{\text{deg}}}{\partial x} = -\frac{1}{m}\frac{\partial V}{\partial x} + \frac{\hbar^2}{2m^2}\frac{\partial}{\partial x}\left(\frac{1}{\sqrt{n}}\frac{\partial^2\sqrt{n}}{\partial x^2}\right),$$

where the last term represents the quantum Bohm potential and ensures phase-space spreading and dispersion.

The competition between nonlinear steepening and quantum dispersion yields a regularized shock front whose width is dictated by λ_F. Numerical simulations of such systems reveal oscillatory tails behind the main shock front analogous to quantum dispersive ripples absent in classical hydrodynamics. These oscillations are robust features in compression zones of quantum ICF implosions and bear signatures observable in spectroscopic diagnostics.

In scenarios where electrostatic forces dominate (e.g., in the early compression phase of spherical implosions), the quantum Vlasov–Poisson system governs the collisionless dynamics:

$$\frac{\partial f_W}{\partial t} + \frac{\mathbf{p}}{m}\cdot\nabla_{\mathbf{r}} f_W - \Theta\left[f_W, \phi\right] = 0, \quad \nabla^2\phi = -\frac{e}{\varepsilon_0}\int f_W d^3p.$$

The self-consistent potential $\phi(\mathbf{r}, t)$ arises from the collective response of the charge-separated system. The system supports quantum Landau damping, modifies collective oscillation spectra, and regularizes phase-space filamentation.

Analytical approaches employ the linearization of the Wigner–Poisson system around a stationary state and Fourier–Laplace transforms, yielding quantum dispersion relations:

$$1 - \frac{e^2}{\varepsilon_0 m}\int \frac{f_0(p)}{(\omega - \mathbf{k}\cdot\mathbf{p}/m)^2 - \hbar^2k^4/\left(4m^2\right)} d^3p = 0.$$

The denominator features a quantum correction $\hbar^2 k^4$ that introduces new modes, modifies phase velocities, and controls shock precursor formation. The presence of these quantum-induced spectral shifts is critical in modeling small-scale electrostatic compression phenomena during laser-driven implosions and stagnation-layer formation.

Shock compression in ICF does not maintain local thermodynamic equilibrium. Accurate modeling requires nonequilibrium quantum kinetic closure schemes, which derive hydrodynamic fields from moment integrals of the Wigner distribution:

$$n(\mathbf{r}, t) = \int f_W(\mathbf{r}, \mathbf{p}, t) d^3p, \quad \mathbf{u}(\mathbf{r}, t) = \frac{1}{n} \int \frac{\mathbf{p}}{m} f_W d^3p.$$

In general, the moment hierarchy in quantum kinetic theory is not closed, necessitating advanced closure techniques. One such method is the use of maximum entropy closures, where the distribution function is approximated by maximizing entropy under given moment constraints. For weakly collisional systems, quantum Chapman–Enskog expansions offer a systematic way to derive hydrodynamic equations. Additionally, quantum lattice Boltzmann methods provide a numerically stable framework for simulating quantum systems, particularly in complex geometries and boundary conditions. These modern approaches aim to bridge microscopic dynamics and macroscopic observables in a consistent and computationally tractable manner. These closures account for anisotropic pressure, heat flux, and higher-order transport phenomena crucial for modeling layered fuel targets and asymmetric implosions.

9.3.2 Laser–Plasma Interaction Under Quantum Dispersion Regimes

The interaction between intense laser radiation and plasma lies at the heart of inertial confinement fusion (ICF), high-intensity laser experiments, and next-generation quantum plasma technologies. In traditional treatments, laser–plasma interactions are described using classical fluid, or kinetic models, where the plasma dielectric function governs wave propagation, absorption, and instability mechanisms. However, in ultra-dense regimes relevant to quantum plasma characterized by high Fermi energies, reduced Debye lengths, and spatial scales comparable to the de Broglie wavelength the inclusion of quantum dispersive effects becomes imperative.

In classical linear response theory, the plasma dielectric function $\varepsilon(\omega, \mathbf{k})$ for electrostatic and electromagnetic perturbations is derived from the Vlasov–Poisson or Vlasov–Maxwell systems. In quantum plasmas, the analogous derivation begins from the linearization of the Wigner–Maxwell system, wherein the Wigner distribution function incorporates the effects of quantum coherence and non-locality.

For a collisionless, degenerate electron plasma, the longitudinal dielectric response $\varepsilon(\omega, \mathbf{k})$ in the linear regime is given by the Lindhard form (at zero temperature) [19, 20]:

$$\varepsilon_L(\omega, k) = 1 + \frac{3\omega_p^2}{k^2 v_F^2}\left[\frac{1}{2} + \frac{1 - \left(\frac{\omega}{kv_F}\right)^2}{4}\ln\left|\frac{1 + \frac{\omega}{kv_F}}{1 - \frac{\omega}{kv_F}}\right|\right],$$

where v_F is the Fermi velocity and ω_p the plasma frequency. In the long-wavelength limit k $\to$ 0, this recovers the quantum plasma oscillation frequency, while for finite k, the response captures Landau damping modified by Pauli blocking and quantum recoil. For transverse electromagnetic waves interacting with the plasma, the quantum dielectric tensor modifies the wave dispersion relation:

$$k^2c^2 = \omega^2\varepsilon_T(\omega, k) = \omega^2 - \omega_p^2 + \hbar^2\frac{k^4}{4m^2}\left(1 - \frac{3}{5}\frac{k^2v_F^2}{\omega^2}\right),$$

where the last term is the leading-order quantum correction to the dispersion, arising from quantum pressure and diffraction. This term becomes dominant in short-wavelength, high-density regimes, leading to deviations from classical electromagnetic propagation, notably increased transparency windows and suppressed cutoff frequencies.

To analyze intense, slowly modulated laser pulses interacting with quantum plasma, a multiple-scale perturbation expansion is employed on the Maxwell–fluid system (with quantum corrections), yielding a quantum nonlinear Schrödinger equation (QNLS) for the envelope of the vector potential $A(x, t)$:

$$i\left(\frac{\partial A}{\partial t} + v_g\frac{\partial A}{\partial x}\right) + \frac{1}{2}\beta\frac{\partial^2 A}{\partial x^2} - \gamma|A|^2A + \sigma_Q\frac{\partial^4 A}{\partial x^4} = 0,$$

where:

$v_g = \partial\omega/\partial k$ is the group velocity,
β is the group velocity dispersion (GVD) coefficient,
γ is the Kerr-type nonlinear coefficient (arising from relativistic or ponderomotive effects),
$\sigma_Q \propto \hbar^2/m^2$ quantifies quantum dispersive corrections.

The presence of the fourth-order derivative term is a hallmark of quantum diffraction, and leads to new families of quantum-modified solitons, modulational stability criteria, and shock-like envelope features.

Analytically, the QNLS admits stable nonlinear envelope solitons when quantum dispersion balances nonlinear self-focusing. This is physically realized when the effective electron pressure (including quantum and degeneracy effects) opposes the laser-induced electron density perturbation.

In classical plasmas, high-intensity laser pulses self-focus via the relativistic or ponderomotive mechanism, wherein the plasma refractive index becomes intensity-dependent:

$$n_{\text{eff}}^2 \approx 1 - \frac{\omega_p^2}{\omega^2 \gamma},$$

with γ representing the relativistic Lorentz factor, or ponderomotive-induced mass correction. In quantum plasmas, quantum pressure, Fermi degeneracy, and quantum recoil contribute additional restoring forces, modifying the refractive index profile.

The modified quantum refractive index becomes:

$$n_{\text{eff}}^2 \approx 1 - \frac{\omega_p^2}{\omega^2} \left(1 + \frac{3}{5} \frac{k_B T_F}{mc^2} + \frac{\hbar^2 k^2}{4m^2 c^2} \right)^{-1},$$

where the Fermi temperature T_F and quantum recoil term regularize the refractive index depression. As a consequence, quantum self-focusing requires a higher threshold intensity and may become fully suppressed in ultra-degenerate regimes. Additionally, filamentation instability where a broad beam breaks into localized filaments undergoes quantum stabilization, reducing transverse modulation growth rates and altering focal spot structures.

These corrections are especially significant in overdense plasma layers created during fuel compression in ICF, where intense short-pulse lasers propagate into partially degenerate electron gases.

In intense laser–plasma interaction, parametric instabilities such as stimulated Raman scattering (SRS) and stimulated Brillouin scattering (SBS) play a critical role in energy depletion, backscatter losses, and electron heating. The three-wave coupling mechanism involves a pump laser decaying into an electron plasma wave and a scattered wave, or an ion-acoustic wave [21].

The quantum-modified dispersion relations for the daughter waves shift the instability thresholds and growth rates. For instance, the SRS growth rate Γ_{SRS} under quantum dispersion is:

$$\Gamma_{\text{SRS}}^2 = \frac{\omega_0^2 \omega_p^2}{4} \left(\frac{k^2 v_{th}^2 + \frac{\hbar^2 k^4}{4m^2}}{\left(\omega_0 - \omega_p\right)^2 - (k v_{th})^2 - \hbar^2 k^4 / 4m^2} \right),$$

where v_{th} is the thermal velocity. The quantum terms in the denominator lead to stabilization of SRS at small scales and suppression of Langmuir wave growth in dense plasmas. Similarly, in SBS, quantum corrections modify the ion-acoustic dispersion and reduce the coupling efficiency in overdense regimes.

From a physical perspective, this reduction in backscatter instabilities is favorable for ICF, as it improves laser energy deposition into the compressed core. However, it also mandates revised models for fast electron generation and laser-plasma energy partitioning.

Although directly measuring quantum corrections in laser–plasma interactions remains challenging, several observable signatures have been theoretically predicted.

These include a blue-shift in scattered light arising from modified dispersion relations, a reduction in hot electron temperatures due to the suppression of Langmuir wave excitation, and noticeable alterations in harmonic generation spectra, especially in the cutoff frequencies. Additionally, quantum effects are expected to contribute to the stabilization of envelope modulations during long-pulse laser propagation, providing indirect, but significant indicators of underlying quantum behaviour in high-intensity plasma environments. Advanced simulations using quantum-corrected particle-in-cell (Q-PIC) codes and quantum hydrodynamic solvers confirm these features. The incorporation of quantum Bohm potential, Fermi pressure, and Wigner-function based current closures is essential for accurate modeling [22].

9.4 Gravitational and Relativistic Confinement Considerations

Gravitational confinement emerges as a natural mechanism in large-scale astrophysical systems where the mass-energy density of the plasma becomes sufficiently high to induce self-gravitational cohesion. In contrast to electromagnetic, or inertial schemes employed in laboratory plasma confinement, gravitational confinement operates independently of external field structures, and instead results from the self-consistent coupling of matter and spacetime curvature. While typically considered in stellar or compact astrophysical environments, gravitational confinement can be analogously studied in scaled laboratory experiments or in theoretical constructs mimicking curved spacetime behaviour in plasmas.

The classical onset criterion for gravitational confinement in an unmagnetized, non-relativistic plasma is governed by the Jeans instability. For a homogeneous plasma of mass density ρ_0, adiabatic sound speed c_s, and gravitational constant G, the linear dispersion relation for a small density perturbation with wavevector $\mathbf{k}$ is given by:

$$\omega^2 = c_s^2 k^2 - 4\pi G \rho_0.$$

Perturbations with wavelength exceeding the *Jeans length* $\lambda_J = c_s\sqrt{\pi/G\rho_0}$ grow exponentially, marking the threshold for gravitational confinement. In astrophysical contexts such as interstellar molecular clouds, or pre-main-sequence stellar cores, this gravitational instability triggers collapse, heating, and subsequent nuclear ignition.

However, in quantum plasmas especially those exhibiting degeneracy pressure or quantum recoil the standard sound speed is modified by both Fermi statistics and the Bohm potential. Including these quantum corrections, the effective pressure becomes:

$$P = P_{\text{thermal}} + P_{\text{deg}} + P_Q,$$

with $P_{\text{deg}} \sim n^{5/3}$ (for non-relativistic degeneracy), and the quantum pressure term associated with the Bohm potential given by:

$$P_Q = -\frac{\hbar^2}{2m_e} n \nabla^2 \ln n.$$

The corresponding modified dispersion relation then reads:

$$\omega^2 = c_s^2 k^2 + \frac{\hbar^2 k^4}{4 m_e m_i} - 4\pi G \rho_0.$$

The critical *quantum-Jeans wavenumber* is obtained by solving $\omega^2 = 0$, leading to:

$$k_{QJ}^2 = \frac{2 m_e m_i}{\hbar^2} \left[-c_s^2 + \sqrt{c_s^4 + \frac{\hbar^2}{m_e m_i} 4\pi G \rho_0} \right].$$

This highlights the stabilizing influence of quantum dispersion at short wavelengths, where the system resists gravitational fragmentation. Notably, at high densities (e.g., $\rho \sim 10^{14} \text{g/cm}^3$, the balance of gravitational potential and quantum degeneracy pressure leads to stable equilibrium configurations, as in white dwarfs, or neutron star crusts.

In compact objects, or high-energy plasma domains where the gravitational potential becomes comparable to the rest energy of the plasma constituents, a general relativistic treatment is necessary. The criterion for relativistic gravitational confinement is encoded in the Tolman–Oppenheimer–Volkoff (TOV) equation, derived from Einstein's field equations under the assumption of spherical symmetry and hydrostatic equilibrium [23]:

$$\frac{dP(r)}{dr} = -\frac{G}{r^2} \left[\rho(r) + \frac{P(r)}{c^2} \right] \left[M(r) + 4\pi r^3 \frac{P(r)}{c^2} \right] \left[1 - \frac{2GM(r)}{rc^2} \right]^{-1},$$

$$\frac{dM(r)}{dr} = 4\pi r^2 \rho(r),$$

where $M(r)$ is the enclosed mass within radius 'r', $\rho(r)$ is the energy density, and $P(r)$ is the pressure. Solutions to the TOV equation yield the maximum gravitationally confined mass for a given equation of state (EoS). For non-interacting, relativistically degenerate fermions, the Chandrasekhar mass $M_{\text{Ch}} \sim 1.44 M_{\text{solar}}$ emerges as the upper limit for electron-degenerate stability.

In quantum plasmas, the relevant EoS includes both degeneracy and plasma interaction effects. For instance, in magnetized compact stars or relativistically hot accretion disks, Landau quantization and relativistic Fermi pressure modify the equilibrium profiles. The inclusion of non-local electromagnetic interactions further alters the pressure tensor, potentially stabilizing configurations otherwise unstable in standard general relativistic hydrodynamics.

While true gravitational confinement cannot be realized at laboratory scales due to the weakness of G, analog systems that emulate gravitational effects through effective field theory or transformation optics offer fertile ground for experimental exploration. Notable among these are:

(i) Optical metric analogs: Utilizing refractive index variations to mimic curved spacetime geometries, e.g., dielectric media with spatially varying permittivity $\epsilon(\mathbf{x})$ to simulate Schwarzschild, or Rindler metrics.
(ii) Bose–Einstein condensates (BECs): Quantum fluid analogs governed by Gross–Pitaevskii equations that reproduce features of horizon formation and acoustic black holes, effectively serving as models for gravitational trapping of perturbations.
(iii) Plasma mirrors and laser-induced spacetimes: High-intensity laser–plasma interactions can generate pressure gradients and accelerations that simulate aspects of relativistic confinement, especially in ultradense plasma sheaths or standing shock regions.

Theoretically, these systems can be analyzed through the correspondence between the stress-energy tensor of the quantum plasma and an effective Einstein–Cartan geometry [24]. The analog Einstein equations:

$$G_{\mu\nu}^{\text{eff}} = 8\pi G_{\text{eff}} T_{\mu\nu}^{\text{plasma}},$$

allow one to compute synthetic curvature from density and flow profiles in the plasma, enabling exploration of confinement thresholds in dynamically evolving field configurations.

To delineate a sharp boundary for gravitational confinement, one can equate the total gravitational potential energy $U_G \sim -GM^2/R$ with the internal energy arising from quantum effects such as degeneracy pressure $U_Q \sim \hbar^2 N^{5/3}/(mR^2)$. This yields the gravitational–quantum equilibrium radius:

$$R_G \sim \frac{\hbar^2}{Gm^3 N^{1/3}}.$$

Below this radius, quantum pressure dominates and halts collapse, while above it, gravity can compress the system further. In astrophysical scenarios, this analysis supports the formation of white dwarfs and neutron stars, in laboratory settings, this condition remains theoretical but provides useful insight into minimum achievable scales for analog confinement systems.

Moreover, in regimes of extreme density and curvature, where spacetime fluctuations and quantum gravitational effects cannot be ignored, the semiclassical Einstein equations:

$$G_{\mu\nu} + \Lambda g_{\mu\nu} = 8\pi G \langle T_{\mu\nu} \rangle,$$

necessitate a quantum field-theoretic treatment of plasma confinement, incorporating renormalized expectation values and effective actions. Although such corrections remain inaccessible to current laboratory systems, their imprint may be observable in ultra-relativistic heavy ion collisions or near horizon-scale plasma structures.

The study of gravitational confinement, though seemingly detached from terrestrial fusion, offers deep insights into the limiting behaviour of high-density plasmas. It frames the fundamental competition between kinetic, quantum, and gravitational energies. In fusion theory, particularly in fast ignition and ultra-compressed target scenarios, gravitational analogs motivate alternative confinement architectures and inspire methods for probing extreme equation-of-state behavior.

As quantum technologies evolve, the possibility of simulating relativistic gravitational confinement in tailored quantum plasmas opens a new frontier in fusion-adjacent research, bridging astrophysical theory with laboratory-scale realizations of gravitational trapping phenomena.

9.4.1 Quantum Plasma Dynamics in Compact Object Analogs

The physics of quantum plasmas in compact astrophysical objects such as white dwarfs, neutron stars, and magnetars provides not only an extreme testing ground for the principles of quantum hydrodynamics, but also a conceptual framework for constructing terrestrial analogs of such phenomena within controlled environments. These analogs, whether realized experimentally, or modeled theoretically, allow one to probe the interplay of quantum degeneracy, strong-field electrodynamics, relativistic corrections, and nonlinear collective behaviour in regimes far beyond those accessible in classical plasma devices. In this section, we develop a detailed mathematical and physical characterization of the governing equations, transport properties, and nonlinear structures relevant to quantum plasma dynamics in compact object analogs.

The dynamics of quantum plasmas in compact object analogs are governed by a coupled set of quantum-corrected hydrodynamic equations that incorporate degeneracy pressure, electromagnetic fields, and relativistic modifications. The foundational equations are:

Continuity Equation:

$$\frac{\partial n}{\partial t} + \nabla \cdot (n\mathbf{v}) = 0,$$

Momentum Conservation with Quantum and Electromagnetic Terms:

$$mn\left(\frac{\partial \mathbf{v}}{\partial t} + \mathbf{v} \cdot \nabla \mathbf{v}\right) = -\nabla P + ne\left(\mathbf{E} + \frac{\mathbf{v} \times \mathbf{B}}{c}\right) + \mathbf{F}_Q,$$

Quantum Force Term (Bohm Potential):

$$\mathbf{F}_Q = \frac{\hbar^2}{2m} n \nabla \left(\frac{\nabla^2 \sqrt{n}}{\sqrt{n}} \right),$$

Maxwell's Equations, with source terms determined self-consistently from the plasma density and current.

The pressure P for a degenerate electron gas follows the polytropic relation:

(a) For non-relativistic degeneracy:

$$P = \frac{\left(3\pi^2\right)^{2/3} \hbar^2}{5m_e} n^{5/3},$$

(b) For ultra-relativistic degeneracy:

$$P = \frac{\left(3\pi^2\right)^{1/3} \hbar c}{4} n^{4/3}.$$

These regimes correspond to compact object analogs of white dwarfs and neutron star crusts, respectively. The competition between pressure gradients and the collective electrostatic and magnetic fields produces highly nonlinear wave structures, shocks, and instabilities.

For fully relativistic treatment, the system is more aptly described using a covariant formalism. The plasma dynamics are encoded in the conservation laws derived from the energy–momentum tensor $T^{\mu\nu}$ and the electromagnetic field tensor $F^{\mu\nu}$:

$$\nabla_\mu T^{\mu\nu} = F^{\nu\lambda} J_\lambda, \quad \nabla_\mu F^{\mu\nu} = 4\pi J^\nu.$$

For a perfect relativistic fluid, the energy–momentum tensor is:

$$T^{\mu\nu} = (\varepsilon + P) u^\mu u^\nu - P g^{\mu\nu},$$

where ε is the energy density, u^μ is the four-velocity, and $g^{\mu\nu}$ is the spacetime metric. The generalized Ohm's law, incorporating quantum corrections, reads:

$$J^\mu = \sigma \left(F^{\mu\nu} u_\nu + \nabla^\mu \mu - \frac{\hbar^2}{2m^2} \nabla^\mu \left(\frac{\Box \sqrt{n}}{\sqrt{n}} \right) \right),$$

where μ is the chemical potential and $\Box$ denotes the d'Alembertian.

In curved spacetime analogs or synthetic laboratory spacetimes, the inclusion of gravitational redshift and frame dragging must be incorporated into the metric $g_{\mu\nu}$. This enables realistic modeling of rotating neutron stars and plasma dynamics near horizons.

A key feature of quantum plasma dynamics in compact object analogs is the emergence of localized, coherent nonlinear structures stabilized by quantum dispersion and magnetic tension. These include:

(i) Quantum Ion-Acoustic Solitons: Balance between quantum pressure and electric field yields stable, non-dispersive solutions of the form:

$$n(x,t) = n_0\left[1 - \frac{\Delta n}{\cosh^2[(x - Vt)/L]}\right],$$

with $L \sim \hbar/\sqrt{m_e e E}$ denoting the quantum-corrected soliton width.

(ii) Magnetized Quantum Vortices: The quantum magnetohydrodynamic (QMHD) equations admit vortex solutions wherein the Bohm potential provides short-scale regularization. These are especially relevant in superconducting neutron star interiors where superfluid and quantum Hall-like structures emerge.

(iii) Spin-Polarization Effects: In highly magnetized environments, spin alignment leads to anisotropic pressure tensors and modifies wave propagation through spin–magnetization coupling:

$$\mathbf{M} = -\mu_B n \tanh\left(\frac{\mu_B B}{k_B T}\right),$$

where μ_B is the Bohr magneton. This introduces a spin-pressure term in the force balance, potentially relevant to magnetars and analog condensed matter systems.

Despite their quantum nature, transport properties such as viscosity, thermal conductivity, and resistivity remain operative in compact plasma analogs. The quantum Boltzmann equation (Uehling–Uhlenbeck form) governs the distribution function $f(\mathbf{x}, \mathbf{p}, t)$ under Pauli exclusion:

$$\frac{\partial f}{\partial t} + \mathbf{v} \cdot \nabla_x f + \mathbf{F} \cdot \nabla_p f = \left(\frac{\partial f}{\partial t}\right)_{\text{coll}},$$

with quantum-corrected collision integrals. In degenerate systems, the thermal conductivity is suppressed, but collective excitations (e.g., zero sound, spin waves) dominate energy transfer. This plays a crucial role in heat redistribution in the neutron star crust and analog experiments using ultracold plasma traps.

The inclusion of strong magnetic fields modifies the transport tensorial structure. The Hall parameter $\beta = \omega_c \tau$, where ω_c is the cyclotron frequency, determines the anisotropy of conduction and viscosity. For $\beta \gg 1$, cross-field transport is quenched, mimicking behaviour observed in strongly magnetized astrophysical plasmas.

Quantum plasma behaviour analogous to compact objects can be approached in the laboratory through:

i. Ultracold Plasma Traps: Laser-cooled neutral atoms ionized just above threshold generate plasmas with well-defined degeneracy and tunable densities ($n \sim 10^8$–10^{10}cm^{-3}).
ii. Dense Laser-Compressed Matter: High-intensity laser pulses induce states of warm dense matter and degenerate plasma with densities up to solid state and beyond ($\rho \gtrsim 1\text{g/cm}^3$), mimicking the physical conditions of compact crustal layers.
iii. Synthetic Curved Spacetime Analogs: Effective metrics can be generated through metamaterials, BECs, or high-gradient laser–plasma interfaces to explore gravitational and relativistic corrections to plasma dynamics.

In such systems, the diagnostic observable such as collective mode frequencies, nonlinear structure formation, or radiation signatures can be directly linked to the theoretical predictions derived from the governing quantum-relativistic equations.

Understanding the dynamics of quantum plasmas in compact object analogs informs fusion research at both theoretical and practical levels. First, it elucidates the ultimate physical limits imposed by degeneracy and quantum nonlocality on matter compression. Second, it inspires novel confinement and compression architectures that exploit quantum stabilization mechanisms. Third, it provides a template for benchmarking extreme equation-of-state models necessary for simulating inertial confinement implosion cores at densities approaching nuclear matter [25].

Moreover, the study of such analogs offers a bridge between controlled fusion research and astrophysical plasma modeling, highlighting universal principles of matter behaviour under quantum statistical, relativistic, and collective constraints. In this regard, compact object analogs serve not merely as pedagogical constructs, but as rigorous testing grounds for high-energy plasma physics and quantum field hydrodynamics.

9.4.2 Application of Relativistic Quantum Plasma Theory to Extreme Confinement

The domain of extreme confinement, where both relativistic and quantum mechanical effects are simultaneously non-negligible constitutes one of the most challenging and revealing regimes of plasma physics. This regime becomes particularly relevant in high-energy–density physics (HEDP), inertial confinement fusion (ICF) cores approaching relativistic degeneracy, and astrophysical environments such as collapsing stellar cores and magneto-gravitational implosions. Relativistic quantum plasma theory provides the essential theoretical infrastructure for modeling the collective behaviour of matter under such extreme compression, wherein the particle velocities approach the speed of light and the de Broglie wavelengths are comparable to interparticle separations.

In relativistically confined quantum plasmas, the proper formalism is the covariant generalization of quantum hydrodynamics (QHD), incorporating both special relativity and quantum statistical mechanics. The plasma is described by a relativistic fluid of fermions governed by a quantum-corrected energy–momentum tensor and coupled to self-consistent electromagnetic and possibly gravitational fields. The principal governing relations include:

i. Energy–Momentum Conservation:

$$\nabla_\mu T^{\mu\nu} = F^{\nu\lambda} J_\lambda,$$

where $T^{\mu\nu}$ is the relativistic energy–momentum tensor, $F^{\mu\nu}$ the electromagnetic field tensor, and J^λ the four-current density.

ii. Relativistic Quantum Pressure for Degenerate Fermions: The pressure P and energy density ε for a relativistically degenerate Fermi gas follow from the Fermi–Dirac statistics:

$$P = \frac{1}{3\pi^2\hbar^3} \int_0^{p_F} \frac{p^4 c^2}{\sqrt{p^2c^2 + m^2c^4}} dp, \quad \varepsilon = \frac{1}{\pi^2\hbar^3} \int_0^{p_F} p^2 \sqrt{p^2c^2 + m^2c^4} dp,$$

where $p_F = \hbar\left(3\pi^2 n\right)^{1/3}$ is the Fermi momentum.

iii. Modified Quantum Bohm Potential in Relativistic Regime: The quantum correction to the momentum equation takes the form of a relativistic Bohm potential:

$$\mathbf{F}_Q^{\text{rel}} = -\frac{\hbar^2}{2m\gamma} \nabla \left(\frac{\Box\sqrt{n}}{\sqrt{n}} \right),$$

where $\gamma = \left(1 - v^2/c^2\right)^{-1/2}$ and $\Box$ denotes the d'Alembertian operator acting in Minkowski space–time.

These relations are supplemented by Maxwell's equations and a continuity equation formulated in a covariant form. The theory accounts for both the relativistic inertia of the plasma constituents and the quantum pressure originating from wavefunction overlap at high densities.

Extreme confinement leads to matter compression to the point where relativistic degeneracy pressure becomes the dominant stabilizing force. To quantify this, one defines a degeneracy parameter χ and a relativistic parameter ξ as:

$$\chi = \frac{E_F}{k_B T}, \quad \xi = \frac{p_F}{mc} = \left(\frac{3\pi^2 n}{n_c} \right)^{1/3}, \quad n_c = \left(\frac{mc}{\hbar} \right)^3.$$

Here, $\chi \gg 1$ signals full degeneracy, while $\xi \gtrsim 1$ indicates the onset of relativistic effects. For instance, in dense inertial confinement cores approaching solid density

($n \sim 10^{30}\mathrm{m}^{-3}$), one finds $\xi \approx 0.1 - 1$, implying a partially relativistic degenerate plasma.

At these thresholds, the classical equations of state become inadequate. Instead, the relativistic pressure scaling $P \propto n^{4/3}$ emerges naturally, modifying the compressibility of the plasma and imposing a limit on achievable core densities. This limit parallels the Chandrasekhar limit in white dwarfs, representing a fusion-relevant analog of astrophysical stability bounds.

Collective oscillations and instabilities in relativistically degenerate quantum plasmas exhibit non-classical dispersion properties. The relativistic quantum-modified Langmuir wave dispersion relation is:

$$\omega^2 = \omega_p^2 + \frac{3P}{nm\gamma^2}k^2 + \frac{\hbar^2 k^4}{4m^2\gamma^2},$$

where $\omega_p = \sqrt{4\pi ne^2/\gamma m}$ is the relativistically corrected plasma frequency. This expression includes several key components: a relativistic correction to the inertial mass represented by the Lorentz factor (γ), a pressure term derived from the relativistic Fermi equation of state, and a quantum dispersion term associated with the Bohm potential. Together, these elements provide a more complete description of plasma behavior under conditions where both quantum and relativistic effects become significant.

As the plasma approaches ultra-relativistic degeneracy, the pressure term dominates, transforming the plasma response from electrostatic oscillations to quasi-acoustic modes governed by the relativistic sound speed:

$$c_s = \sqrt{\frac{\partial P}{\partial \varepsilon}} \rightarrow \frac{c}{\sqrt{3}}(\text{ultra - relativistic limit}).$$

Such collective modes are critical in the energy transport and symmetry preservation in rapidly imploding plasma shells, especially in high-gain ICF capsules.

At extreme densities and temperatures, the interaction of plasma with radiation fields becomes nonlinear, and quantum electrodynamic (QED) effects modify the effective dynamics. Notable corrections include:

a. Vacuum Polarization Effects: Under extreme fields, the photon propagator acquires an effective mass, modifying the dielectric response:

$$\epsilon(\omega, k) = 1 - \frac{\omega_p^2}{\omega^2 - \delta\Pi(\omega, k)},$$

where $\delta\Pi$ is the one-loop QED polarization tensor correction.

b. Nonlinear Compton Scattering and Radiation Reaction: In ultra-intense laser-driven compression scenarios, relativistic electrons emit photons via nonlinear Compton scattering, leading to significant energy loss and a backreaction on dynamics. The radiation reaction force, in the Landau–Lifshitz approximation, is:

$$\mathbf{F}_{\mathrm{RR}} = \frac{2e^4}{3m^2c^5}[\mathbf{E} + \mathbf{v} \times \mathbf{B}]^2\mathbf{v}.$$

c. Photon–Photon Interactions and Pair Creation: At field strengths approaching the Schwinger limit ($E_s \sim 10^{18}\mathrm{V/m}$), virtual photons interact nonlinearly, potentially giving rise to electron–positron pairs. Although such intensities exceed current fusion devices, planned exawatt-class laser facilities and astrophysical analogs provide fertile ground for exploration.

Insights from relativistic quantum plasma theory have a direct impact on the design of fusion confinement systems, particularly in advanced approaches such as fast ignition and shock-compression schemes. In these paradigms, dense plasma layers must be compressed beyond classical thermodynamic thresholds without triggering instabilities, or stagnation driven by quantum degeneracy. Relativistic effects significantly influence key physical properties: they alter the sound speed, thereby affecting shock wave propagation, they suppress electron mean free paths, leading to reduced thermal conductivity, they limit compressibility through quantum-statistical stiffening; and they introduce magneto-inertial anisotropies that impact the symmetry of implosions. Moreover, relativistic degeneracy can stabilize the plasma against hydrodynamic instabilities such as Rayleigh–Taylor modes, promoting greater uniformity during implosion if carefully managed. As a result, relativistic quantum plasma theory not only enhances diagnostic modeling, but also supports predictive control strategies in cutting-edge fusion technologies, including direct-drive inertial confinement fusion (ICF) and laser-plasma compression systems operating near relativistic intensity regimes [26, 27].

9.4.3 *Coupling of Curvature Effects with Quantum Corrections in Plasma Flows*

The study of quantum plasma dynamics under confinement geometries where curvature effects become non-negligible either due to spatial constraints, externally imposed geometrical conditions, or underlying spacetime curvature presents a novel domain of theoretical and experimental relevance. The incorporation of curvature modifies both the quantum statistical properties and collective dynamics of plasma species, giving rise to modified dispersion relations, new forms of quantum stress, and curvature-induced instabilities or coherence phenomena. We employ a differential geometric framework to elucidate this coupling systematically and embed it within a generalized quantum hydrodynamic and field-theoretic context.

To account for curvature effects, one considers a plasma embedded in a curved Riemannian, or pseudo-Riemannian manifold $(\mathcal{M}, g_{\mu\nu})$, where $g_{\mu\nu}$ is the local metric tensor. The fundamental dynamical equations of quantum hydrodynamics are then expressed in covariant form, with all derivatives replaced by covariant derivatives ∇_μ, and physical quantities expressed as geometric tensors, or scalar densities. The key governing relations are:

(i) Continuity Equation:

$$\nabla_\mu (mu^\mu) = 0,$$

where n is the proper number density and u^μ the four-velocity of the plasma fluid.

(ii) Energy–Momentum Conservation:

$$\nabla_\nu T^{\mu\nu} = qF^{\mu\nu} J_\nu,$$

with $T^{\mu\nu}$ incorporating both classical and quantum contributions, and $F^{\mu\nu}$ being the electromagnetic field tensor on the curved manifold.

(iii) Quantum Bohm Tensor on Curved Manifolds: The quantum Bohm potential generalizes to a curvature-sensitive tensorial form:

$$Q^{\mu\nu} = -\frac{\hbar^2}{4m^2}\left[\nabla^\mu \nabla^\nu \ln n + \nabla^\mu \ln n \nabla^\nu \ln n - \frac{1}{2} g^{\mu\nu} \Box \ln n\right],$$

where $\Box = g^{\mu\nu}\nabla_\mu \nabla_\nu$ is the covariant d'Alembert operator. This tensor represents the stress induced by quantum diffraction and density gradients in curved space.

The total energy–momentum tensor thus includes classical ideal fluid contributions, pressure due to degeneracy, and quantum geometric stress:

$$T^{\mu\nu} = (nh + P + Q)u^\mu u^\nu + (P + Q)g^{\mu\nu} + Q^{\mu\nu}.$$

The curvature modifies wave propagation and collective excitations by introducing geometric terms into the dispersion relations. In a spherically symmetric, positively curved geometry (such as the interior of a compressing spherical shell), the Laplacian operator becomes:

$$\nabla^2 \to \frac{1}{r^2}\frac{\partial}{\partial r}\left(r^2 \frac{\partial}{\partial r}\right) - \frac{L^2}{\hbar^2 r^2},$$

where L^2 represents the generalized angular momentum operator on a curved manifold. The quantization of modes in such a geometry leads to curvature-corrected eigen frequencies.

For Langmuir-type electrostatic oscillations in curved space, the dispersion relation modifies to:

$$\omega^2 = \omega_p^2 + \frac{3v_s^2}{r^2}\ell(\ell + 1) + \frac{\hbar^2}{4m^2 r^4}\ell^2(\ell + 1)^2 + \mathcal{O}(\mathcal{R}),$$

where $\mathcal{R}$ is the Ricci scalar curvature and the last term accounts for explicit curvature corrections to the quantum potential. These corrections increase in magnitude with confinement and compression, thus altering the threshold for phase mixing and mode coupling in imploding plasma shells.

In toroidal or helical confinement devices, the electromagnetic field tensor $F^{\mu\nu}$ itself inherits geometric dependence due to the curvature of field lines and the topology of the system. The modified Maxwell equations on curved manifolds read:

$$\nabla_\mu F^{\mu\nu} = \mu_0 J^\nu, \quad \nabla_{[\lambda} F_{\mu\nu]} = 0,$$

where square brackets denote antisymmetrization. The curvature of spacetime or spatial embedding manifests through Christoffel symbols in the covariant derivatives, affecting charge conservation and electromagnetic wave propagation [28]. In such settings, the plasma permittivity tensor becomes position-dependent:

$$\varepsilon^{ij}\left(x^k\right) = \delta^{ij} + \chi^{ij}\left(g_{\mu\nu}, R^\alpha_{\beta\mu\nu}\right),$$

with χ^{ij} denoting the nonlocal response function depending on curvature tensors such as the Riemann curvature $R^\alpha_{\beta\mu\nu}$. These modifications lead to curvature-induced birefringence, spatially dependent refractive indices, and topologically induced current flows, with implications for energy transport and magnetic reconnection in curved plasmas.

In strongly compressed inertial fusion systems, local energy densities may transiently reach values sufficient for metric back reaction to become non-negligible. While full general relativity is not typically invoked in terrestrial fusion research, analog gravity models where effective curved metrics emerge from plasma excitations offer a conceptual bridge. The effective acoustic metric for quantum ion-acoustic waves, for instance, is given by:

$$ds^2_{\text{eff}} = \frac{n_0}{c_s}\left[-\left(c_s^2 - v^2\right)dt^2 - 2v_i dx^i dt + \delta_{ij} dx^i dx^j\right],$$

where c_s is the ion-acoustic speed and v_i the fluid velocity. The curvature of this emergent metric affects quantum field propagation in the plasma, allowing the modeling of analogue Hawking radiation, mode trapping, and effective particle production under fusion-like gradients. Further, the inclusion of spacetime curvature directly modifies the thermodynamic relations through the Tolman-Ehrenfest effect:

$$T(x)\sqrt{g_{00}(x)} = \text{const},$$

implying that thermal equilibrium in curved plasma configurations requires spatially varying temperatures. This gradient becomes relevant in assessing thermalization and energy distribution in imploding geometries.

The interplay between curvature effects and quantum corrections fundamentally transforms the behavior, operational regimes, and diagnostic approaches of confined plasma systems. In scenarios such as inertial confinement fusion (ICF) capsules, or magnetic nozzles, compression-induced curvature during implosion significantly influences transport properties, stability thresholds, and the structure of evolving modes. Additionally, geometric constraints can give rise to topologically protected states, quantized flow patterns and coherent plasma structures stabilized by curvature-induced Berry phases, which offer promising avenues for enhancing confinement. In laser–plasma interactions, spatially structured laser fields can effectively imprint curvature onto the plasma metric, thereby modulating quantum responses such as high-harmonic generation and enabling the control of anomalous transport phenomena. Altogether, the geometric embedding of quantum plasma flows is not a mere mathematical abstraction, but a physically impactful element with direct implications for plasma compression, confinement, and diagnostics. These insights underscore the growing necessity of incorporating geometric fluid dynamics and quantum field theory in curved space as essential frameworks for advancing controlled fusion research and quantum plasma engineering [29].

References

1. Qin, H., Tang, W. M., & Rewoldt, G. (1999). Linear Gyrokinetic theory for kinetic magnetohydrodynamic eigenmodes in tokamak plasmas. *Physics of Plasmas, 6*(6), 2544–2562.
2. Rinderknecht, H. G., Amendt, P. A., Wilks, S. C., & Collins, G. (2018). Kinetic physics in ICF: Present understanding and future directions. *Plasma Physics and Controlled Fusion, 60*(6), Article 064001.
3. Cai, B. J., Li, B. A., & Zhang, Z. (2023). Central speed of sound, the trace anomaly, and observables of neutron stars from a perturbative analysis of scaled Tolman-Oppenheimer-Volkoff equations. *Physical Review D, 108*(10), Article 103041.
4. Oz, E., Myers, C. E., Yamada, M., Ji, H., Kulsrud, R. M., & Xie, J. (2011). Experimental verification of the Kruskal-Shafranov stability limit in line-tied partial-toroidal plasmas. *Physics of Plasmas, 18*(10).
5. Shi, T., Zheng, L. J., Wan, B. N., Sun, Y., Shen, B., & Qian, J. P. (2016). Ideal magnetohydrodynamic theory for localized interchange modes in toroidal anisotropic plasmas. *Physics of Plasmas, 23*(8).
6. Kalis, J. (2024). Characterization of edge modes in different confinement regimes of fusion plasmas.
7. Helander, P., Beidler, C. D., Bird, T. M., Drevlak, M., Feng, Y., Hatzky, R., Jenko, F., Kleiber, R., Proll, J. H. E., Turkin, Y., & Xanthopoulos, P. (2012). Stellarator and tokamak plasmas: a comparison. *Plasma Physics and Controlled Fusion, 54*(12), 124009.
8. Martinez, B., d'Humières, E., & Gremillet, L. (2020). Synchrotron radiation from ultrahigh-intensity laser-plasma interactions and competition with Bremsstrahlung in thin foil targets. *Physical Review Research, 2*(4), Article 043341.
9. Efimkin, D. K., & Kargarian, M. (2021). Topological spin-plasma waves. *Physical Review B, 104*(7), Article 075413.
10. Wang, W., Gao, W., Cao, L., Xiang, Y., & Zhang, S. (2020). Photonic topological fermi nodal disk in non-Hermitian magnetic plasma. *Light: Science and Applications, 9*(1), 40.
11. Mahajan, S. M., & Asenjo, F. A. (2011). Vortical dynamics of spinning quantum plasmas: Helicity conservation. *Physical Review Letters, 107*(19), Article 195003.
12. Liang, Y., & Lin, S. (2025). Spin alignment of quarkonia in vortical quark-gluon plasma. *Chinese Physics C.*

13. Li, T. C., Howes, G. G., Klein, K. G., Liu, Y. H., & TenBarge, J. M. (2019). Collisionless energy transfer in kinetic turbulence: Field–particle correlations in Fourier space. *Journal of Plasma Physics, 85*(4), 905850406.
14. Kim, J., Glier, T. E., Grimm-Lebsanft, B., Buchenau, S., Teubner, M., Biebl, F., Kim, N. J., Kim, H., Yi, G. C., Rübhausen, M., & Yoon, S. (2019). Quantum confinement induced excitonic mechanism in zinc-oxide-nanowalled microrod arrays for UV–Vis surface-enhanced Raman scattering. *The Journal of Physical Chemistry C, 123*(40), 24957–24962.
15. Al-Yousef, H. A., & Khalil, S. M. (2020). Degeneracy in magneto-active dense plasma. *Advances in Mathematical Physics, 2020*(1), 6495807.
16. Hay, M. J., & Fisch, N. J. (2015). Ignition threshold for non-Maxwellian plasmas. *Physics of Plasmas, 22*(11).
17. Bredow, R., Bornath, T., Kraeft, W. D., Dharma-wardana, M. W. C., & Redmer, R. (2015). Classical-map hypernetted chain calculations for dense plasmas. *Contributions to Plasma Physics, 55*(2–3), 222–229.
18. Dulat, A., Lad, A. D., Aparajit, C., Choudhary, A., Ved, Y. M., Veisz, L., & Ravindra Kumar, G. (2024). Single-shot, spatio-temporal analysis of relativistic plasma optics. *Optica, 11*(8), 1077–1084.
19. Akbari-Moghanjoughi, M. (2023). Quasiparticle approach to collective quantum dielectric response. *Physics of Plasmas, 30*(10).
20. Akbari-Moghanjoughi, M. (2015). Hydrodynamic limit of Wigner-Poisson kinetic theory: revisited. *Physics of Plasmas, 22*(2).
21. Shekhanov, S. A., & Tikhonchuk, V. T. (2021). SRS-SBS competition and nonlinear laser energy absorption in a high temperature plasma. *Plasma Physics and Controlled Fusion, 63*(11), Article 115016.
22. Figueiredo, J. L., Mendonça, J. T., & Terças, H. (2024). Quantum kinetic theory of light-matter interactions in degenerate plasmas. *Physical Review A, 110*(6), Article 063519.
23. Nashed, G. G. L., & Bamba, K. (2022). Realistic compact stars in conformal teleparallel gravity. *Progress of Theoretical and Experimental Physics, 2022*(10), 103E01.
24. Shukla, P. K., & Eliasson, B. (2010). Recent developments in quantum plasma physics. *Plasma Physics and Controlled Fusion, 52*(12), Article 124040.
25. Yang, Y. (2025). The role of quantum computing in advancing plasma physics simulations for fusion energy and high-energy. *Frontiers in Physics, 13*, 1551209.
26. Misra, A. P., & Sahu, B. (2015). Multidimensional ion acoustic solitary waves and shocks in quantum plasmas. *Physica A: Statistical Mechanics and its Applications, 421*, 269–278.
27. Joseph, I., Shi, Y., Porter, M. D., Castelli, A. R., Geyko, V. I., Graziani, F. R., Libby, S. B., & DuBois, J. L. (2023). Quantum computing for fusion energy science applications. *Physics of Plasmas*, 30(1).
28. Alcubierre, M. (2008). *Introduction to 3+ 1 numerical relativity* (vol. 140). OUP Oxford.
29. Chi, H., Ou, Y., Eldred, T. B., Gao, W., Kwon, S., Murray, J., Dreyer, M., Butera, R. E., Foucher, A. C., Ambaye, H., Keum, J., & Moodera, J. S. (2023). Strain-tunable Berry curvature in quasi-two-dimensional chromium telluride. *Nature Communications, 14*(1), 3222.

Chapter 10
Quantum Plasmas in High Energy Contexts

In the landscape of contemporary physics, the interface between plasma physics and quantum field theory becomes sharply prominent when plasma systems are driven into regimes characterized by relativistic velocities, ultra-strong fields, or quantum-statistical coherence. Quantum plasmas, which at their core involve collective behaviours of charged particles governed by quantum mechanical principles, acquire a distinct and elevated role in high energy physics (HEP) due to their capacity to mediate, amplify, or restructure field–matter interactions under extreme conditions. These quantum plasmas are no longer dilute ensembles of semi-classical particles; rather, they embody a dynamically rich medium where quantized field excitations, relativistic kinetics, and nonlinear electrodynamic processes are inseparably interwoven.

Conventionally, quantum plasmas are defined by the presence of at least one of the following: quantum degeneracy effects (e.g., Fermi pressure in dense electron gases), wavefunction interference effects (e.g., Bohm potential contributions to force balance), or quantum radiation effects (e.g., photon emission, recoil, and pair creation). While such definitions hold in low-to-moderate energy settings, their extrapolation to the high-energy regime necessitates deeper structural modifications. In high energy physics, the plasma constituents electrons, positrons, ions, and photons are treated not merely as wavepackets with quantum statistical behavior, but as excitations of underlying quantum fields, interacting via gauge-invariant couplings. Hence, the formulation of a high-energy quantum plasma rests upon the synthesis of quantum electrodynamics (QED) with plasma kinetic theory.

Thus, high-energy quantum plasma is a relativistic, many-body system of charged particles and photons where collective dynamics are governed by a quantum field-theoretic treatment of matter and gauge fields, subject to strong-field, high-intensity, or ultrafast temporal modulations, such that nonperturbative quantum electrodynamic effects significantly modify plasma behavior.

The energy scales relevant to quantum plasmas in high-energy physics (HEP) differ significantly from those encountered in fusion or astrophysical contexts. These

P. Kumar, *Quantum Plasma*, https://doi.org/10.1007/978-981-95-0481-7_10

regimes typically range from the moderate relativistic domain where electron energies are often accessible through tabletop ultra-intense laser–plasma interactions to the highly relativistic regime, characterized by energies reaching up to the GeV scale, as observed in laser-wakefield accelerators or beam–plasma interactions within collider environments. At the extreme end lies the ultra-relativistic, or critical QED regime, where field strengths approach, or surpass the Schwinger limit, and the vacuum itself begins to act as an active medium.

Quantum plasmas in these regimes cannot be treated within the perturbative hierarchy of standard QED. Instead, they demand resummation techniques, the inclusion of higher-loop diagrams in effective actions, and real-time kinetic equations for quasi-particles that include quantum recoil and radiation reaction. Moreover, these systems often exhibit non-equilibrium dynamics, requiring treatments that account for decoherence, thermalization, and quantum correlations far from equilibrium.

10.1 Thermal Quantum Plasmas Versus Strong-Field Relativistic Plasmas

The realm of quantum plasma physics bifurcates into two fundamentally distinct but occasionally overlapping regimes, thermal quantum plasmas and strong-field relativistic plasmas. Both emerge from the incorporation of quantum mechanical principles into plasma dynamics, yet their governing parameters, dominant interactions, and underlying physical phenomena are sharply differentiated. While thermal quantum plasmas are predominantly governed by quantum statistical effects at finite temperature and density, strong-field relativistic plasmas are dominated by extreme external fields, relativistic particle dynamics, and nonperturbative quantum electrodynamic (QED) processes.

The fundamental distinction between thermal quantum plasmas and strong-field relativistic plasmas begins with the nature of their controlling parameters. In thermal quantum plasmas, the quantum nature of particles arises due to high number densities and low temperatures such that the de Broglie wavelengths of particles begin to overlap significantly. The relevant condition is:

$$n\lambda_{\mathrm{dB}}^{3} \gtrsim 1, \quad \text{where} \quad \lambda_{\mathrm{dB}} = \frac{\mathrm{h}}{\sqrt{2\pi \mathrm{m} \mathrm{k}_{\mathrm{B}} \mathrm{T}}}$$

Here, n is the particle number density, m the mass of the species (typically electrons), T the temperature, and λ_{dB} is the thermal de Broglie wavelength. This inequality signals the onset of quantum degeneracy, such that the plasma must be treated with quantum statistics Fermi–Dirac for electrons and Bose–Einstein for bosonic excitations (e.g., photons or plasmon-like modes). The system is therefore dominated by quantum pressure, exchange–correlation effects, and tunneling terms, such as the quantum Bohm potential.

By contrast, in strong-field relativistic plasmas, the central role is played not by temperature, but by external field strength and relativistic energy input. Here, the dominant parameters are the normalized vector potential a_0 and the quantum nonlinearity parameter χ_e, defined respectively as:

$$a_0 = \frac{eE_0}{m_e c\omega_0}, \quad \chi_e = \frac{e\hbar}{m_e^3 c^4}\sqrt{-\left(F_{\mu\nu}p^\nu\right)^2}$$

where E_0 and ω_0 are the amplitude and frequency of the applied field, $F_{\mu\nu}$ is the electromagnetic field tensor, and p^ν is the four-momentum of the particle. The regime $a_0 \gtrsim 1$ marks the transition to relativistic motion under oscillatory fields, while $\chi_e \gtrsim 1$ denotes the onset of QED-dominated dynamics, including recoil, radiation reaction, and pair production.

10.1.1 Kinetic Descriptions and Field-Particle Interactions

Thermal quantum plasmas are best described by kinetic equations that incorporate quantum statistics and coherence effects, often via the Wigner function $f_W(\mathbf{r}, \mathbf{p}, t)$, which generalizes the classical distribution function:

$$\frac{\partial f_W}{\partial t} + \frac{\mathbf{p}}{m} \cdot \nabla_r f_W - \mathcal{Q}\left[f_W, \phi\right] = 0$$

Here, $\mathcal{Q}\left[f_W, \phi\right]$ represents the quantum potential term, a non-local pseudo-differential operator involving gradients of the electrostatic potential ϕ. In this regime, quantum corrections manifest as dispersive terms in the dielectric function, and collective modes such as Langmuir and ion-acoustic waves acquire quantum dispersion corrections due to the Fermi pressure and Bohm term. In strong-field relativistic plasmas, however, the system cannot be described by a single-particle Wigner function alone. One must adopt covariant QED kinetic theory, where the evolution of particle distributions is governed by the Dirac–Heisenberg–Wigner formalism, or real-time Schwinger-Keldysh techniques [1]. The quantum Vlasov equation now includes radiative and pair production processes via collision integrals $C[f]$ derived from first principles:

$$\left(\frac{\partial}{\partial t} + \mathbf{v} \cdot \nabla_\mathbf{r} + \mathbf{F}_{\text{eff}} \cdot \nabla_\mathbf{p}\right) f(\mathbf{r}, \mathbf{p}, t) = \mathcal{C}_{\text{QED}}\left[f, F_{\mu\nu}\right]$$

where $\mathbf{F}_{\text{eff}}$ incorporates the Lorentz force as well as radiation reaction, and $\mathcal{C}_{\text{QED}}$ includes nonlinear Compton scattering, Breit–Wheeler pair creation, and stochastic photon emission.

10.1.2 Role of Temperature Versus Field Strength

In thermal quantum plasmas, temperature albeit low remains the fundamental energy scale. It governs degeneracy pressure, thermal excitations, and screening lengths. One typically observes Fermi–Dirac equilibrium, with chemical potential μ evolving with density and temperature. The plasma frequency acquires quantum corrections as:

$$\omega_p^2 = \frac{ne^2}{\varepsilon_0 m_e}\left(1 + \frac{3}{5}\frac{T}{T_F}\right)$$

with T_F the Fermi temperature. Conversely, in strong-field relativistic plasmas, the notion of thermal equilibrium often breaks down. The system is typically in a transient, non-equilibrium state, driven by field-induced acceleration rather than collisional relaxation. The effective 'temperature' becomes ill-defined, instead energy spectra are described by power laws, stochastic broadening, or field-induced spectral hardening [2]. For example, the photon spectrum emitted from nonlinear Compton scattering follows a non-thermal distribution:

$$\frac{dN}{d\omega} \propto \left(\frac{\omega}{\omega_c}\right)^{-2/3} \exp\left(-\frac{2\omega}{3\omega_c}\right), \quad \omega_c = \frac{3}{2}\gamma^3\omega_0$$

This is a direct manifestation of the strong-field driven dynamics, unattainable from any thermally equilibrated quantum plasma.

10.1.3 Quantum Coherence and Radiation Effects

A crucial aspect distinguishing the two regimes is the nature and role of coherence. In thermal quantum plasmas, coherence emerges as statistical i.e., related to indistinguishability and exchange rather than phase coherence. Wavefunctions overlap, but decoherence times are short due to collisions and the statistical ensemble.

In strong-field relativistic plasmas, quantum coherence can be externally induced and preserved by ultrashort laser pulses, or field configurations with specific symmetry. Such coherence enables the observation of quantum interference in particle emission, photon entanglement, and higher harmonic generation, forming the basis of phenomena such as radiation reaction suppression, coherence-enhanced pair production, and vacuum polarization. Moreover, in the thermal regime, radiation is often treated as a perturbative loss term or is thermally reabsorbed. In contrast, in the relativistic regime, radiation becomes a driving mechanism it shapes particle orbits, modifies energy–momentum balance, and introduces stochastic quantum jumps via discrete photon emissions. Radiation reaction in this context must be modeled using either the Landau–Lifshitz prescription, or full QED Monte Carlo methods embedded in particle-in-cell (PIC) simulations [3, 4].

Table 10.1 Comparative features of thermal quantum plasmas and strong-field relativistic plasmas, highlighting their distinct physical regimes and governing mechanisms

Feature	Thermal quantum plasma	Strong-field relativistic plasma
Dominant parameter	Temperature TT, density nn	Field strength E, laser amplitude a_0
Governing theory	Quantum kinetic theory, wigner formalism	Quantum field theory, QED kinetic equations
Particle dynamics	Non-relativistic or mildly relativistic	Ultra-relativistic
Radiation	Weak, thermal, collisional	Strong, coherent, quantum recoil dominant
Coherence	Statistical (Fermi–Dirac)	Field-induced quantum coherence
Pair production	Negligible or thermally suppressed	Field-induced (Breit–Wheeler, Schwinger)
Applicable systems	Dense astrophysical cores, degenerate semiconductors	Laser-plasma accelerators, QED cascades, vacuum breakdown

This rigorous demarcation between thermal and strong-field relativistic quantum plasmas provides an essential conceptual and mathematical framework for the subsequent sections of the chapter. It is only through such a distinction that one can systematically explore the physics of quantum plasma in high energy physics without conflating effects of fundamentally different origin. While both regimes are quantum in nature, they inhabit disparate sectors of the parameter space, demand different theoretical treatments, and give rise to distinct phenomena. The interplay between these regimes especially during transitions such as laser-induced heating or field-induced compression will be addressed in later sections, where hybrid models and nonequilibrium formalisms bridge the two domains [5] (Table 10.1).

10.1.4 Overview of Energy Scales: From keV–MeV (Laser-Induced) to GeV–TeV

The study of quantum plasma in high-energy physics necessitates a precise understanding of the energy scales involved, particularly due to their profound influence on collective plasma behavior, particle-field interactions, and the applicability of different theoretical frameworks. The transition from the keV–MeV domain typical of laser-driven and fusion-scale plasmas to the GeV–TeV domain relevant for astrophysical jets and particle collider environments introduces new phenomena including vacuum polarization, non-linear QED effects, and quantum degeneracy under extreme relativistic conditions.

Energy Scales and Physical Contexts

The characteristic energy scale E of a plasma particle is typically related to temperature by the Boltzmann relation:

$$E = k_B T,$$

where k_B is the Boltzmann constant. This yields:

i. keV (10^3eV) scale: Corresponds to temperatures on the order of $T \sim 10^7$K, typical in magnetic fusion plasmas, inertial confinement targets, and X-ray generation zones in high-intensity laser experiments.
ii. MeV (10^6eV) scale: Appears in high-intensity laser-plasma interactions where relativistic electrons and positron pairs are produced, or in gamma-ray bursts.
iii. GeV (10^9eV) scale: Characteristic of high-energy astrophysical jets, synchrotron sources, and laser wakefield acceleration (LWFA) in long-distance plasma channels.
iv. TeV (10^{12}eV) scale: Achieved in particle colliders (e.g., LHC) and models for early-universe plasma states where electroweak unification becomes relevant.

Each scale reflects a hierarchy of physical phenomena, and correspondingly, the governing theoretical models must be adapted transitioning from classical or semi-classical kinetic theory at keV scales to quantum field theoretical (QFT) approaches at the GeV–TeV frontier.

The table summarizes the transition across energy scales:

Energy scale	Physical domain	Dominant theory	Key phenomena
keV	Laser plasma, ICF	Quantum kinetics, Schrödinger, Dirac	Ponderomotive acceleration, tunneling ionization
MeV	Relativistic laser-plasma, GRBs	Dirac-Maxwell, radiation reaction	Pair creation, nonlinear Compton, Breit-wheeler
GeV	Wakefield, astrophysical jets	QED, Vlasov-Fokker–Planck	Vacuum birefringence, trapping, gamma rays
TeV	Collider, early universe	Finite-T QFT, QCD plasma	Electroweak symmetry breaking, Schwinger effect

The study of quantum plasma at these varying energy scales not only offers insights into plasma-specific phenomena but also bridges connections across high-energy astrophysics, cosmology, and collider physics. The mathematical frameworks evolve from semiclassical field-particle treatments to fully quantized gauge field theories, reflecting the increasing role of fundamental interactions, symmetry principles, and vacuum structure in plasma behaviour at extreme energies.

10.1.5 Relevance to High-Intensity Laser Physics, Particle Colliders, and Astrophysical Observations

Quantum plasma phenomena at high energy scales are increasingly central to three distinct yet converging domains of contemporary physics, high-intensity laser physics, particle collider experiments, and astrophysical observations. Each of these domains provides a unique natural laboratory to probe the interplay of collective plasma behavior, quantum field theoretical effects, and extreme-field electrodynamics.

High-Intensity Laser Physics and Strong-Field Quantum Plasma

The advent of ultra-short, high-power laser systems enabled by chirped pulse amplification (CPA) has made it feasible to attain intensities exceeding $10^{22}\mathrm{W/cm}^2$. In such regimes, laser–plasma interactions cease to be adequately described by classical electrodynamics or standard fluid models; instead, they invoke relativistic quantum plasma dynamics where nonlinear QED processes and collective quantum coherence become dominant. The effective coupling is governed by the dimensionless laser amplitude a_0, which, for an electron of charge 'e' and mass m_e, and for laser electric field E, is expressed as:

$$a_0 = \frac{eE}{m_e c\omega}.$$

When $a_0 \gg 1$, electron motion becomes relativistic, and radiation reaction (RR) must be incorporated via the Landau–Lifshitz, or quantum-corrected models. These effects significantly influence energy distribution, angular emission spectra, and particle trapping in plasma waves.

Moreover, the interaction of intense lasers with overdense or near-critical plasmas results in the formation of electron–positron–photon cascades. These cascades are modeled using Boltzmann-like kinetic equations for the distribution functions $f_\varepsilon(p, t), f_\gamma(k, t), f_{\varepsilon^+}(p, t)$, each coupled by nonlinear QED processes including nonlinear Compton scattering and multiphoton Breit–Wheeler pair production. The evolution equations take the schematic form:

$$\left(\frac{\partial}{\partial t} + \mathbf{v}\cdot\nabla_\mathbf{x} + \mathbf{F}_{\text{eff}}\cdot\nabla_\mathbf{p}\right)f_i = \sum_j \mathcal{C}_{ij}[f_j],$$

where $\mathcal{C}_{ij}$ represents the collision operators derived from strong-field QED matrix elements.

Laser wakefield acceleration (LWFA) further exemplifies the relevance of quantum plasma, where intense laser pulses excite relativistic plasma waves capable of trapping and accelerating electrons to GeV-scale energies over centimeter-scale distances. Theoretical modeling employs a combination of Maxwell–Vlasov equations, envelope approximation for laser evolution, and quantum-corrected ponderomotive forces [6].

Particle Colliders and Quantum Field Theoretic Plasma Dynamics

In the realm of high-energy colliders such as the Large Hadron Collider (LHC), quantum plasma emerges as an essential medium in two key contexts: quark–gluon plasma (QGP) formation and beam–plasma instabilities in accelerator environments. The QGP, generated in heavy-ion collisions at center-of-mass energies exceeding a few hundred GeV per nucleon pair, represents a deconfined state of matter where color charges are screened, and collective behaviour is governed by non-Abelian gauge theories at finite temperature and chemical potential.

The description of such plasmas is rooted in the formalism of thermal quantum chromodynamics (QCD), with the grand canonical partition function given by:

$$Z = \int \mathcal{D}A\mathcal{D}\overline{\psi}\mathcal{D}\psi e^{-S_{QCD}[A,\psi,\overline{\psi};T,\mu]},$$

where S_{QCD} incorporates the gluon and quark fields, temperature T, and chemical potential μ. Perturbative techniques are limited due to the strong coupling in the QGP near the critical temperature, necessitating non-perturbative approaches such as lattice QCD, or holographic duality models (e.g., AdS/CFT correspondence) to evaluate transport coefficients like shear viscosity η, bulk viscosity, and the jet quenching parameter $\hat{q}$.

Further, beam–plasma interactions in particle colliders particularly in plasma-based acceleration schemes such as plasma wakefield accelerators (PWFA) invoke quantum corrections to collective beam dynamics, such as quantum emittance, Landau damping in the quantum regime, and synchrotron radiation reaction in ultra-relativistic bunches. These effects demand an extension of conventional envelope equations and collective modes analysis using quantum Vlasov–Maxwell systems, augmented by spin–orbit coupling and Pauli corrections in spin-polarized beams.

Astrophysical Observations and Quantum Plasmas under Extreme Conditions

Astrophysical plasmas span energy densities and field strengths vastly exceeding terrestrial capabilities, rendering them fertile grounds for quantum plasma phenomena. In environments such as magnetars, neutron stars, active galactic nuclei (AGN), and gamma-ray bursts (GRBs), one encounters magnetic field strengths $B \gtrsim 10^{13} - 10^{15}\text{G}$, approaching or exceeding the quantum critical field:

$$B_{\text{cr}} = \frac{m_e^2 c^3}{e\hbar} \approx 4.41 \times 10^{13}\text{G}.$$

Under such conditions, vacuum birefringence, photon splitting ($\gamma \rightarrow \gamma + \gamma$), and pair production ($\gamma \rightarrow e^+ + e^-$) occur spontaneously. These processes are captured via the effective Heisenberg–Euler Lagrangian $\mathcal{L}_{\text{HE}}$, which introduces nonlinear corrections to Maxwell's equations in vacuum. The physical consequences include anisotropic dispersion relations, nonlinear wave propagation, and polarization-dependent transparency.

In accretion disks and relativistic jets, quantum plasma processes dictate radiation transport, magnetohydrodynamic (MHD) stability, and energy dissipation. For instance, Compton drag, inverse Compton scattering, and stimulated pair cascades alter radiation spectra observable in X-ray and gamma-ray frequencies. Such systems are mathematically modeled through relativistic radiative transport equations coupled to quantum kinetic plasma models:

$$\left(\frac{\partial}{\partial t} + \mathbf{v} \cdot \nabla_{\mathbf{x}} + \mathbf{F}_{\text{rad}} \cdot \nabla_{\mathbf{p}}\right) f = \mathcal{C}_{\text{QED}}[f],$$

with $\mathbf{F}_{\text{rad}}$ representing radiation reaction and QED-modified Lorentz forces.

Astroseismology of neutron stars further necessitates the inclusion of quantum magnetohydrodynamics (QMHD), where quantum corrections to pressure, magnetization, and spin alignment (through Pauli paramagnetism and spin–spin interactions) lead to novel wave modes and dispersion characteristics.

Unifying Perspective and Observational Implications

The convergence of high-intensity laser physics, collider-based high-energy plasmas, and extreme astrophysical systems highlights the growing need for a unified theoretical framework grounded in quantum plasma theory. This unified approach draws upon several foundational elements, including quantum kinetic theory, which governs phase-space evolution under the influence of quantum statistics and coherence, quantum electrodynamics (QED) in plasma, which accounts for nonlinear self-interactions, pair production, and vacuum corrections, and strong-field and dense-matter physics, which describes matter under conditions of extreme pressure, electromagnetic fields, and energy density. Observationally, quantum plasma effects manifest in various distinct signatures, harmonic generation and spectral broadening in laser–plasma experiments, jet suppression and elliptic flow in collider-produced quark-gluon plasmas, and anomalous transparency, spectral cutoffs, and polarization shifts in high-energy astrophysical emissions such as X-rays and gamma rays. In all these cases, the integration of theoretical models with numerical simulations and precise experimental or observational diagnostics is essential for achieving a robust and quantitative understanding of quantum plasma dynamics.

This triangulation of theory, experiment, and observation defines the frontier of high-energy quantum plasma physics and demands a cross-disciplinary approach that incorporates modern quantum field methods, classical–quantum correspondence techniques, and novel experimental platforms.

10.2 Relativistic Quantum Plasma

In high-energy environments where particle velocities approach the speed of light, the conventional treatment of plasma parameters becomes insufficient. The incorporation of relativistic effects into the characterization of plasma particularly the Debye length, thermal velocity, and inertial mass marks a critical extension of plasma theory necessary for accurately modeling the behaviour of matter under extreme conditions such as those encountered in high-intensity laser–plasma interactions, relativistic beam–plasma systems, and early-universe plasmas.

Relativistic Thermal Motion and Distribution Functions

The starting point for relativistic generalization lies in the redefinition of the equilibrium particle distribution. For a plasma species of mass m, charge q, and inverse temperature $\beta = 1/(k_B T)$, the relativistic Maxwell–Jüttner distribution replaces the classical Maxwell–Boltzmann form and is expressed as:

$$f(\mathbf{p}) = \frac{n_0}{4\pi m^3 c^3 K_2\left(\beta mc^2\right)} \exp\left(-\beta\gamma(\mathbf{p})mc^2\right),$$

where $\gamma(\mathbf{p}) = \sqrt{1 + \frac{p^2}{m^2c^2}}$ is the Lorentz factor, and K_2 is the modified Bessel function of the second kind. This distribution respects Lorentz invariance and encapsulates the energy–momentum relationship of relativistic particles, yielding modifications to both the thermal velocity and inertia.

The relativistic mean energy per particle is obtained as:

$$\langle\mathcal{E}\rangle = mc^2\left(\frac{K_1\left(\beta mc^2\right)}{K_2\left(\beta mc^2\right)} + \frac{3}{\beta mc^2}\right),$$

which smoothly recovers the classical limit $\langle\mathcal{E}\rangle \approx \frac{3}{2}k_B T$ as $\beta mc^2 >> 1$, and grows as $\sim 3k_B T$ in the ultrarelativistic limit $\beta mc^2 << 1$.

Relativistic Debye Length

In non-relativistic plasma physics, the Debye length λ_D characterizes the exponential screening of electrostatic potentials and is given by:

$$\lambda_D = \sqrt{\frac{\varepsilon_0 k_B T}{n_0 q^2}}.$$

In the relativistic regime, the screening length must be re-derived from first principles by analyzing the static limit of the longitudinal component of the polarization tensor in the quantum field theoretic description. One approach involves computing the dielectric response function $\epsilon_L(k, \omega = 0)$ from the relativistic Vlasov equation, or from the one-loop photon self-energy in finite-temperature QED [7].

For a collisionless, isotropic, relativistic plasma, the relativistic Debye length λ_D^{rel} generalizes as:

$$\lambda_D^{\text{rel}} = \left[\frac{\varepsilon_0}{q^2}\int \frac{d^3p}{(2\pi\hbar)^3}\frac{f_0(\mathbf{p})}{\gamma(\mathbf{p})}\left(\frac{\partial f_0}{\partial \mathcal{E}}\right)\right]^{-1/2},$$

where $f_0(\mathbf{p})$ is the equilibrium distribution (e.g., Jüttner distribution), and the factor $\gamma(\mathbf{p})$ arises from time dilation effects in the Lorentz transformation of the force equation. Physically, this implies that relativistic motion suppresses the effectiveness of electrostatic screening due to increased inertial mass and delayed collective response.

In the ultrarelativistic limit, where $k_BT >> mc^2$, the thermal velocity asymptotically approaches "c", and the energy distribution becomes dominated by momentum, resulting in:

$$\lambda_D^{\text{ur}} \approx \sqrt{\frac{\varepsilon_0 k_B T}{n_0 q^2}}\left(\frac{K_2(\beta mc^2)}{K_1(\beta mc^2)}\right)^{1/2} \rightarrow \sqrt{\frac{\varepsilon_0 k_B T}{n_0 q^2}},$$

which formally resembles the classical form but arises under a different thermodynamic scaling.

Inertial Mass Corrections and Effective Mass

The inertia of particles in a relativistic quantum plasma cannot be solely equated to the rest mass mm. Instead, the effective inertial response arises from the full energy–momentum tensor and is manifest in the momentum transport equations and in wave propagation phenomena.

The relativistic momentum equation for the fluid velocity $\mathbf{u}$ incorporates the relativistic enthalpy $h = \rho + p$, yielding:

$$\frac{d}{dt}(h\gamma\mathbf{u}) = qn(\mathbf{E} + \mathbf{u} \times \mathbf{B}) - \nabla p,$$

where $\gamma = (1 - u^2/c^2)^{-1/2}$, and ρ is the energy density. This equation shows that the effective inertial mass is no longer constant, but instead dynamically varies with the energy and pressure of the plasma medium. The inertia per unit volume is effectively $\gamma^2 h$, leading to inertial mass corrections in wave dispersion and momentum exchange.

Further, in quantum kinetic theory, the effective mass can be derived from the energy dispersion relation modified by collective interactions. For instance, the electron effective mass in a hot dense plasma is given by the pole of the propagator:

$$m_{\text{eff}}^2 = m^2 + \Pi_T(\omega = 0, k \rightarrow 0),$$

where Π_T is the transverse part of the photon polarization tensor. At one-loop level in finite-temperature QED, this gives rise to:

$$m_{\text{eff}}^2 \approx m^2 + \frac{e^2T^2}{8},$$

in the ultrarelativistic limit. This shows that even in the absence of rest mass, the thermal bath confers an effective inertial mass due to medium-induced self-energy corrections. Analogous behaviour is encountered in neutrino and gluon propagation through the early universe plasma and in quark–gluon plasma contexts, highlighting the universality of inertial mass renormalization in relativistic quantum media.

The relativistic modifications to Debye screening and inertia play a pivotal role in the behaviour of high-energy plasmas. In laser–plasma accelerators operating at relativistic intensities, reduced screening alters the structure of plasma wakes and enhances field gradients, affecting beam quality and energy gain. In astrophysical contexts, relativistic Debye shielding determines the charge separation scales in pair plasmas near pulsars and black hole magnetospheres. Moreover, inertial mass corrections influence the propagation of magnetosonic and Langmuir modes in strongly coupled relativistic media, thereby affecting radiation emission, instabilities, and energy transport. From a theoretical standpoint, these modifications necessitate the deployment of covariant kinetic theories, effective field theories at finite temperature and density, and numerical models that incorporate relativistic corrections in both particle and field evolution [8].

10.2.1 Dirac–Maxwell and Klein–Gordon–Maxwell Systems: Relativistic Field-Theoretic Descriptions of Quantum Plasmas

In regimes of extreme energy density, such as those found in relativistic laser-plasma interactions, compact astrophysical environments, and early-universe cosmology, the dynamics of charged particles and electromagnetic fields must be described within a unified relativistic and quantum field-theoretic framework. Traditional fluid and kinetic models become inadequate when the Compton wavelength becomes comparable to interparticle spacing, and particle creation/annihilation processes become non-negligible. In this context, two foundational systems govern the theoretical modeling of relativistic quantum plasmas: the Dirac–Maxwell system, which describes spin-½ fermionic matter fields interacting with gauge fields, and the Klein–Gordon–Maxwell system, which applies to spin-0 bosonic charged fields.

Both systems constitute self-consistent, covariant, nonlinear field theories where the matter field dynamics (obeying Dirac, or Klein–Gordon equations) are coupled to the electromagnetic field dynamics governed by Maxwell's equations, with source terms constructed from conserved Noether currents [9].

The Dirac–Maxwell System

The Dirac–Maxwell system provides a first-principles description of a relativistic quantum plasma composed of spin-½ particles (electrons, positrons) interacting via the quantized electromagnetic field. The coupled dynamics derive from the total action functional:

$$S[\overline{\psi}, \psi, A^\mu] = \int d^4x \left[\overline{\psi}(i\hbar\gamma^\mu D_\mu - mc)\psi - \frac{1}{4\mu_0} F_{\mu\nu}F^{\mu\nu} \right],$$

where $\psi(x)$ is the Dirac spinor field, $\overline{\psi} = \psi^\dagger\gamma^0$, A^μ is the electromagnetic four-potential, $F_{\mu\nu} = \partial_\mu A_\nu - \partial_\nu A_\mu$ is the field strength tensor, and $D_\mu = \partial_\mu + i\frac{q}{\hbar}A_\mu$ is the gauge-covariant derivative. The gamma matrices γ^μ satisfy the Clifford algebra $\{\gamma^\mu, \gamma^\nu\} = 2\eta^{\mu\nu}$. Varying the action with respect to $\overline{\psi}$ yields the Dirac equation in an external field:

$$(i\hbar\gamma^\mu D_\mu - mc)\psi = 0,$$

while variation with respect to A^μ gives Maxwell's equations with a quantum current source:

$$\partial_\nu F^{\nu\mu} = \mu_0 J^\mu, \quad J^\mu = q\overline{\psi}\gamma^\mu\psi.$$

These two equations together form the Dirac–Maxwell system. The current J^μ is conserved by virtue of Noether's theorem: $\partial_\mu J^\mu = 0$, and it serves as the source for the dynamical electromagnetic field. This coupling introduces nonlinearity, backreaction, and relativistic quantum effects into plasma modeling. The Dirac–Maxwell system accommodates key relativistic phenomena such as Zitterbewegung (rapid oscillatory motion of wavepackets), electron–positron pair creation in supercritical fields (e.g., via Schwinger mechanism), spin polarization effects in strong fields, and self-consistent field generation. Moreover, solutions of the Dirac–Maxwell equations include solitary wave structures, such as Dirac solitons and relativistic quantum vortices, which are of interest in high-intensity laser–plasma interactions and pair-plasma astrophysical jets.

For semiclassical analyses, one often adopts a mean-field approach, replacing field operators by classical expectation values and using Wigner-function formulations, or quantum hydrodynamic moments derived from the Dirac equation. These approaches retain spinor structure and allow investigation of collective modes, instability growth rates, and spin–orbit coupling in relativistic quantum plasmas [10].

The Klein–Gordon–Maxwell System

While fermionic plasmas are dominant in physical realizations, bosonic charged fields can model scalar charge distributions, heavy ion beams, or condensed quasi-particle systems (e.g., charged scalar condensates in early-universe models). The

Klein–Gordon–Maxwell (KGM) system describes the relativistic dynamics of spin-0 complex scalar fields interacting with electromagnetic fields and provides a mathematically simpler, but physically rich counterpart to the Dirac–Maxwell theory.

The KGM action functional is:

$$S[\phi, \phi^*, A^\mu] = \int d^4x \left[(D^\mu \phi)^* D_\mu \phi - \frac{m^2 c^2}{\hbar^2} |\phi|^2 - \frac{1}{4\mu_0} F_{\mu\nu} F^{\mu\nu} \right],$$

where $\phi(x)$ is the complex scalar field and $D_\mu = \partial_\mu + i\frac{q}{\hbar} A_\mu$ is the gauge-covariant derivative as before. Variation with respect to ϕ^* and A^μ yields:

$$D^\mu D_\mu \phi + \frac{m^2 c^2}{\hbar^2} \phi = 0, \quad \partial_\nu F^{\nu\mu} = \mu_0 J^\mu,$$

with the associated current:

$$J^\mu = \frac{iq\hbar}{2m} \left(\phi^* D^\mu \phi - \phi (D^\mu \phi)^* \right).$$

The Klein–Gordon–Maxwell system thus encodes the covariant dynamics of a charged scalar plasma interacting with electromagnetic fields. Although lacking spinor degrees of freedom, the scalar field formulation permits soliton-like solutions, wave collapse phenomena, and nonlinear plasma wave interactions in relativistic regimes.

A significant feature of the KGM system is the emergence of bosonic collective modes and quantum coherent field structures. In analogy with Bose–Einstein condensates, scalar fields in certain parameter regimes exhibit macroscopic quantum coherence, leading to plasma analogues of quantum vortices and domain walls. Moreover, in strong fields or high-density regimes, pair production and charge–density wave instabilities become dynamically relevant.

Mathematically, the KGM system supports a rich spectrum of solutions, including spherically symmetric bound states, breathers, and topological structures in curved spacetime generalizations (e.g., charged boson stars or scalar field collapse in black hole formation scenarios). In plasma physics, its simplified structure enables analytical insight into nonlinear field–plasma interactions, especially in studies of ultra-intense scalar laser–plasma analogues [11].

Gauge Invariance, Conserved Quantities, and Quantization

Both the Dirac–Maxwell and Klein–Gordon–Maxwell systems exhibit local U(1) gauge invariance, ensuring charge conservation and electromagnetic consistency. The Lagrangian formulations guarantee manifest Lorentz covariance, and their canonical quantization leads to the foundations of quantum electrodynamics (QED). In practical modeling of quantum plasmas, however, one frequently employs a mean-field semiclassical truncation, in which the matter fields are treated as classical wave functions or expectation-valued operators, and electromagnetic fields evolve classically or within effective quantum corrections.

The Hamiltonian densities for both systems can be derived via Legendre transformation, and conserved quantities such as the total energy–momentum tensor and total charge can be expressed as functionals of the matter and field variables. These provide tools for stability analysis and numerical simulations of plasma field dynamics. In contemporary research, these systems also serve as effective field theories describing emergent phenomena in high-energy density laboratory plasmas, relativistic beam–plasma interactions, and exotic astrophysical plasmas. They provide a framework for modeling quantum turbulence, anomalous transport phenomena, and nonlinear wave–particle interactions in plasmas where conventional models fail.

The Dirac–Maxwell and Klein–Gordon–Maxwell systems form the backbone of relativistic quantum plasma theory. They are indispensable for describing matter–radiation interactions in the quantum regime, understanding strong-field QED processes such as vacuum polarization and photon splitting, and simulating next-generation plasma-based accelerators operating near, or beyond the Schwinger limit. Moreover, their coupling to gravitational fields in curved spacetime generalizations opens avenues for studying quantum plasma effects in black hole accretion disks, early cosmology, and relativistic compact objects [12].

As quantum technologies evolve to manipulate ultrafast, high-field, relativistic plasmas such as in quantum laser-plasma interfaces, relativistic quantum computing platforms, or controlled pair plasma generation these field-theoretic models will become increasingly relevant for both foundational understanding and experimental design.

10.2.2 Wave Propagation and Collective Excitations in High-Energy Quantum Regimes

In the domain of high-energy quantum plasmas, the dispersion relation the relation between wave frequency and wave vector encodes the fundamental spectrum of permissible collective excitations. These include longitudinal and transverse plasma waves, quantum-modified Langmuir modes, electromagnetic and electrostatic instabilities, and even nonlinear wave structures in relativistic regimes. In contrast to classical plasma theory, where dispersion arises from fluid, or kinetic models incorporating thermal motion and electromagnetic response, relativistic quantum plasmas require a field-theoretic or quantum kinetic treatment that simultaneously incorporates quantum statistics, relativistic kinematics, spin effects, and vacuum polarization phenomena.

Linear Response in Relativistic Quantum Plasmas

The derivation of dispersion relations typically begins with linearizing the governing equations around a homogeneous equilibrium background. In relativistic quantum plasmas, the appropriate dynamical equations may be taken as:

1. The quantum Vlasov–Maxwell system for semiclassical fermionic distributions,
2. The Wigner–Maxwell system for quantum phase-space analysis,
3. or, in fully field-theoretic treatments, the one-loop polarization tensor from quantum electrodynamics in medium.

In all cases, the key idea is to compute the dielectric response function $\varepsilon^{\mu\nu}(k)$, which relates induced currents $J^{\mu}(k)$ to applied field perturbations $A^{\nu}(k)$, and whose poles determine the allowed dispersion modes. The general wave equation in momentum space follows from Maxwell's equations:

$$\left[k^2\eta^{\mu\nu} - k^{\mu}k^{\nu} - \mu_0\Pi^{\mu\nu}(k)\right]A_{\nu}(k) = 0,$$

where $\Pi^{\mu\nu}(k)$ is the polarization tensor, encoding the linear electromagnetic response of the quantum plasma medium. The dispersion relation emerges from the condition that the determinant of the coefficient matrix vanishes:

$$\det\left[k^2\eta^{\mu\nu} - k^{\mu}k^{\nu} - \mu_0\Pi^{\mu\nu}(k)\right] = 0.$$

The detailed structure of $\Pi^{\mu\nu}(k)$ depends on the underlying particle distribution (degenerate, thermal, isotropic, magnetized), the quantum-statistical regime (Fermi–Dirac or Bose–Einstein), and the energy scale (non-relativistic, semi-relativistic, or ultra-relativistic).

Dispersion from Quantum Kinetic Theory

In the semi-relativistic quantum kinetic framework, one may begin with the covariant Vlasov equation for a spin-½ plasma, linearized about an equilibrium distribution $f_0(p)$:

$$\left(p^{\mu}\frac{\partial}{\partial x^{\mu}} + qF^{\mu\nu}p_{\nu}\frac{\partial}{\partial p^{\mu}}\right)f(x, p) = 0.$$

For quantum corrections, this distribution can be replaced by the Wigner function W(x, p), whose dynamics include both phase-space evolution and quantum interference terms via the Moyal bracket expansion. Linearization yields an equation for the perturbation $\delta W(k, p)$, whose solution leads to an induced current $J^{\mu}(k)$ proportional to $A^{\nu}(k)$, defining $\Pi^{\mu\nu}(k)$ explicitly [13]. For instance, in an unmagnetized, isotropic relativistic electron plasma with equilibrium Fermi–Dirac distribution:

$$f_0(p) = \frac{1}{e^{(E_p-\mu)/k_BT} + 1}, \quad E_p = \sqrt{p^2c^2 + m^2c^4},$$

the longitudinal dielectric function is given by:

$$\varepsilon_L(\omega, k) = 1 + \frac{1}{\epsilon_0 k^2} \int \frac{q^2 \vec{k} \cdot \partial f_0/\partial \vec{p}}{\omega - \vec{k} \cdot \vec{v}} d^3 p,$$

where $\vec{v} = \vec{p}c^2/E_p$ is the relativistic velocity. Quantum recoil and wave–particle resonances enter through the denominator $\omega - \vec{k} \cdot \vec{v}$, especially in degenerate regimes where Landau damping and quantum oscillations are prominent.

Dispersion in Degenerate and Relativistic Regimes

Let us consider a fully degenerate relativistic Fermi gas at zero temperature. The equilibrium distribution becomes a step function up to the Fermi momentum p_F related to the density via:

$$n = \frac{g}{(2\pi\hbar)^3} \int_{|\vec{p}| \le p_F} d^3 p, \quad g = 2 \text{ (spin)}.$$

In this limit, longitudinal oscillations correspond to quantum-relativistic Langmuir waves, with dispersion relation (at leading order):

$$\omega^2 = \omega_p^2 + \frac{3}{5} v_F^2 k^2 + \mathcal{O}(k^4),$$

where $v_F = p_F / \sqrt{p_F^2 + m^2 c^2}$ is the relativistic Fermi velocity and $\omega_p = \sqrt{ne^2/\varepsilon_0 m_{\text{eff}}}$ is the effective plasma frequency, with m_{eff} incorporating relativistic mass corrections. Higher-order quantum corrections include the Bohm potential term and spin–orbit contributions in polarized plasmas. In the ultra-relativistic $p_F >> mc$, the dispersion modifies to:

$$\omega^2 = \frac{4}{3} \pi \alpha c^2 k^2 + \mathcal{O}(k^4),$$

where α is the fine-structure constant, and the behaviour approaches that of massless quantum plasmas relevant to early-universe conditions or graphene-like relativistic media [14].

Magnetic Field Effects and Anisotropic Dispersion

In strongly magnetized relativistic quantum plasmas, the dispersion relations acquire anisotropy due to the quantization of transverse motion into Landau levels. The equilibrium distribution becomes discretized:

$$E_{n,p_z} = \sqrt{p_z^2 c^2 + m^2 c^4 + 2n\hbar q B c}, \quad n \in \mathbf{N},$$

and wave propagation perpendicular to $\vec{B}$ must be treated using tensorial dielectric functions. The longitudinal and transverse components of $\varepsilon^{ij}(\omega, \vec{k})$ split into magnetized modes:

i. Ordinary (O) and extraordinary (X) modes for electromagnetic waves,
ii. Upper hybrid and Bernstein modes for electrostatic oscillations.

The resulting dispersion relations exhibit mode splitting, cutoff frequencies, and cyclotron resonances. Quantum corrections modify cyclotron frequencies via anomalous magnetic moments and spin polarization, yielding shifts in electron Bernstein modes and opening possibilities for quantum magnetosonic waves and quantum cyclotron harmonics, important in neutron star magnetospheres and ultraintense laser–plasma experiments.

Vacuum Polarization and Quantum Electrodynamic Corrections

At field strengths approaching or exceeding the Schwinger limit $E_c = m^2c^3/e\hbar$, vacuum becomes nonlinear and birefringent due to quantum fluctuations. The dispersion relations then acquire additional terms from vacuum polarization effects. These are captured by the Heisenberg–Euler effective action, which modifies the electromagnetic field Lagrangian:

$$\mathcal{L}_{\text{eff}} = -\frac{1}{4}F_{\mu\nu}F^{\mu\nu} + \frac{\alpha^2}{90m^4}\left[(F_{\mu\nu}F^{\mu\nu})^2 + \frac{7}{4}(F_{\mu\nu}\tilde{F}^{\mu\nu})^2\right] + \cdots,$$

leading to nonlinear wave equations. As a result, the refractive index becomes field-dependent, yielding photon splitting, birefringence, and vacuum Cherenkov radiation. These effects imprint themselves onto plasma dispersion relations and may be probed in laser–plasma facilities operating at high intensities [15].

Dispersion relations in relativistic quantum plasmas capture the complex interaction between quantum mechanics, relativistic dynamics, and collective field behaviours. Unlike classical plasmas, where dispersion is largely governed by charge inertia and electromagnetic field propagation, relativistic quantum plasmas reveal a richer array of physical phenomena. These include quantum recoil and coherence effects at microscopic scales, spin–field interactions that result in helicity-dependent wave propagation, and modifications in wave velocities due to degeneracy pressure. Additionally, Landau quantization introduces discrete oscillatory responses, while vacuum nonlinearities give rise to unusual refractive behaviors. Together, these effects influence key processes such as wave–particle resonance, plasma-based particle acceleration, high-harmonic generation, photon–photon scattering, and the development of instabilities in pair plasmas. Thus, precise formulations of dispersion relations are vital for accurately modeling and interpreting experimental results in high-intensity laser–plasma interactions, astrophysical radiation environments, and emerging quantum plasma technologies.

10.2.3 Applications—Gamma-Ray Bursts, Relativistic Shocks, and Quark-Gluon Plasma Precursors

In the domain of high-energy physics, quantum plasmas manifest in astrophysical transients, extreme relativistic environments, and the early-universe conditions. This section elaborates on three critical scenarios gamma-ray bursts (GRBs), relativistic shocks, and the formation of quark-gluon plasma (QGP) precursors where quantum plasma dynamics become indispensable. Each of these applications exhibits a confluence of quantum field theoretical effects, relativistic kinetic theory, and collective plasma behaviour under extreme conditions of energy density and temperature [16].

Gamma-ray bursts, the most luminous electromagnetic events in the universe, result from catastrophic stellar collapses or compact object mergers, producing outflows with Lorentz factors $\gamma \gg 1$. The physical conditions in the relativistic jet cores and emission regions entail ultra-high temperatures (T $\gtrsim$ 1011 K), intense magnetic fields (B ~ 1015 G), and dense pair plasma formation, thereby demanding a quantum electrodynamical plasma description.

In these environments, pair production is driven by vacuum polarization and thermal QED processes. The one-loop Heisenberg–Euler effective Lagrangian for an electromagnetic field in a plasma medium becomes relevant:

$$\mathcal{L}_{\text{eff}} = -\frac{1}{4}F_{\mu\nu}F^{\mu\nu} + \frac{e^4}{360\pi^2 m_e^4}\left[(F_{\mu\nu}F^{\mu\nu})^2 + \frac{7}{4}(F_{\mu\nu}\tilde{F}^{\mu\nu})^2\right],$$

where $F_{\mu\nu}$ is the electromagnetic field tensor and $\tilde{F}^{\mu\nu}$ its dual. This nonlinear term governs vacuum birefringence and photon splitting processes, essential in describing radiative transport in the GRB plasma.

Relativistic quantum Vlasov equations govern the phase-space evolution of electron–positron plasmas with strong field coupling:

$$\left[p^\mu \partial_\mu - eF^{\mu\nu}p_\nu \frac{\partial}{\partial p^\mu}\right] f(x, p) = \mathcal{C}[f],$$

where $\mathcal{C}[f]$ denotes the quantum Boltzmann collision integral including annihilation and Compton scattering processes. The Debye screening length is suppressed relativistically and given by:

$$\lambda_D \sim \left(\frac{T}{ne^2}\right)^{1/2}\left(\frac{m_e}{T}\right)^{1/2} \sim \frac{1}{T},$$

in the ultra-relativistic limit $T >> m_e$, revealing that electrostatic shielding is minimal and long-range QED effects dominate.

Relativistic shocks are ubiquitous in GRBs, pulsar wind nebulae, and active galactic nuclei jets. These shocks form when supersonic relativistic flows are abruptly decelerated by ambient media, leading to compression and particle acceleration. The plasma here is collisionless, magnetized, and highly anisotropic, often modeled using quantum kinetic and fluid equations.

The standard magnetohydrodynamic description is modified in the relativistic quantum regime by including quantum pressure and spin effects:

$$\partial_\mu T^{\mu\nu} = F^{\nu\lambda} J_\lambda, \quad T^{\mu\nu} = (\varepsilon + P)u^\mu u^\nu - P g^{\mu\nu} + \Pi^{\mu\nu}_{\text{quantum}},$$

where $\Pi^{\mu\nu}_{\text{quantum}}$ includes terms arising from Bohm potential and spin-vorticity coupling. The quantum-corrected Rankine-Hugoniot conditions across the shock front yield altered entropy generation and compression ratios, particularly significant in degenerate or pair plasmas [17].

Quantum effects influence shock-driven acceleration mechanisms such as diffusive shock acceleration (DSA). The synchrotron cooling length in the quantum regime is governed by the radiation reaction force:

$$F^\mu_{\text{rad}} = \frac{2e^4}{3m^2}\left(F^{\mu\lambda} F_{\lambda\nu} u^\nu + (F_{\alpha\beta} u^\alpha u^\beta)^2 u^\mu\right),$$

leading to the Landau-Lifshitz prescription for radiation backreaction, which dominates for electrons in ultra-strong magnetic fields (approaching the critical QED field $B_c \sim 4.4 \times 10^{13}$G) [18].

Quark-Gluon Plasma Precursors and Pre-Hadronic Quantum Plasmas

The initial stage of relativistic heavy-ion collisions creates a non-equilibrium plasma of deconfined quarks and gluons, governed by quantum chromodynamics (QCD). Before thermalization into the strongly coupled quark-gluon plasma (sQGP), the system exhibits properties of a quantum non-Abelian plasma with both chromoelectric and chromomagnetic instabilities.

The color Wigner function formalism generalizes quantum kinetic theory to QCD plasmas:

$$\left[p^\mu D_\mu - g F^{\mu\nu}_a T^a p_\nu \frac{\partial}{\partial p^\mu}\right] W(x, p) = \mathcal{C}_{\text{QCD}}[W],$$

where $D_\mu = \partial_\mu - igA^a_\mu T^a$ is the gauge covariant derivative and T^a the generators of SU(3). Color instabilities (e.g., Weibel-type modes) emerge due to anisotropic parton distributions [19], with growth rates given by:

$$\gamma_k \sim gT\sqrt{\frac{\xi}{1+\xi}},$$

where ξ quantifies the momentum anisotropy. These instabilities can drive rapid isotropization and induce chromoelectric field amplification.

Longitudinal chromoelectric oscillations are described by gauge-covariant generalizations of the dielectric tensor $\epsilon_{\mu\nu}^{ab}(k)$. In the Abelian limit, collective excitations follow a modified dispersion relation:

$$\omega^2 = k^2 + m_D^2\left(1 - \frac{i\pi\omega}{2k}\right), \quad m_D^2 = \frac{g^2T^2N_c}{3},$$

indicating Landau damping and color Debye screening. The dynamical screening length regulates parton scattering and energy loss, crucial for understanding jet quenching [20].

The pre-equilibrium evolution is modeled via quantum field-theoretical approaches such as the Keldysh-Schwinger formalism. The two-particle irreducible (2PI) effective action for gauge fields and quarks permits real-time simulations of decoherence and hydrodynamization:

$$\Gamma_{\text{2PI}}[A, G] = S_{\text{YM}}[A] + i\text{Tr}\ln G^{-1} + \frac{i}{2}\text{Tr}(D^{-1}G) + \Phi[G],$$

where $\Phi[G]$ encodes non-perturbative self-interactions. This allows the extraction of transport coefficients and entropy production rates without assuming near-equilibrium hydrodynamics.

The interplay of relativistic quantum field theory, plasma kinetics, and collective effects in these high-energy phenomena reveals the ubiquity of quantum plasmas as a unifying framework. From gamma-ray bursts to early-universe QCD matter, the phenomena span scales and interactions that require an inherently quantum, relativistically covariant, and gauge-invariant plasma theory. These scenarios provide not only natural laboratories for probing fundamental interactions, but also demand a synthesis of theoretical approaches ranging from effective actions to quantum kinetic equations that form the foundation of modern quantum plasma physics in high-energy regimes [21].

10.3 Laser–Plasma Interaction in the Quantum Regime

The interaction of ultra-intense laser fields with quantum plasmas represents a frontier in both theoretical modeling and experimental diagnostics, where relativistic dynamics, quantum coherence, and strong-field electrodynamics converge. When laser intensities surpass $10^{22}\,\text{W/cm}^2$, the associated electric field strength E_L approaches and exceeds the Schwinger critical field $E_S = m_e^2c^3/e\hbar \sim 1.32 \times 10^{18}\,\text{V/m}$, thereby probing a regime where vacuum polarization, pair production, and quantum radiation reaction become non-negligible.

The fundamental description of quantum plasma in this regime requires coupling the quantum Vlasov equation, or Wigner-Maxwell formalism to quantum field equations. In high-intensity regimes, the plasma particles (primarily electrons) are accelerated to relativistic velocities, and the quantum corrections such as the Bohm potential, spin terms, and relativistic mass variations must be retained explicitly.

Let the laser be modeled as an external electromagnetic field:

$$A^{\mu}_{\text{ext}}(x) = \left(0, \frac{E_0}{\omega} \sin(k \cdot x)\hat{\mathbf{e}}\right),$$

with amplitude E_0, frequency ω, and wavevector $k^{\mu} = (\omega, \mathbf{k})$, in the Lorentz gauge. The one-particle Wigner function $f(x, p)$ evolves according to the relativistic quantum Vlasov equation in the collisionless limit:

$$\left[p^{\mu}\partial_{\mu} + eF^{\mu\nu}p_{\nu}\partial^{p}_{\mu} + \mathcal{O}(\hbar^2)\right]f(x, p) = 0,$$

where $F^{\mu\nu} = \partial^{\mu}A^{\nu} - \partial^{\nu}A^{\mu}$ includes the total field, and $\mathcal{O}(\hbar^2)$ denotes higher-order quantum corrections, including spin–orbit coupling and vacuum dispersion.

The self-consistent evolution of the field is governed by Maxwell's equations with quantum currents:

$$\partial_{\mu}F^{\mu\nu} = j^{\nu} = e\int \frac{d^3p}{p^0} p^{\nu} f(x, p),$$

with possible inclusion of quantum radiative corrections (e.g., Heisenberg–Euler terms) when field intensities exceed E_S.

At such high intensities, the electron dynamics is best described through solutions of the Dirac equation in strong backgrounds:

$$\left(i\gamma^{\mu}\partial_{\mu} - e\gamma^{\mu}A_{\mu} - m\right)\psi(x) = 0.$$

For a plane-wave laser field, Volkov solutions provide exact expressions for the wavefunctions:

$$\psi_p(x) = \left[1 + \frac{e\, \not{k}\, \not{A}}{2(k \cdot p)}\right]u(p)e^{iS(x)},$$

where $S(x)$ is the classical action including field-dependent phase, and $u(p)$ is a free Dirac spinor. These solutions reveal how particle wavefunctions acquire ponderomotive shifts and nonlinearities, underpinning phenomena such as nonlinear Compton scattering and laser-induced pair production [22].

A central parameter controlling quantum effects is the dimensionless quantum nonlinearity parameter:

$$\chi_e = \frac{e\hbar}{m^3 c^4}\sqrt{-(F_{\mu\nu}p^\nu)^2},$$

which quantifies the ratio of the field in the electron's rest frame to the Schwinger field. For $\chi_e \gtrsim 1$, quantum recoil, radiation reaction, and even pair cascades become prominent. For a laser-electron collision at angle θ,

$$\chi_e \approx \frac{\gamma E_L}{E_S}(1-\cos\theta),$$

revealing the key role of Lorentz factor and geometry in accessing the quantum radiation regime.

Inclusion of radiation reaction is crucial in this regime. The quantum-corrected Landau–Lifshitz equation, or alternatively, the stochastic emission model based on quantum probabilities $dP/d\omega$, must be embedded into the particle dynamics. The radiation spectrum becomes broadened and redshifted due to quantum recoil. The expected photon emission rate in nonlinear Compton scattering, at leading order in χ_e, is given by:

$$\frac{d^2N_\gamma}{d\omega dt} \sim \alpha\frac{m^2}{\omega}\left[\frac{2}{\chi_e^2} + \frac{\omega^2}{m^2\gamma^2}\right]\exp\left(-\frac{2\omega}{3\chi_e m}\right),$$

indicating suppression of high-frequency emission due to quantum coherence loss.

Collective behaviour persists even in ultra-relativistic regimes. The formation of Langmuir waves, wakefields, and parametric instabilities is now influenced by quantum degeneracy and relativistic mass shift. For instance, the dispersion relation of Langmuir oscillations is modified as:

$$\omega^2 = \omega_{pe}^2\left(1 + \frac{3k^2v_F^2}{\omega^2} + \frac{\hbar^2k^4}{4m_e^2\omega^2}\right) + \delta\omega_{\mathrm{rel}},$$

where v_F is the Fermi velocity and $\delta\omega_{\mathrm{rel}}$ denotes relativistic corrections due to mass dilation and Doppler shifts. This equation unifies quantum statistical effects with relativistic kinematics.

In spatially inhomogeneous settings, laser filamentation and self-channeling manifest modified growth rates due to quantum pressure and spin alignment. These nonlinear structures are analyzed using envelope equations derived from the Wigner–Maxwell system, or the Dirac–Heisenberg–Wigner formalism.

Several laser facilities, such as ELI, Apollon, and future upgrades to XFEL systems, are approaching or have exceeded the $10^{22}\,\mathrm{W/cm^2}$ threshold. Quantum signatures such as hard gamma emission, nonlinear Breit–Wheeler pair production, and radiation-dominated wakefield acceleration are under active investigation. The interplay of laser polarization, plasma density, and quantum recoil defines optimal parameters for observing these effects. Notably, quantum suppression of ponderomotive force and nonlocal response functions modify energy deposition and particle trapping.

The interaction of ultra-intense lasers with quantum plasmas necessitates a synthesis of relativistic quantum electrodynamics, kinetic plasma theory, and nonlinear optics. It opens the avenue to not only test fundamental physics, such as vacuum breakdown and radiation reaction, but also enables compact accelerators, bright gamma-ray sources, and analogues of early universe conditions in the laboratory. The formulation presented here serves as a foundational pillar for deeper investigations into nonlinear QED processes, quantum hydrodynamics, and stochastic field-particle coupling in extreme plasma environments.

10.3.1 Quantum Corrections to Ponderomotive Force

The ponderomotive force constitutes a fundamental nonlinear mechanism in the dynamics of plasmas under oscillating electromagnetic fields. In classical electrodynamics, this force represents a time-averaged effect whereby a charged particle is pushed away from regions of high-intensity electromagnetic fields. In the context of quantum plasmas, especially under high-energy or high-frequency excitation, this classical picture becomes inadequate due to the emergence of quantum statistical and dispersive effects.

For a classical charged particle of charge q and mass m, subject to an oscillating electric field $\mathbf{E}(\mathbf{r}, t) = \mathbf{E}_0(\mathbf{r})\cos(\omega t)$, the classical ponderomotive force $\mathbf{F}_p$ is derived from the cycle-averaged Newton–Lorentz equation:

$$\mathbf{F}_p = -\frac{q^2}{4m\omega^2}|\nabla \mathbf{E}_0|^2,$$

which reveals that the particle is expelled from regions with higher field intensities. However, this expression omits quantum corrections that arise due to the wave nature of particles, the uncertainty principle, and quantum degeneracy pressure, especially at length scales comparable to the de Broglie wavelength.

To incorporate quantum corrections, one resorts to the quantum hydrodynamic (QHD) model or the Wigner–Moyal formalism. We begin with the QHD momentum equation for a single-fluid plasma system:

$$m\left(\frac{\partial \mathbf{u}}{\partial t} + (\mathbf{u}\cdot\nabla)\mathbf{u}\right) = q(\mathbf{E} + \mathbf{u}\times\mathbf{B}) - \nabla P + \mathbf{F}_Q,$$

where $\mathbf{u}$ is the fluid velocity, P the pressure (which may be thermal or degeneracy pressure), and $\mathbf{F}_Q$ denotes the quantum force term arising from the Bohm potential:

$$\mathbf{F}_{\mathrm{Q}} = \frac{\hbar^2}{2m}\nabla\left(\frac{\nabla^2\sqrt{n}}{\sqrt{n}}\right),$$

with n the number density. This term encapsulates the quantum pressure associated with the Heisenberg uncertainty principle.

To obtain the quantum-corrected ponderomotive force, one must introduce the oscillating laser field into the QHD framework and apply a multi-timescale perturbation analysis. The electric field is introduced as a rapidly oscillating component superimposed on a slowly varying background.

By expanding the fluid variables as $n = n_0 + n_1 + \cdots$, $\mathbf{u} = \mathbf{u}_1 + \mathbf{u}_2 + \cdots$, and retaining terms to second order, one obtains a quantum-corrected force per unit mass:

$$\mathbf{F}_{\mathrm{qp}} = -\frac{q^2}{4m^2\omega^2}\nabla|\mathbf{E}_0|^2 + \frac{\hbar^2 q^2}{8m^3\omega^4}\nabla\left(\nabla^2|\mathbf{E}_0|^2\right) + \cdots \sqrt{a^2+b^2},$$

where the first term is the classical ponderomotive force, while the second term represents a leading-order quantum correction derived from the second-order expansion of the Bohm potential in response to field gradients. This correction becomes significant in the high-frequency (short-wavelength) limit, where spatial gradients of the field energy density are large, or when the electron temperature is low enough for quantum effects to dominate.

Physically, the additional quantum term implies that particles can be pushed not only by gradients in the field intensity, but also by the curvature (i.e., Laplacian) of that intensity, reflecting nonlocal quantum dispersive effects [23].

The Wigner–Moyal kinetic framework offers a fully quantum and phase-space-resolved description of the ponderomotive force, avoiding the assumption of fluid closure. The Wigner function $f_W(\mathbf{r}, \mathbf{p}, t)$ satisfies the Wigner equation:

$$\frac{\partial f_W}{\partial t} + \frac{\mathbf{p}}{m}\cdot\nabla_{\mathbf{r}} f_W + \frac{q}{i\hbar}\int d^3s\, e^{i\mathbf{s}\cdot\mathbf{p}/\hbar}\left[\phi\left(\mathbf{r}+\frac{\mathbf{s}}{2}\right) - \phi\left(\mathbf{r}-\frac{\mathbf{s}}{2}\right)\right] f_W(\mathbf{r}, \mathbf{p}, t) = 0,$$

where $\phi(\mathbf{r})$ is the electrostatic potential. Expanding the potential term in powers of $\hbar$, one recovers the classical Vlasov equation in the limit $\hbar \to 0$. However, retaining terms up to $\hbar^2$ allows us to extract nonlocal quantum contributions to the force. From this kinetic formalism, the ponderomotive force density can be expressed as:

$$\mathbf{F}_{\mathrm{W}} = -\int \frac{\mathbf{p}}{m}\left(\frac{\partial f_W}{\partial t}\right)_{\text{ponderomotive}} d^3p,$$

with corrections involving $\hbar^2\nabla(\nabla^2\phi)$ explicitly appearing, consistent with the Bohm-type dispersive structure.

Quantum corrections to the ponderomotive force become significant under specific extreme physical conditions. In regimes with high field gradients where the intensity of the electromagnetic field varies rapidly over spatial scales comparable to the electron's de Broglie wavelength, quantum dispersion effects play a crucial role. Similarly, in ultracold dense plasmas, such as those found in solid-state systems, or white dwarfs, quantum effects like degeneracy pressure and quantum recoil can dominate over classical thermal contributions. High-frequency lasers also contribute to pronounced quantum corrections, when the laser frequency is sufficiently large, the induced oscillations lead to strong quantum kinetic effects, particularly under conditions where quantum coherence becomes non-negligible. Further, in ultra-intense laser fields, relativistic effects come into play. The relativistic increase in electron mass alters the effective plasma frequency and introduces higher-order nonlinearities in the relativistic quantum hydrodynamics (QHD) framework. These relativistic corrections, when combined with quantum effects, give rise to complex hybrid ponderomotive behaviours that depart significantly from classical predictions.

Quantum corrections to the ponderomotive force signify a crucial extension to classical plasma theory, capturing the influence of quantum statistical pressure, spatial nonlocality, and high-frequency field effects. Their derivation from both fluid and kinetic perspectives illuminates their robustness across modeling frameworks. In high-intensity laser–plasma interactions, especially those approaching or surpassing the Schwinger limit, these corrections become indispensable for accurately predicting electron dynamics, wavebreaking thresholds, and particle acceleration mechanisms. Subsequent sections will explore how these corrections influence quantum wakefield generation and the modulation of laser energy absorption in overcritical plasmas.

10.3.2 Nonlinear Compton Scattering and Electron Acceleration

In high-intensity quantum plasma environments, the interaction of relativistic electrons with ultra-intense electromagnetic fields results in profound non-perturbative phenomena. Among these, nonlinear Compton scattering constitutes a cornerstone process wherein an electron absorbs multiple photons from the laser field and emits a single high-energy photon, effectively transferring momentum and energy in a nonlinearly enhanced quantum electrodynamical regime.

To analyze nonlinear Compton scattering, one begins by recognizing that in the presence of an intense electromagnetic field, the conventional plane-wave solutions of the Dirac equation no longer suffice. Instead, the exact electron wavefunctions must be described by Volkov states, which are solutions to the Dirac equation in a background classical plane-wave field $A^\mu(\phi)$ [24], where $\phi = k_\mu x^\mu$ denotes the laser phase:

$$(i\gamma^\mu \partial_\mu - e\gamma^\mu A_\mu - m)\Psi(x) = 0.$$

The Volkov solution for a spinor in a monochromatic plane-wave field is given by:

$$\Psi_p(x) = \left(1 + \frac{e\, \not{k}\, \not{A}}{2k \cdot p}\right) \frac{u(p)}{\sqrt{2p^0 V}} e^{-ip\cdot x - i\int^{\phi}\left[\frac{ep\cdot A}{k\cdot p} - \frac{e^2 A^2}{2k\cdot p}\right] d\phi'}.$$

Here, $u(p)$ is the free Dirac spinor, $\not{A} = \gamma_\mu A^\mu$, and V is the normalization volume. The presence of the term $e^2 A^2$ signals the nonlinear nature of the interaction: the electron experiences an effective mass increase due to the laser field, a phenomenon often referred to as the 'mass shift' or 'dressing':

$$m_*^2 = m^2\left(1 + a_0^2\right), \quad \text{with} \quad a_0 = \frac{eE_0}{m\omega}.$$

Here, a_0 is the dimensionless laser intensity parameter, with E_0 the electric field amplitude and ω the laser frequency. For $a_0 \gtrsim 1$, the interaction enters the nonlinear regime, characterized by multi-photon processes and radiation reaction effects.

The differential scattering probability per unit volume and time, derived from the S-matrix element involving initial and final Volkov spinors and an emitted photon [25], includes contributions from harmonics of the laser field. The emitted photon's energy ω' and momentum k'^μ are subject to energy–momentum conservation:

$$p + n\, k = p' + k',$$

where 'n' denotes the number of absorbed photons from the external field. The quantum nature of the interaction is encapsulated in the quantum nonlinearity parameter χ, defined as:

$$\chi = \frac{e\sqrt{-(F^{\mu\nu} p_\nu)^2}}{m^3} = \frac{\gamma E_\perp}{E_S},$$

where $E_S = m^2 c^3/(e\hbar)$ is the Schwinger field and $E_\perp$ is the component of the electric field perpendicular to the electron's motion. For $\chi \gtrsim 1$, the quantum recoil during photon emission becomes significant, necessitating fully quantum kinetic models.

In nonlinear Compton scattering, high-energy photon emission leads to recoil on the emitting electron, referred to as radiation reaction. Quantum-corrected models such as the Landau–Lifshitz equation with stochastic photon emission or the quantum-corrected Vlasov–Maxwell system must be employed. The effective equation of motion for the electron in a strong field, incorporating quantum recoil, reads:

$$\frac{dp^\mu}{d\tau} = \frac{e}{m} F^{\mu\nu} p_\nu + \tau_0 \left[\frac{d^2 p^\mu}{d\tau^2} - \frac{p^\mu}{m^2}\left(\frac{dp^\nu}{d\tau}\frac{dp_\nu}{d\tau}\right)\right],$$

with $\tau_0 = \frac{2e^2}{3m}$ the characteristic radiation reaction time. In the quantum regime, one replaces the continuous recoil with a probabilistic emission spectrum governed by the quantum synchrotron function $R(\chi, \eta)$, $\eta = \hbar\omega'/(\gamma m)$.

In quantum plasma environments driven by laser intensities exceeding $10^{22}\,\mathrm{W/cm^2}$, the interaction is dominated by radiation reaction-limited regimes, wherein electrons emit high-energy photons during acceleration and deceleration cycles. The energy gain of an electron is limited by both dephasing and quantum recoil [26], and the final electron energy $\mathcal{E}_f$ is constrained by:

$$\mathcal{E}_f \lesssim \left(\frac{\mathrm{m^2c^4}}{\mathrm{e^2}}\right)^{1/3}\left(\frac{\mathrm{E_0}}{\mathrm{E_S}}\right)^{-2/3}.$$

This saturation mechanism defines the quantum radiation-dominated regime (QRDR), and necessitates full QED particle-in-cell (QED-PIC) simulations for accurate modeling. Within this regime, the acceleration of electrons approaches the radiation-reaction limit, with stochasticity in photon emission introducing beam broadening and spectral features.

Experiments at facilities such as SLAC, ELI, and the upcoming LUXE project have begun probing the nonlinear Compton regime. In inverse Compton setups, relativistic electron beams collide with counter-propagating laser pulses, and the emitted spectrum exhibits harmonic structures and nonlinear redshifts. Moreover, these processes are crucial for the generation of attosecond gamma-ray bursts and for advancing compact laser-based particle accelerators.

Ultimately, nonlinear Compton scattering in quantum plasmas exemplifies the rich interplay between classical radiation theory and quantum electrodynamics under extreme field conditions. Its proper understanding not only bridges theoretical gaps in strong-field QED, but also enables next-generation applications in ultrafast photon sources, radiation therapy, and laboratory astrophysics.

10.3.3 Importance in Laser Wakefield Acceleration and Next-Generation Accelerators

Laser Wakefield Acceleration (LWFA) has emerged as a transformative approach in the field of particle acceleration, offering acceleration gradients several orders of magnitude higher than those achievable with conventional radio-frequency (RF) accelerators. By leveraging the interaction between ultra-intense laser pulses and plasma, LWFA facilitates the generation of compact, high-energy electron beams. In LWFA, an ultra-short, high-intensity laser pulse propagates through a plasma, displacing electrons via the ponderomotive force and creating a trailing region of positive charge due to the heavier, relatively immobile ions. This charge separation establishes a plasma wakefield with electric fields capable of accelerating electrons to relativistic energies over centimeter-scale distances.

At extremely high laser intensities and plasma densities, quantum effects become significant, necessitating corrections to the classical Laser Wakefield Acceleration (LWFA) model. Among these, quantum recoil and radiation reaction play a critical role: as electrons are accelerated and emit high-energy photons, they experience a recoil force that alters their motion. This quantum radiation reaction is incorporated into the momentum equation through additional terms, often modeled using the Landau-Lifshitz formulation or other quantum-corrected equations of motion. Another important quantum phenomenon is quantum degeneracy pressure, which arises in dense plasmas due to the Pauli exclusion principle. This pressure resists further compression and influences the plasma's response to the laser pulse, particularly in degenerate plasmas where electron thermal energy is lower than the Fermi energy. Additionally, quantum effects modify the dispersion relation of plasma waves; for instance, in cold plasmas, the quantum-corrected dispersion relation for longitudinal waves includes the reduced Planck constant and affects the plasma frequency and wave number. This directly impacts the phase velocity of the wakefield and, consequently, the mechanisms of electron trapping and acceleration.

These quantum-informed modifications have significant implications for the development of next-generation accelerators. One of the main advantages of LWFA is compactness, the extremely high acceleration gradients (on the order of gigavolts per meter) allow for much shorter accelerator lengths compared to conventional radio-frequency (RF) accelerators, which are typically limited to gradients around 100 megavolts per meter. This reduction in size also translates to greater cost efficiency, lowering both construction and operational expenses and expanding accessibility for research and medical uses. In terms of performance, LWFA has demonstrated the ability to produce electron beams with low emittance and narrow energy spread, qualities that are desirable for applications in free-electron lasers and high-energy physics experiments. Moreover, advancements in staging, linking multiple LWFA units in sequence are enabling the achievement of even higher energies, provided that efficient inter-stage coupling is maintained to preserve beam quality.

Despite these benefits, LWFA still faces several technical challenges. Beam stability is one such issue, as fluctuations in laser parameters and plasma density can degrade beam quality, this necessitates active feedback and precise control systems. Achieving controlled electron injection into the wakefield is another key concern, as it determines the production of monoenergetic beams. Various injection techniques, such as density transition injection and ionization injection, are currently being explored. Further, as LWFA systems operate at ultra-high intensities, it becomes increasingly important to manage quantum effects like radiation reaction and even pair production, which can otherwise disrupt beam dynamics. Finally, integrating LWFA modules with existing accelerator infrastructure poses practical difficulties, including challenges in beam transport, timing synchronization, and compatibility with traditional accelerator components. Addressing these issues will be vital for fully realizing the potential of LWFA in future accelerator technologies.

10.3.4 Applications in Attosecond Pulse Generation and Extreme Ultraviolet (EUV) Sources

The advent of attosecond pulse generation and *extreme ultraviolet (EUV)* sources has revolutionized ultrafast science, enabling the observation and control of electron dynamics on their natural timescales. Quantum plasmas, characterized by high-density electron populations and significant quantum effects, play a pivotal role in these advancements. This section delves into the mechanisms by which quantum plasmas facilitate attosecond pulse generation and EUV emission, highlighting the underlying physics and potential applications.

High-harmonic generation (HHG) is a nonlinear optical process wherein the interaction of intense laser fields with matter leads to the emission of radiation at integer multiples of the fundamental frequency. In quantum plasmas, the collective behaviour of electrons and the presence of strong electromagnetic fields enhance the efficiency and coherence of HHG.

The HHG process in quantum plasmas can be described by the time-dependent Schrödinger equation (TDSE) coupled with Maxwell's equations. The TDSE accounts for the quantum dynamics of electrons under the influence of the laser field, while Maxwell's equations govern the propagation of electromagnetic waves within the plasma. The interplay between these equations captures the essential physics of HHG in quantum plasmas [27].

Quantum plasmas exhibit phenomena such as tunneling ionization, where electrons penetrate potential barriers due to quantum tunneling, and quantum recoil, where the momentum exchange between photons and electrons becomes significant. These effects contribute to the generation of high-order harmonics and the formation of attosecond pulses.

Plasma Mirrors and Relativistic Effects

Plasma mirrors are formed when intense laser pulses ionize a solid surface, creating a dense plasma layer that reflects subsequent laser pulses. At relativistic intensities, the oscillating plasma surface can upshift the frequency of reflected light through the Doppler effect, leading to the generation of high-order harmonics and attosecond pulses.

In the relativistic regime, the plasma surface oscillates with velocities approaching the speed of light. The reflected electric field $Er(t)$ can be expressed as:

$$E_r(t) = E_0 \cos[\omega_0 t + \phi(t)],$$

where ω_0 is the fundamental frequency, and $\phi(t)$ represents the phase modulation due to the moving plasma surface. The time-dependent phase $\phi(t)$ leads to the generation of high-order harmonics in the reflected spectrum.

Attosecond Pulse Generation Mechanisms

Attosecond pulses in quantum plasmas can be generated through several advanced mechanisms, each exploiting unique plasma-laser interactions. One such mechanism is Coherent Wake Emission (CWE), which occurs when the plasma frequency aligns with a harmonic of the driving laser, leading to efficient generation of high-order harmonics. Another method is the Relativistic Oscillating Mirror (ROM) effect, where the Doppler upshift of light reflected from a plasma surface oscillating at relativistic speeds results in the production of high-frequency radiation. Additionally, resonant harmonic generation utilizes specific resonances within the plasma to selectively enhance certain harmonic orders, thereby amplifying the strength of the resulting attosecond pulses.

The generation of attosecond pulses and extreme ultraviolet (EUV) radiation has opened transformative possibilities in ultrafast science. In time-resolved spectroscopy, these pulses provide the temporal resolution necessary to observe electron dynamics in atoms, molecules, and solid-state systems, shedding light on ultrafast phenomena such as photoionization and electron correlation. In the fields of imaging and metrology, EUV radiation serves as a powerful tool for achieving high-resolution techniques like coherent diffractive imaging and holography, which have important applications in both material science and biological research. Further, attosecond pulses offer a valuable diagnostic tool for probing the ultrafast dynamics of plasmas, contributing to a deeper understanding of complex processes in inertial confinement fusion, laser-plasma interactions, and astrophysical events.

10.4 Critical Fields and the Schwinger Limit

In the realm of quantum electrodynamics (QED), the vacuum is not an inert backdrop but a dynamic medium capable of exhibiting nonlinear phenomena under extreme conditions. One such phenomenon is the spontaneous creation of electron–positron pairs from the vacuum when subjected to an immensely strong electric field. This critical field strength, known as the Schwinger critical field E_e, signifies the threshold at which the vacuum becomes unstable and real particle-antiparticle pairs emerge.

The Schwinger critical field can be derived by equating the work done by an electric field over a characteristic quantum distance to the rest energy of an electron–positron pair.

i. Work done by the Electric Field: Consider an electric field E acting over a distance equal to the reduced Compton wavelength of the electron $\lambda_C = \frac{\hbar}{m_e c}$. The work done W is: $W = eE\lambda_C = eE\frac{\hbar}{m_e c}$
ii. Energy Required for Pair Production: The energy needed to create an electron–positron pair is twice the rest energy of an electron: $\Delta E = 2m_e c^2$

iii. Equating Work and Energy: Setting $W = \Delta E$:

$$eE\frac{\hbar}{m_e c} = 2m_e c^2$$

Solving for E: $E = \frac{2m_e^2 c^3}{e\hbar}$. The factor of 2 is often omitted in literature, leading to the standard expression for the Schwinger critical field: $E_c = \frac{m_e^2 c^3}{e\hbar}$. Numerically, this evaluates to approximately: $E_c \approx 1.3 \times 10^{18}$ V/m This field strength is enormous, highlighting the extreme conditions required for vacuum instability.

The Schwinger critical field represents the electric field strength at which the vacuum becomes unstable, leading to spontaneous electron–positron pair production. This phenomenon, known as the Schwinger effect, is a non-perturbative prediction of QED and illustrates the nonlinear nature of the vacuum under strong fields.

In classical electrodynamics, the vacuum is linear and does not exhibit such behavior. However, QED predicts that under fields approaching E_c, virtual particle-antiparticle pairs can be 'pulled apart' by the field, becoming real particles. This process results in the decay of the electric field's energy into matter, a direct manifestation of energy-mass equivalence.

While the Schwinger effect has profound theoretical implications, observing it experimentally poses significant challenges due to the immense field strengths required. Current and planned high-intensity laser facilities aim to approach these conditions, potentially enabling the observation of vacuum pair production. Such experiments would not only confirm a fundamental prediction of QED but also open avenues for exploring new physics in extreme field regimes.

The Schwinger critical field $E = \frac{2m_e^2 c^3}{e\hbar}$ delineates the boundary between stable and unstable vacuum states under the influence of an electric field. Its derivation and implications underscore the rich interplay between quantum mechanics and electromagnetism, revealing the vacuum as a dynamic entity capable of manifesting matter under extreme conditions. Understanding and potentially observing the Schwinger effect remains a frontier in high-energy physics and quantum field theory [28].

10.4.1 Quantum Plasma Behaviour in Sub-critical, Critical, and Supercritical Regimes

The behaviour of quantum plasmas under varying electromagnetic field strengths reveals a rich tapestry of phenomena, particularly as fields approach, or exceed the Schwinger critical field (E_c). This section delves into the distinct characteristics of quantum plasmas in sub-critical ($E < E_c$), critical ($E \approx E_c$), and supercritical ($E > E_c$) regimes, emphasizing the interplay between quantum electrodynamics (QED) effects and plasma dynamics.

Sub-critical Regime

In the sub-critical regime, electromagnetic fields are insufficient to induce significant vacuum polarization, or spontaneous pair production. Quantum plasmas in this domain exhibit behaviour predominantly governed by collective interactions and quantum statistical effects. In this regime, the plasma exhibits a linear response to external perturbations, allowing its dielectric properties to be accurately described using conventional quantum plasma models. Although quantum electrodynamics (QED) effects are negligible, important quantum corrections, such as the inclusion of the Bohm potential and Fermi pressure significantly influence the plasma's behavior. Despite the presence of these quantum effects, the plasma remains stable, with no indications of vacuum decay or pair production phenomena.

Critical Regime

As the field strength approaches E_c, the plasma enters a transitional phase where QED effects become pronounced, leading to notable modifications in plasma properties. In this regime, the vacuum itself begins to exhibit nonlinear behavior, leading to effects such as vacuum polarization, which manifests as birefringence and alterations in refractive indices. Concurrently, the onset of electron–positron pair production becomes possible, though typically occurring at low rates, thereby introducing additional charged particles into the plasma environment. As a result, the plasma's response transitions from linear to nonlinear, requiring the incorporation of quantum electrodynamics (QED) corrections into theoretical models to accurately capture its complex dynamics.

Supercritical Regime

In the supercritical regime, electromagnetic fields surpass E_c, leading to profound alterations in plasma dynamics driven by intense QED effects. In this extreme regime, electron–positron pair production occurs at a dramatically increased rate, often resulting in the formation of dense pair plasmas. The intense electromagnetic fields cause accelerated particles to emit substantial radiation, giving rise to significant radiation reaction effects that alter particle trajectories and energy distributions. Additionally, complex cascade processes, such as Breit–Wheeler pair production and nonlinear Compton scattering become dominant, further enriching the plasma dynamics. As these interactions progress, the energy of the electromagnetic field is rapidly depleted through radiation and particle creation, leading to field attenuation and a self-regulating behaviour within the plasma.

The transition of quantum plasmas through sub-critical, critical, and supercritical regimes underscores the intricate interplay between quantum mechanics and plasma physics. Understanding these behaviours is pivotal for advancing high-intensity laser technologies, exploring astrophysical phenomena, and probing the fundamental limits of QED.

10.5 Inverse Compton Scattering

Inverse Compton Scattering (ICS) is a pivotal process in high-energy astrophysics and laboratory plasma physics, wherein low-energy photons gain energy by scattering off high-energy electrons. This mechanism is instrumental in generating high-energy photons, such as X-rays and gamma rays, in various environments, including astrophysical sources like pulsars and active galactic nuclei, as well as in laboratory settings utilizing laser-plasma interactions. Understanding the distinctions between classical and quantum descriptions of ICS is crucial for accurately modeling these phenomena.

In the classical regime, ICS is treated using the Thomson scattering framework, applicable when the photon energy in the electron rest frame is much less than the electron rest mass energy, i.e., $\hbar\omega' \ll m_e c^2$. Under this approximation, the scattering is elastic, and the energy gain of the photon is modest.

Consider a photon of energy $\hbar\omega$ scattering off an electron with Lorentz factor γ. In the electron rest frame, the photon energy is Doppler shifted:

$$\omega' = \gamma\omega(1 - \beta\cos\theta),$$

where $B = v/c$ and θ is the angle between the photon and electron velocities. After scattering, the photon energy in the lab frame becomes:

$$\omega_s = \gamma\omega'\left(1 + \beta\cos\theta'\right),$$

where θ is the scattering angle in the electron rest frame. For head-on collisions $\theta = \pi$) and backscattering ($\theta' = 0$), the maximum photon energy is:

$$\omega_{s,\max} = \frac{4\gamma^2\omega}{1 + 4\gamma\omega/m_e c^2}.$$

In the classical limit, where $\gamma\omega \ll m_e c^2$, this simplifies to $\omega_{s,\max} \approx 4\gamma^2\omega$.

The total power radiated by an electron due to ICS in the Thomson regime is given by:

$$P = \frac{4}{3}\sigma_T c\gamma^2\beta^2 U_{\text{ph}},$$

where σ_T is the Thomson cross-section, and U_{ph} is the energy density of the photon field.

When the photon energy in the electron rest frame approaches or exceeds $m_e c^2$, quantum effects become significant, necessitating a quantum electrodynamics (QED) treatment. This regime is characterized by the Klein-Nishina formula for the differential cross-section and includes phenomena such as recoil and multiple photon emissions [29].

The differential cross-section for ICS in the quantum regime is given by the Klein-Nishina formula:

$$\frac{d\sigma}{d\Omega} = \frac{r_e^2}{2}\left(\frac{\omega_s}{\omega}\right)^2\left(\frac{\omega_s}{\omega} + \frac{\omega}{\omega_s} - \sin^2\theta\right),$$

where r_e is the classical electron radius, and θ is the scattering angle. This expression accounts for the decrease in cross-section at high photon energies due to recoil effects.

The significance of quantum effects is quantified by the dimensionless parameter χ_e:

$$\chi_e = \frac{e\hbar}{m_e^3 c^4}\sqrt{-(F^{\mu\nu}p_\nu)^2},$$

where $F^{\mu\nu}$ is the electromagnetic field tensor, and p_ν is the four-momentum of the electron. When $\chi_e \gtrsim 1$, quantum effects such as radiation reaction and pair production become prominent.

In intense electromagnetic fields, such as those produced by high-intensity lasers, nonlinear effects become significant. Here, the electron can absorb multiple photons simultaneously before emitting a single high-energy photon, a process known as nonlinear inverse Compton scattering (NICS) [30].

The strength of the nonlinear interaction is characterized by the dimensionless parameter a_0:

$$a_0 = \frac{eE}{m_e c\omega},$$

where E is the electric field amplitude of the laser, and ω is its angular frequency. When $a_0 \gg 1$, the interaction enters the nonlinear regime, and the electron's motion becomes highly relativistic in the laser field.

In the nonlinear regime, the emitted photon spectrum exhibits harmonics of the fundamental frequency, and the emission is no longer confined to a narrow cone. The spectrum becomes broadened, and the angular distribution widens due to the complex electron trajectories in the intense field.

In plasma environments, collective effects and the presence of background fields can modify the ICS process. Plasma dispersion can alter the phase velocity of electromagnetic waves, affecting the resonance conditions for scattering. Additionally, plasma-induced fields can influence electron trajectories, modifying the emission characteristics.

In laser-plasma interactions, wakefield acceleration can produce relativistic electron beams that subsequently undergo ICS with the driving laser or with secondary laser pulses. The combination of plasma acceleration and ICS enables the generation of high-brightness, tunable X-ray and gamma-ray sources in compact setups.

The transition from classical to quantum regimes in inverse Compton scattering is governed by the energy of the interacting photons and electrons, as well as the intensity of the electromagnetic fields involved. Understanding these regimes is essential for interpreting high-energy astrophysical phenomena and for developing advanced radiation sources in laboratory settings. Plasma environments introduce additional complexities, but also offer opportunities for novel applications of ICS in generating high-energy photons.

10.6 Quantum Recoil, Nonlinear Compton Shift, and Spectral Broadening

In high-intensity quantum plasma environments, particularly in regimes governed by extreme field strengths and relativistic particle dynamics, the classical picture of radiation emission becomes insufficient. In such regimes, the interplay between strong-field quantum electrodynamics (QED) effects and plasma collective interactions gives rise to complex radiative behaviors. Central among these are the phenomena of quantum recoil, the nonlinear Compton shift, and the spectral broadening of high-energy photons emitted during electron–laser interactions.

Quantum Recoil in Strong-Field Electron-Photon Interactions

Quantum recoil refers to the momentum transfer experienced by a charged particle, such as an electron, upon emitting or absorbing a photon whose energy is non-negligible compared to the rest energy of the particle. While negligible in the low-energy (Thomson) regime, quantum recoil becomes significant when the photon energy approaches or exceeds the electron's kinetic or rest energy.

Let ω_i and ω_f be the initial and final photon energies, and let $\gamma \gg 1$ denote the Lorentz factor of the relativistic electron in a laser-driven plasma wakefield or a strong external field. The recoil parameter 'r' is defined as:

$$r = \frac{\hbar\omega_i}{\gamma m_e c^2}.$$

When $r \ll 1$, recoil effects are negligible and radiation can be treated classically (Thomson scattering). However, when $r \sim 1$, quantum recoil must be incorporated, marking the transition to the nonlinear Compton regime.

In covariant form, the four-momentum conservation condition for single-photon emission by an electron interacting with an intense electromagnetic wave can be written as:

$$p^{\mu} + sk^{\mu} = p'^{\mu} + k'^{\mu},$$

where p^μ and p'^μ are the initial and final electron four-momenta, k^μ is the four-wavevector of the external field (e.g., laser), k'^μ is the four-momentum of the emitted photon, and $s \in \mathbf{Z}^+$ corresponds to the harmonic number in multiphoton processes.

The recoil alters the kinematics of Compton scattering and affects both the energy spectrum and angular distribution of emitted photons, necessitating quantum corrections to classical radiation formulas. These corrections are encoded in the differential emission probability per unit time (in the Furry picture) via strong-field QED, where the quantum nonlinear Compton cross-section is given in terms of modified Bessel functions involving the nonlinearity parameter χ_e [31].

Nonlinear Compton Shift and Harmonic Structure

In the classical linear Compton process, the scattered photon energy is given by:

$$\omega' = \frac{\omega}{1 + \frac{\omega}{m_e c^2}(1 - \cos\theta)},$$

where ω is the incident photon frequency and θ the scattering angle. In contrast, the nonlinear Compton process occurring in a plasma or laser field of high intensity $I \gtrsim 10^{20} \mathrm{W/cm^2}$ entails the simultaneous absorption of multiple photons from the driving field. The process is governed by the dimensionless field strength a_0, defined as:

$$a_0 = \frac{eE_0}{m_e c\omega},$$

where E_0 is the peak electric field of the laser. For $a_0 \gtrsim 1$, the electron motion becomes relativistically nonlinear, leading to harmonic radiation at frequencies $\omega_n \approx n\omega$, with red-shifted Compton edges due to recoil.

The energy of the emitted photon in the nth harmonic under nonlinear Compton scattering is given approximately by:

$$\hbar\omega_{n'} = \frac{n\hbar\omega}{1 + \frac{n\hbar\omega}{\gamma m_e c^2}(1 - \cos\theta) + a_0^2/2},$$

where the denominator incorporates both recoil and ponderomotive effects. This shift is termed the nonlinear Compton shift, and it modifies the location of spectral peaks from the predictions of linear theory.

Spectral Broadening in Quantum Plasma Environments

In quantum plasmas, the radiation spectra of electrons interacting with collective or external fields are inherently broadened due to a combination of quantum recoil, stochastic photon emission, and fluctuating plasma backgrounds. The interaction of a relativistic electron with a spatially and temporally varying electromagnetic field such as in laser-plasma accelerators leads to a broadened emission line, characterized by both Doppler and recoil-induced widening.

Mathematically, the spectral distribution of emitted radiation is described via the quantum synchrotron function $\mathcal{F}(\xi)$, defined as:

$$\mathcal{F}(\xi) = \xi \int_{\xi}^{\infty} K_{5/3}(z)dz,$$

where $\xi = \omega/\omega_c$, $K_{5/3}$ is the modified Bessel function of the second kind, and ω_c is the characteristic frequency, modified in quantum plasma settings by field-dependent corrections:

$$\omega_c = \frac{3}{2}\gamma^2 \frac{eB_\perp}{m_e c}\big(1 + \delta_q + \delta_p\big).$$

Here, $\delta_q \sim \chi_e$ accounts for quantum corrections, and $\delta_p \sim \omega_p^2/\omega^2$ encapsulates plasma dispersion effects. In strongly magnetized quantum plasmas, such as those found in magnetars, or laser-driven solid-density plasmas, these contributions are non-negligible, leading to complex spectral profiles.

Moreover, quantum stochasticity in photon emission induces additional broadening, well described by stochastic differential equations for the radiation reaction force in the quantum regime:

$$\frac{dp^\mu}{d\tau} = -\frac{2}{3}\frac{e^2}{m_e^2 c^3}\big(F^{\mu\nu}F_{\nu\lambda}p^\lambda + \ldots\big) + \text{stochastic noise}.$$

Incorporating these quantum Langevin-type corrections is essential for accurate modeling of radiation spectra, particularly in experiments at facilities such as ELI or XFELs.

The rich phenomenology of high-energy photon emission in quantum plasma environments manifested through quantum recoil, nonlinear Compton shifts, and spectral broadening reflects the interplay between strong-field QED and collective plasma effects. These features are central to next-generation high-intensity laser experiments and astrophysical diagnostics, such as inverse Compton x-ray sources and gamma-ray bursts. Future extensions must integrate real-time quantum kinetic descriptions, incorporating Wigner–Moyal or Dirac-Heisenberg-Wigner formalisms, to model these phenomena in fully non-equilibrium and dynamically evolving quantum plasmas.

10.7 Application in Compact Gamma-Ray Sources and Diagnostics

The advent of high-intensity laser-plasma interactions and advancements in accelerator technologies have ushered in a new era of compact gamma-ray sources. These sources, rooted in the principles of quantum plasma physics, offer unprecedented capabilities for generating and diagnosing high-energy photons.

At the heart of compact gamma-ray sources lies the interaction between relativistic electrons and intense electromagnetic fields. One primary mechanism is the nonlinear inverse Compton scattering (NICS), where an electron absorbs multiple photons from a laser field and emits a single high-energy photon. The energy of the emitted photon $\hbar\omega'$ in the head-on collision geometry can be approximated by:

$$\hbar\omega' \approx \frac{4\gamma^2\hbar\omega}{1 + a_0^2 + \frac{4\gamma\hbar\omega}{m_e c^2}}$$

where:

γ is the Lorentz factor of the electron,
ω is the frequency of the incident laser photon,
$a_0 = \frac{eE}{m_e c\omega}$ is the normalized vector potential,
E is the electric field amplitude of the laser,
m_e is the electron mass,
c is the speed of light.

The term a_0^2 accounts for the nonlinear effects due to the intense laser field. As a_0 increases, the emission spectrum broadens, and higher harmonics become significant, leading to the generation of gamma-rays with energies extending into the MeV range.

Laser-plasma accelerators (LPAs) have emerged as compact alternatives to conventional accelerators, capable of producing GeV-scale electron beams over centimeter-scale distances. In LPAs, an intense laser pulse propagates through an underdense plasma, driving a wakefield that accelerates electrons to relativistic energies. When these electrons interact with a counter-propagating laser pulse or a plasma mirror, they can emit gamma-rays via NICS [32].

The brilliance $\mathcal{B}$ of the emitted gamma-ray beam, a critical parameter for applications, is given by:

$$\mathcal{B} = \frac{N_\gamma}{\sigma_x \sigma_y \sigma_\theta^2 \Delta\omega/\omega},$$

where:

N_γ is the number of photons,
σ_x and σ_y are the transverse beam sizes,
σ_θ is the divergence angle,
$\Delta\omega/\omega$ is the relative bandwidth.

High-brilliance gamma-ray beams are essential for applications requiring high spatial and temporal resolution, such as nuclear resonance fluorescence and medical imaging.

Compact gamma-ray sources have emerged as powerful diagnostic tools across various high-energy physics applications. One key technique is Nuclear Resonance Fluorescence (NRF), which exploits the resonant absorption and re-emission of gamma-rays by atomic nuclei, enabling precise isotope-specific identification. The effectiveness of NRF relies heavily on the narrow linewidth and tunability of gamma-ray sources, which are essential for resolving closely spaced nuclear energy levels. In the realm of High-Energy Density Physics (HEDP), gamma-ray diagnostics offer valuable insights into plasma properties such as temperature and density by analyzing spectral signatures from nuclear reactions and bremsstrahlung radiation. These sources also play a critical role in material characterization, where gamma-ray radiography provides non-destructive, high-resolution imaging of dense materials due to their deep penetration capabilities. In medical applications, particularly in oncology, gamma-ray beams are utilized in radiotherapy to target tumors with high precision, underscoring the need for finely controlled beam parameters to minimize collateral damage to healthy tissues.

Designing compact gamma-ray sources requires careful optimization of several interrelated factors. Laser parameters, such as intensity, pulse duration, and spot size directly impact the acceleration gradient and the energy of the resulting electron beams. Plasma density plays a crucial role in shaping the wakefield structure and defining the dephasing length, which limits the maximum electron energy achievable. Additionally, producing high-quality electron beams with low emittance and narrow energy spread is essential for generating collimated, monochromatic gamma-ray outputs. Synchronization between the electron beam and the scattering laser pulse must be precisely maintained to ensure effective overlap and interaction. To support these design efforts, advanced simulation tools like particle-in-cell (PIC) codes are widely used to model laser-plasma interactions and optimize experimental configurations [33, 34].

Looking ahead, compact gamma-ray sources hold great potential, though several challenges remain to be addressed. Ensuring beam stability on a shot-to-shot basis is vital for consistent and reproducible diagnostic performance. Scaling these systems to reach higher gamma-ray energies while maintaining a compact footprint demands innovations in laser technology and advanced plasma channeling methods. Further, integrating these sources with robust detection systems capable of handling high photon fluxes and energies is essential for realizing their full diagnostic capabilities. Continued interdisciplinary research in laser physics, plasma science, and accelerator technology is crucial for overcoming these challenges and advancing the development of next-generation compact gamma-ray systems.

References

1. Jana, C., Loganayagam, R., & Rangamani, M. (2020). Open quantum systems and Schwinger-Keldysh holograms. *Journal of High Energy Physics, 2020*(7), 1–58.
2. Lumb, S., Lumb, S., Munjal, D., & Prasad, V. (2015). Intense field induced excitation and ionization of an atom confined in a dense quantum plasma. *Physica Scripta, 90*(9), Article 095603.
3. Fedeli, L., Zaïm, N., Sainte-Marie, A., Thévenet, M., Huebl, A., Myers, A., Vay, J.L., & Vincenti, H. (2022). PICSAR-QED: a Monte Carlo module to simulate strong-field quantum electrodynamics in particle-in-cell codes for exascale architectures. *New Journal of Physics, 24*(2), 025009.
4. Moritaka, T., Baiotti, L., Lin, A., Weiwu, L., Sakawa, Y., Kuramitsu, Y., Morita, T., & Takabe, H. (2013). Plasma particle-in-cell simulations with QED reactions for pair production experiments using a high-Z solid target. *Journal of Physics: Conference Series, 454*(1):012016. IOP Publishing.
5. Dharma-Wardana, M. W. C., & Perrot, F. (1998). Energy relaxation and the quasiequation of state of a dense two-temperature nonequilibrium plasma. *Physical Review E, 58*(3), 3705.
6. Tajima, T., Yan, X. Q., & Ebisuzaki, T. (2020). Wakefield acceleration. *Reviews of Modern Plasma Physics, 4*(1), 7.
7. Mottola, E., & Szép, Z. (2010). Systematics of high temperature perturbation theory: The two-loop electron self-energy in QED. *Physical Review D—Particles, Fields, Gravitation, and Cosmology, 81*(2), 025014.
8. Shaposhnikov, M. E. (1997). Finite temperature effective theories. *Subnucl. Ser, 34*, 360–397.
9. Al-Naseri, H., & Brodin, G. (2023). Applicability of the Klein-Gordon equation for pair production in vacuum and plasma. *Physical Review E, 108*(5), Article 055205.
10. Manfredi, G., Hervieux, P. A., & Hurst, J. (2019). Phase-space modeling of solid-state plasmas: A journey from classical to quantum. *Reviews of Modern Plasma Physics, 3*(1), 13.
11. Shi, Y., Qin, H., & Fisch, N. J. (2021). Plasma physics in strong-field regimes: Theories and simulations. *Physics of Plasmas, 28*(4).
12. Gavrus, C. D. (2017). *Global well-posedness and parametrices for critical Maxwell-Dirac and massive Maxwell-Klein-Gordon equations with small Sobolev data* (Doctoral dissertation, UC Berkeley).
13. Brodin, G., & Zamanian, J. (2022). Quantum kinetic theory of plasmas. *Reviews of Modern Plasma Physics, 6*(1), 4.
14. Jakovac, J. (2025). *Theoretical simulations of quasi-particle spectra in doped graphene* (Doctoral dissertation, University of Rijeka. Faculty of Physics).
15. Macleod, A. J., Noble, A., & Jaroszynski, D. A. (2019). Cherenkov radiation from the quantum vacuum. *Physical Review Letters, 122*(16), Article 161601.
16. Drake, R. P. (2009). Perspectives on high-energy-density physics. *Physics of Plasmas, 16*(5).
17. Gedalin, M. (2023). Rankine-Hugoniot relations in turbulent shocks. *Frontiers in Physics, 11*, 1325995.
18. Amano, T., Matsumoto, Y., Bohdan, A., Kobzar, O., Matsukiyo, S., Oka, M., Niemiec, J., Pohl, M., & Hoshino, M. (2022). Nonthermal electron acceleration at collisionless quasi-perpendicular shocks. *Reviews of Modern Plasma Physics, 6*(1), 29.
19. Hauksson, S., Jeon, S., & Gale, C. (2022). Momentum broadening of energetic partons in an anisotropic plasma. *Physical Review C, 105*(1), Article 014914.
20. Chandra, V., & Sreekanth, V. (2017). Impact of momentum anisotropy and turbulent chromo-fields on thermal particle production in quark–gluon-plasma medium. *The European Physical Journal C, 77*, 1–12.
21. Gelis, F., & Hauksson, S. (2024). Study of a longitudinally expanding plasma with the 2PI effective action. In: *EPJ web of conferences* (vol. 296, p. 13003). EDP Sciences.
22. Marklund, M., & Shukla, P. K. (2006). Nonlinear collective effects in photon-photon and photon-plasma interactions. *Reviews of Modern Physics, 78*(2), 591–640.

23. Fedele, R., Tanjia, F., De Nicola, S., Jovanović, D., & Shukla, P. K. (2012). Quantum ring solitons and nonlocal effects in plasma wake field excitations. *Physics of Plasmas, 19*(10).
24. Mendonca, J. T., & Serbeto, A. (2011). Volkov solutions for relativistic quantum plasmas. *Physical Review E—Statistical, Nonlinear, and Soft Matter Physics, 83*(2), 026406.
25. Seipt, D., Del Sorbo, D., Ridgers, C. P., & Thomas, A. G. R. (2018). Theory of radiative electron polarization in strong laser fields. *Physical Review A, 98*(2), Article 023417.
26. Lee, M. J., & Jung, Y. D. (2018). Resonant streaming instability in a non-degenerate quantum plasma including the electron quantum recoil effect. *Physics of Plasmas, 25*(9).
27. Ganeev, R. A. (2023). High-order harmonic generation in laser-induced low-density plasma: Past and recent achievements. *Applied Physics B, 129*(1), 17.
28. Zhang, P., Bulanov, S. S., Seipt, D., Arefiev, A. V., & Thomas, A. G. R. (2020). Relativistic plasma physics in supercritical fields. *Physics of Plasmas, 27*(5).
29. Mehlhaff, J., Werner, G., Cerutti, B., Uzdensky, D., & Begelman, M. (2024). Kinetic simulations and gamma-ray signatures of Klein-Nishina relativistic magnetic reconnection. *Monthly Notices of the Royal Astronomical Society, 527*(4), 11587–11626.
30. Galbiati, M., Formenti, A., Grech, M., & Passoni, M. (2023). Numerical investigation of non-linear inverse Compton scattering in double-layer targets. *Frontiers in Physics, 11*, 1117543.
31. Sampath, A. (2020). *Strong-field QED and collisional effects in electron beam-plasma interaction* (Doctoral dissertation, Ruprecht-Karls-Universität Heidelberg).
32. Arber, T. D., Bennett, K., Brady, C. S., Lawrence-Douglas, A., Ramsay, M. G., Sircombe, N. J., Gillies, P., Evans, R. G., Schmitz, H., Bell, A. R., & Ridgers, C. P. (2015). Contemporary particle-in-cell approach to laser-plasma modelling. *Plasma Physics and Controlled Fusion, 57*(11), 113001.
33. Haugbølle, T., Frederiksen, J. T., & Nordlund, Å. (2013). Photon-plasma: A modern high-order particle-in-cell code. *Physics of Plasmas, 20*(6).
34. Wu, D., Yu, W., Fritzsche, S., & He, X. T. (2020). Particle-in-cell simulation method for macroscopic degenerate plasmas. *Physical Review E, 102*(3), Article 033312.

Chapter 11
Experimental Techniques and Simulations

While theoretical developments have matured substantially across various subfields of quantum plasma research, ranging from compact astrophysical objects to ultracold laboratory setups, the empirical grounding and numerical benchmarking remain sparse and fragmented. This chapter serves to unify disparate efforts across distinct domains where quantum plasma manifests in complementary forms. In laboratory settings, the drive to realize quantum plasmas is propelled by advances in ultra-intense laser systems, optical and magnetic trapping technologies, and nanostructured materials. Systems such as ultracold neutral plasmas, dense laser-compressed targets, and quantum wells in semiconductors offer platforms wherein Fermi statistics, strong coupling, and collective excitations intersect in measurable ways. In these environments, nonlinear collective modes, Landau quantization, and tunneling-induced instabilities become accessible to diagnostic scrutiny and simulation validation.

11.1 Laboratory Platforms

The experimental realization of quantum plasma regimes necessitates precision-controlled platforms wherein quantum statistical effects, wavefunction coherence, and particle indistinguishability dominate collective dynamics.

11.1.1 Ultracold Plasma Experiments

Ultracold plasmas are generated by photoionizing laser-cooled neutral atoms or Rydberg gases, resulting in partially ionized plasmas with electron temperatures in the range $T_e \sim 1$ K and ion temperatures near $T_i \sim 1$ mK. The initial neutral

P. Kumar, *Quantum Plasma*, https://doi.org/10.1007/978-981-95-0481-7_11

atomic gas is typically trapped and cooled via magneto-optical traps, and a two-photon ionization scheme is employed to liberate electrons with controlled excess kinetic energy. The ensuing plasma is quasi-neutral and spatially confined, and its lifetime is limited by expansion and recombination.

From a physical standpoint, the de Broglie wavelength of electrons,

$$\lambda_{de} = \frac{h}{\sqrt{2\pi m_e k_B T_e}},$$

can exceed the interparticle spacing for $T_e \lesssim 1\text{K}$, indicating the relevance of quantum diffraction. Further, the Coulomb coupling parameter for ions,

$$\Gamma_i = \frac{Z^2 e^2}{4\pi \epsilon_0 a k_B T_i},$$

where a is the Wigner–Seitz radius, can exceed unity $\left(\Gamma_i \gtrsim 1\right)$ in these conditions, leading to strongly coupled ion dynamics, potentially approaching Coulomb crystallization. Electron–electron interactions, however, remain in the weakly coupled domain due to their higher kinetic energy. The platform is uniquely suited for studying wave-particle duality, disorder-induced heating, ambipolar expansion, and non-equilibrium relaxation in quantum kinetic regimes.

A salient feature is the possibility of observing quantum many-body phenomena on classical plasma length scales, thus bridging cold atom physics and plasma collective behavior. The theoretical modeling often incorporates quantum Vlasov and Wigner–Moyal equations, with particular emphasis on quantum diffusion and Landau damping in a partially degenerate regime.

11.1.2 Laser-Produced Dense Plasmas

High-intensity laser systems, including femtosecond pulse lasers and extreme ultraviolet (XUV) sources, can create dense quantum plasmas via irradiation of solid, or near-solid density targets. At intensities exceeding $10^{15}\,\text{W/cm}^2$, the target surface is rapidly ionized, forming a dense electron–hole plasma with electron temperatures ranging from several eV to keV and densities on the order of $n_e \sim 10^{21} - 10^{24}\,\text{cm}^{-3}$. The characteristic Fermi temperature $T_F = (\hbar^2/2m_e)(3\pi^2 n_e)^{2/3}/k_B$ may be comparable to, or exceed T_e, leading to degeneracy effects and Pauli blocking of collision processes.

The quantum electrodynamics of this regime is further enriched by the rapid nonthermal electron heating and the formation of transient population inversions. These features necessitate a quantum statistical treatment of distribution functions, employing non-equilibrium Green's functions, or quantum hydrodynamic models that include quantum pressure and exchange–correlation effects. The laser-induced plasma evolves through ultrafast nonequilibrium phases, with the possibility of observing time-resolved phenomena such as transient Fermi surface formation, high-harmonic generation, and quantum electrodynamic nonlinearity [1].

Further, laser-solid interactions often lead to the formation of steep density gradients and ultrathin plasmas wherein tunneling and field emission become prominent, making this an ideal testbed for verifying relativistic quantum kinetic models and field-theoretic descriptions involving semiclassical Boltzmann-Vlasov frameworks with quantum corrections.

11.1.3 High-Pressure Shock Tubes and Diamond Anvil Cells

Quantum plasmas at high densities and moderate temperatures collectively referred to as warm dense matter (WDM) can be accessed using dynamic compression techniques such as shock tubes and static methods like diamond anvil cells (DACs). In these experiments, material densities can exceed solid density by a factor of $2 - 10$, and temperatures can lie in the range $1\ \text{eV} < T < 100\ \text{eV}$. The defining feature is the simultaneous significance of degeneracy ($T \sim T_F$) and strong coupling ($\Gamma \sim 1$), placing the system outside the validity of classical, or ideal quantum models [2].

In this regime, electrons are partially degenerate, and ion structure is partially liquid-like, with quantum diffraction and exchange–correlation contributing non-negligibly to the equation of state. The experimental observable includes X-ray absorption spectra, which are interpreted via models incorporating quantum statistical potentials and density functional theory (DFT)-informed opacities.

Physically, such platforms allow exploration of collective modes in partially degenerate environments and testing of sum rules for quantum dielectric functions under high-pressure conditions. They provide insight into the role of quantum kinetic effects in screening, compressibility, and conductivity. Theoretical interpretations often utilize quantum Langevin dynamics, orbital-free DFT, and path-integral Monte Carlo simulations that resolve both the thermodynamic and transport properties of WDM.

11.1.4 Inertial Confinement Fusion (ICF) Experiments

Inertial confinement fusion schemes, such as those at the National Ignition Facility, provide an avenue for the creation of extremely high-density, high-temperature plasmas over ultrashort timescales. The imploding fuel core, typically deuterium–tritium (D-T), experiences a compression that brings the plasma density to $n \sim 10^{26}\text{cm}^{-3}$ and the temperature to $T \sim 10^3\text{eV}$, placing it well within the quantum plasma domain.

The inner core exhibits electron degeneracy ($T/T_F << 1$) and significant Fermi pressure, which modifies both the hydrodynamic stability and the energy transport properties. Electrons are no longer treated as a classical Maxwellian bath, but require quantum statistical formulations that include the Fermi–Dirac distribution and temperature-dependent screening length:

$$\lambda_{TF} = \left(\frac{\pi}{4}\right)^{1/2} \frac{\hbar v_F}{e\sqrt{n_e}},$$

where v_F is the Fermi velocity. The screening plays a crucial role in modifying fusion cross-sections, energy deposition from alpha particles, and thermal conductivity in the dense core.

Moreover, the multiscale nature of ICF from laser-plasma interaction at the ablation front to quantum-degenerate physics in the core necessitates hybrid simulation approaches combining classical radiation hydrodynamics with quantum kinetic modules. This platform offers a unique opportunity to test theoretical models of degenerate radiation transport, nonlinear dielectric response, and the stability of quantum-shaped implosions under the influence of hydrodynamic instabilities.

11.1.5 Quantum Plasmas in Condensed Systems

Beyond high-temperature ionized media, quantum plasma-like phenomena appear in condensed matter systems characterized by collective electronic excitations and quantum statistical correlations. Plasmons in graphene, for example, manifest as collective oscillations of Dirac electrons governed by relativistic quantum kinetics. The dispersion relation of two-dimensional plasmons,

$$\omega_p(k) \sim \sqrt{\frac{2\pi e^2 n_s k}{\epsilon m^*}},$$

exhibits scale-dependent behaviour, with quantum capacitance and tunneling effects modifying the spectrum at sub-micron wavelengths. The interplay between quantum coherence, spin–orbit coupling, and many-body interactions renders these systems ideal for exploring nonlocal and anisotropic dielectric responses.

Similarly, two-dimensional electron gases (2DEGs) in semiconductor heterostructures provide a low-dimensional realization of quantum plasma dynamics with tunable density and degeneracy. The experimental probes, including inelastic light scattering, angle-resolved photoemission spectroscopy (ARPES), and terahertz reflectometry, enable direct measurement of plasmon lifetimes, dispersion, and damping mechanisms. These systems allow direct observation of quantum corrections to the dielectric function, including exchange–correlation potentials and dynamic local-field effects [3, 4].

Superconducting qubits and circuit quantum electrodynamics platforms also manifest collective electronic dynamics under extreme coherence. While not plasmas in the traditional sense, the collective charge and phase degrees of freedom in these systems admit an effective quantum hydrodynamic description with analogs of plasmonic oscillations. These platforms offer unmatched control over decoherence, making them ideal quantum simulators for exploring nonperturbative regimes of quantum kinetic theory.

11.2 Diagnostics and Measurement Tools

The diagnostic methodologies employed in quantum plasma research must transcend classical paradigms, incorporating tools and techniques sensitive not only to thermal and collective phenomena, but also to quantum statistical effects, wavefunction coherence, Fermi degeneracy, and nonlocal field fluctuations. In quantum plasmas where the de Broglie wavelength of electrons becomes comparable to, or exceeds the interparticle spacing, and where Pauli exclusion, quantum tunneling, and exchange–correlation interactions are non-negligible diagnostic instruments must probe both microscopic and macroscopic observables with commensurate spatial, temporal, and spectral resolution.

11.2.1 Thomson Scattering: Probing Electron Density, Temperature, and Quantum Effects

Thomson scattering in a plasma corresponds to the elastic scattering of electromagnetic radiation by free electrons. In quantum plasmas, the scattering cross-section becomes sensitive to quantum statistical properties through modifications in the dynamic structure factor $S(\mathbf{k}, \omega)$, which governs the scattered intensity:

$$\frac{d^2\sigma}{d\Omega d\omega} = r_e^2\left(\frac{\omega_s}{\omega_0}\right)S(\mathbf{k}, \omega),$$

where r_e is the classical electron radius, ω_0 and ω_s are the incident and scattered frequencies respectively, and $\mathbf{k} = \mathbf{k}_s - \mathbf{k}_0$ is the momentum transfer vector. In the quantum regime, $S(\mathbf{k}, \omega)$ incorporates the quantum dielectric function $\epsilon(\mathbf{k}, \omega)$, typically modeled using the Lindhard, or Mermin formulations:

$$S(\mathbf{k}, \omega) = \frac{-\hbar}{\pi}\frac{1}{1 - e^{-\hbar\omega/k_B T}}\text{Im}\left[\frac{-1}{\epsilon(\mathbf{k}, \omega)}\right].$$

Thomson scattering thus serves as a non-invasive probe of electron density via the position of the plasma resonance peak, and temperature via the spectral broadening. Crucially, quantum degeneracy manifests as asymmetric broadening of the spectral line shape due to Pauli blocking and Fermi velocity constraints, while quantum diffraction effects are encoded in the modification of the electron response function. Low-k behaviour allows access to collective modes, while high-k scattering captures individual particle dynamics, enabling simultaneous diagnosis of classical and quantum behaviours [5].

Modern laser Thomson scattering setups achieve temporal resolutions on the order of femtoseconds and spatial resolutions below microns, making them apt for ultracold plasmas, solid-density laser-produced plasmas, and warm dense matter. The comparison between experimentally measured spectra and quantum statistical models enables direct inference of the degeneracy parameter $\chi = T/T_F$, confirming the quantum nature of the plasma.

11.2.2 X-ray Absorption and Emission Spectroscopy: Detecting Ionization Potential Depression and Quantum Screening

X-ray diagnostics provide access to core-level transitions in partially ionized, high-density plasmas where quantum effects are profound. The X-ray absorption coefficient $\mu(\omega)$ is directly tied to the frequency-dependent opacity and is influenced by bound–free and free–free transitions, which are in turn modified by quantum screening and degeneracy. The absorption edge is shifted and broadened due to ionization potential depression (IPD), which, under quantum conditions, must incorporate non-classical screening lengths and exchange–correlation corrections [6].

A representative model is the Stewart-Pyatt or Ecker-Kröll approach [7], which gives the IPD as:

$$\Delta E_{\text{IPD}} \approx \frac{Ze^2}{4\pi\epsilon_0 r_D}, \quad r_D^{-2} = \sum_s \frac{n_s q_s^2}{\epsilon_0 k_B T_s},$$

modified in the quantum regime by replacing the Debye length with the Thomas–Fermi screening length:

$$\lambda_{\text{TF}} = \left(\frac{\pi}{4}\right)^{1/2} \frac{\hbar v_F}{e\sqrt{n_e}}.$$

In a quantum plasma, the ionization energy is not a fixed quantity, but is dynamically altered by the dense electronic environment. Fermi pressure, degeneracy-induced screening, and quantum diffraction reduce the effective potential well depth, thereby lowering the ionization threshold. This is experimentally visible as a redshift in K-edge and L-edge absorption features, and the broadening correlates with collisional and many-body quantum effects. The line ratios in X-ray emission (e.g., Ly-α to He-α transitions) encode information on electron temperature and ionization equilibrium, enabling non-perturbative temperature and density diagnostics [8].

Free-electron laser (FEL) sources and X-ray free-electron lasers (XFELs) provide temporally coherent, high-brilliance sources that can interrogate dense plasmas on attosecond timescales. These techniques are indispensable in warm dense matter studies and inertially confined plasmas where direct optical probing is infeasible.

11.2.3 Interferometry and Reflectometry: Reconstructing Quantum-Scale Density Profiles

Interferometry determines electron density profiles by measuring phase shifts of a probe laser beam traversing the plasma. The phase shift $\Delta\phi$ is related to the electron density $n_e(x)$ via:

$$\Delta\phi = \frac{e^2}{8\pi^2\epsilon_0 m_e c\omega} \int n_e(x)\,dx.$$

In quantum plasmas, the refractive index becomes a function of the quantum-modified dielectric function $\varepsilon(\omega)$, which deviates from its classical form due to collective Fermi motion, quantum polarization, and dispersive corrections:

$$n(\omega) = \sqrt{\epsilon(\omega)} \quad \text{with} \quad \epsilon(\omega) = 1 - \frac{\omega_p^2}{\omega^2 + i\nu\omega - \Pi_q(\omega)},$$

where $\Pi_q(\omega)$ encodes quantum polarization corrections.

Reflectometry, in turn, measures the amplitude and phase of a reflected wave and provides spatially resolved profiles near the critical surface (where $\omega = \omega_p$), enabling determination of density scale lengths and density fluctuations with sub-skin-depth resolution.

Quantum diffraction leads to spatial smearing of density gradients, and this manifests in interferometry as non-monotonic or broadened phase profiles. Moreover, nonlocal conductivity and quantum tunneling give rise to partial transmission through classically opaque regions, an effect detectable through modified reflectivity spectra. These observations signal breakdowns of the local approximation in the dielectric response, demanding full quantum treatments such as the Wigner–Poisson formalism or hydrodynamic models with Bohm potential terms.

High-resolution Mach–Zehnder, or Nomarski interferometers are integrated into optical access systems of laser plasma chambers and magneto-optical traps. In dense plasma contexts, femtosecond time-resolved reflectometry using ultrashort pulses maps density fronts and quantum pre-plasma layers preceding main plasma formation [9].

11.2.4 Collective Oscillation Probes: Langmuir Frequency Shifts and Quantum Dispersion Relations

The collective dynamics of quantum plasmas are encoded in their dielectric response function and collective excitation spectra. The dispersion relation for longitudinal modes, in the linearized quantum hydrodynamic (QHD) approximation, reads:

$$\omega^2 = \omega_p^2 + \frac{3}{5} v_F^2 k^2 + \frac{\hbar^2 k^4}{4 m_e^2},$$

where $\omega_p = \sqrt{n_e e^2/\epsilon_0 m_e}$ is the classical plasma frequency, the second term accounts for Fermi pressure, and the third term originates from the Bohm quantum potential reflecting quantum diffraction.

In high-density regimes, the Langmuir oscillation peak is shifted upward from the classical ω_p value, and the k-dependence becomes non-negligible due to the quantum pressure and nonlocal effects. Measurement of this frequency shift and its k-dispersion thus directly reflects quantum corrections.

Langmuir probes and microwave backscattering can detect this frequency shift. Quantum degeneracy raises the effective pressure, thereby stiffening the plasma and increasing the restoring force for electron oscillations. Simultaneously, wavevector-dependent dispersion reveals nonlocal response and finite-size corrections in bounded quantum systems. This is critical in systems such as nanoscale plasmas or ultrathin films where boundary conditions and quantum confinement alter the collective mode spectrum.

Ultrafast pump–probe techniques, in which a strong pump pulse excites plasma oscillations and a time-delayed probe measures the response, are used to observe temporal oscillations at quantum-modified Langmuir frequencies. Time-resolved electron energy loss spectroscopy (EELS) and terahertz reflectometry also allow measurement of dispersion relations in solid-state quantum plasmas and 2D materials [10].

11.3 Simulation Techniques for Quantum Plasma

The study of quantum plasma systems requires simulation frameworks that incorporate essential quantum mechanical effects while respecting the multi-scale and often nonlinear nature of plasma dynamics. Unlike classical plasma models, which rely on Vlasov–Maxwell, or fluid equations devoid of quantum corrections, quantum plasma simulations demand the inclusion of phenomena such as wavefunction interference, tunneling, Fermi–Dirac statistics, and spin dynamics. Three major theoretical frameworks dominate the simulation landscape: Quantum Hydrodynamics (QHD), Quantum Kinetic Theory (QKT), and Multi-Fluid Models.

11.3.1 Particle-In-Cell (PIC) with Quantum Corrections

The Particle-in-Cell (PIC) method has long served as a foundational tool for simulating classical and weakly coupled plasmas. In its standard formulation, macroparticles representing charged species evolve under electromagnetic fields that are updated on a spatial grid via Maxwell's equations. However, in quantum plasma regimes particularly in high-density, ultra-short scale systems where quantum coherence and relativistic inertia become relevant PIC methodologies must be appropriately extended [11].

A quantum-corrected PIC (QPIC) scheme introduces quantum force terms and relativistic corrections into the Lorentz force equations governing particle trajectories. The modified equations of motion for a particle with species index α read:

$$\frac{d\mathbf{p}_\alpha}{dt} = q_\alpha(\mathbf{E} + \mathbf{v}_\alpha \times \mathbf{B}) + \mathbf{F}^{(\alpha)}_{\text{quantum}},$$

$$\frac{d\mathbf{r}_\alpha}{dt} = \mathbf{v}_\alpha = \frac{\mathbf{p}_\alpha}{\gamma_\alpha m_\alpha}, \quad \gamma_\alpha = \left(1 + \frac{|\mathbf{p}_\alpha|^2}{m_\alpha^2 c^2}\right)^{1/2}.$$

The quantum correction $\mathbf{F}^{(\alpha)}_{\text{quantum}}$ typically includes:

Bohm Force: Approximated for PIC via smoothed density fields:

$$\mathbf{F}_{\text{Bohm}} = \frac{\hbar^2}{2m_\alpha}\nabla\left(\frac{\nabla^2\sqrt{n_\alpha}}{\sqrt{n_\alpha}}\right).$$

Spin Force: For spin-$\frac{1}{2}$ particles in magnetized plasmas:

$$\mathbf{F}_{\text{spin}} = \nabla\left(\boldsymbol{\mu}_\alpha \cdot \mathbf{B}\right), \quad \boldsymbol{\mu}_\alpha = -g\mu_B\mathbf{S}_\alpha.$$

In practice, implementing the Bohm force requires careful treatment of numerical noise due to finite particle statistics. Smoothing procedures and finite-difference approximations to the quantum potential are often used on the mesh-resolved densities. The time-step is constrained not only by the Courant condition, but also by quantum phase evolution scales particularly in high-density or strongly confined systems.

Relativistic quantum PIC codes have become indispensable in simulating laser-plasma interactions near solid densities and in describing quantum wakefield acceleration, where both quantum recoil and radiation reaction must be included.

This framework proves highly effective for a range of advanced plasma phenomena. It is particularly well-suited for modeling laser-driven quantum plasma jets, where intense laser fields interact with dense quantum media to produce high-velocity plasma flows. Additionally, it supports the study of ultra-short laser pulse propagation in dense plasma targets, capturing the intricate dynamics influenced by quantum effects. Further, the framework is ideal for investigating field-induced tunneling and quantum reflection, enabling detailed analysis of particle behaviourunder extreme electromagnetic fields in quantum regimes.

Such simulations capture the interplay between nonlinear plasma oscillations, relativistic beam dynamics, and quantum coherence, thereby providing insights into laboratory realizations of pair plasmas, quantum beam–plasma instabilities, and coherent radiation emission [12].

11.3.2 Monte Carlo Simulations in Quantum Plasma

The Monte Carlo (MC) methodology occupies a distinguished place in computational quantum plasma physics, owing to its capacity to simulate the stochastic behavior of many-body systems with intrinsic quantum and statistical fluctuations. While deterministic approaches such as the spectral, or grid-based solutions of kinetic, or fluid equations provide valuable insights into structured, large-scale dynamics, they often fall short when addressing the microscopic and thermodynamic complexities inherent to strongly correlated quantum plasmas. Monte Carlo methods, by contrast, allow for a direct statistical sampling of phase-space configurations and field-theoretic ensembles, providing a foundational route for addressing equilibrium and non-equilibrium properties beyond the reach of mean-field approximations.

From a statistical physics standpoint, a quantum plasma may be viewed as an ensemble of charged particles typically electrons, positrons, and ions interacting via quantum-modified electromagnetic fields. At equilibrium, this ensemble is described by the grand canonical partition function incorporating fermionic and bosonic degrees of freedom, often regulated by Fermi–Dirac and Bose–Einstein statistics, respectively. In this context, the Monte Carlo approach operates by evaluating multidimensional integrals over many-particle configurations to approximate ensemble averages of observables such as internal energy, pressure, transport coefficients, and correlation functions. These integrals, generally intractable by analytical means for large systems, become numerically accessible via stochastic importance sampling [13].

Path Integral Monte Carlo

One of the most powerful formulations applicable to quantum plasma is the Path Integral Monte Carlo (PIMC) method. In this framework, the quantum statistical density matrix is re-expressed as a functional integral over imaginary-time paths, enabling a full accounting of quantum diffraction, exchange effects, and zero-point motion. The discretized imaginary-time axis allows the quantum system to be mapped onto an effective classical polymer-like system, where each quantum particle is represented by a chain of beads connected via thermal and quantum links. This mapping facilitates the simulation of systems with significant quantum degeneracy, including dense hydrogen, electron–hole plasmas, and quark–gluon analogs under astrophysical conditions [14].

Physically, PIMC captures the interplay between Coulomb interactions and quantum uncertainty by enabling the visualization of spatial and temporal path entanglements. This is critical in regimes where the thermal de Broglie wavelength becomes comparable to the interparticle distance, such as in warm dense matter, or laser-compressed plasmas. It is precisely in such conditions that classical molecular dynamics or mean-field treatments lose validity, while PIMC offers a first-principles route to understanding electron screening, bound state formation, and collective excitations under quantum constraints [15].

Fermionic Sign Problem

A major limitation in applying Monte Carlo methods to fermionic quantum plasmas particularly at low temperatures is the so-called "fermionic sign problem." This problem arises due to the antisymmetrization requirement on the fermionic wavefunction, leading to rapidly oscillating integrands and exponentially vanishing signal-to-noise ratios. Physically, this reflects the challenge of capturing quantum exchange interactions and Pauli exclusion effects within a stochastic framework. To mitigate this, various strategies have been adopted, including fixed-node approximations, restricted path integrals, and trial wavefunction approaches that introduce variational principles to guide the sampling procedure toward physically relevant configurations.

These variational Monte Carlo (VMC) schemes incorporate trial wavefunctions parameterized by Slater determinants, Jastrow factors, or more advanced machine-learned forms, allowing the quantum plasma to be probed across a variety of strongly coupled regimes. Variational optimization of energy, or entropy functionals then yields insight into ground and excited states, metastable phases, and entropic barriers relevant to thermalization and phase transitions. In particular, for magnetized quantum plasmas, the inclusion of vector potential couplings and spin-field correlations within these variational schemes allows the simulation of spin-polarized currents, Zeeman-shifted excitations, and vortex-antivortex lattice formation [16].

Kinetic Monte Carlo

While traditional MC methods are well-suited for equilibrium thermodynamic studies, non-equilibrium phenomena in quantum plasma such as relaxation, diffusion, and current-driven instabilities necessitate the use of Kinetic Monte Carlo (KMC) approaches. KMC algorithms simulate the temporal evolution of systems by modeling discrete transitions between quantum states, driven by scattering processes, tunneling events, or external field perturbations. The stochastic rate constants for these events are typically derived from Fermi's Golden Rule, or non-equilibrium

Green's function formalism, making this approach well-suited for investigating electron transport across quantum wells, barrier tunneling in high-field regimes, and energy redistribution in plasma after rapid perturbations.

Physically, KMC captures the probabilistic unfolding of quantum dynamics in situations where coherence is partially suppressed, or where the timescale of interest is much larger than the decoherence time. In magnetized or spin–orbit coupled quantum plasmas, this method has been adapted to model spin-flip scattering, Landau level transitions, and gyroresonant heating mechanisms, allowing for an in-depth statistical understanding of dissipation, confinement breakdown, and stochastic heating [17].

Modern Advancements and Machine-Learning Enhanced Monte Carlo

Recent advances in computational physics have led to the emergence of machine learning-enhanced Monte Carlo algorithms, which harness neural networks or Gaussian processes to accelerate convergence and reduce computational overhead. These developments have been instrumental in extending the applicability of MC simulations to larger quantum plasmas, higher-dimensional configuration spaces, and more complex interaction kernels, including those derived from non-Abelian gauge theories.

Such hybrid approaches allow the plasma physicist to not only simulate, but also infer hidden structures, such as effective quantum potentials, emergent symmetries, or critical points associated with quantum phase transitions. For instance, unsupervised clustering techniques can reveal topological structures in the sampled configuration space, indicating the presence of solitonic excitations, vortex sheets, or domain boundaries in spin-polarized plasmas [18].

Monte Carlo simulations thus provide a foundational pillar in the computational exploration of quantum plasmas, uniquely capable of resolving microscopic quantum fluctuations, collective behaviours, and thermal-statistical observables in highly nonlinear regimes. They reveal how quantum coherence competes with Coulomb disorder, how entanglement propagates in ionized media, and how temperature, magnetic fields, and spin interactions conspire to shape the plasma's macroscopic properties. As experimental access to ultracold plasmas, warm dense matter, and magneto-optical traps continues to improve, Monte Carlo methods remain indispensable for interpreting data, validating theoretical predictions, and designing novel quantum devices and controlled plasma configurations.

11.3.3 Molecular Dynamics Simulations in Quantum Plasma

Molecular dynamics (MD) simulations provide a foundational computational approach to study the microscopic, time-dependent behavior of particle ensembles under mutual interactions and external fields. In the domain of quantum plasma characterized by particle degeneracy, quantum coherence, and collective excitations MD simulations offer critical insights into regimes that defy analytical treatment and are often beyond the reach of direct experimentation. Unlike field-based formulations that describe plasma through continuum variables, MD simulations resolve the individual trajectories of particles, capturing their mutual interactions with remarkable spatiotemporal resolution. When equipped with appropriate quantum corrections, MD methods reveal non-classical kinetic, structural, and thermodynamic phenomena inherent to quantum plasma systems.

The classical variant of MD begins by considering charged particles, such as electrons and ions, as discrete entities evolving in time under deterministic laws of motion. The inter-particle forces are typically derived from electrostatic Coulomb interactions, possibly modified by magnetic field contributions or boundary effects. In this framework, the particle ensemble is evolved step-by-step, with each configuration determined by the instantaneous positions and velocities of all constituents. While this approach has proven highly effective in capturing the dynamics of high-temperature, weakly coupled plasmas, it becomes physically inadequate when applied to quantum plasma environments where quantum statistical effects become prominent.

Quantum plasma environments exhibit conditions where the thermal de Broglie wavelength of particles is comparable to, or exceeds, the inter-particle separation. This regime entails several important consequences: the indistinguishability of particles (particularly fermions), the delocalization of wavefunctions, and the emergence of quantum pressure and exchange–correlation forces. These phenomena are fundamentally incompatible with the assumptions of classical mechanics and necessitate modifications to the MD formalism. Consequently, classical MD must be systematically extended to account for quantum effects, either by embedding them into effective interaction potentials or by coupling the MD engine to quantum field solvers that capture nonlocal coherence [19].

Effective Quantum Corrections and Semi-Classical Strategies

One of the most widely adopted strategies for incorporating quantum effects into MD is the use of quantum-statistical effective potentials. These potentials are constructed to reproduce quantum mechanical pair correlations without explicitly solving the many-body Schrödinger equation. They regulate short-range divergences that arise in classical Coulomb interactions by including repulsive core terms that mimic the Pauli exclusion principle and quantum diffraction. This effectively models the spread of the wavefunction and the suppression of close encounters between identical fermions. Physically, such corrections lead to more accurate representations of screening effects, thermodynamic stability, and collision dynamics in dense, partially degenerate plasmas.

From a physical viewpoint, the use of quantum effective potentials enables simulations to probe thermodynamic transitions such as Wigner crystallization, degenerate screening, and correlated tunneling. The resulting dynamics reveal the critical role of quantum coherence in regulating the stability of multi-particle configurations, particularly in systems with strong inter-particle coupling. Moreover, the interplay between classical kinetic motion and quantum repulsion introduces a layer of complexity in collision processes, where effective scattering cross-sections and relaxation times deviate markedly from their classical counterparts [20].

Coupling with Electromagnetic Field Evolution

In strongly magnetized or optically active plasma environments, the charged particle motion becomes inextricably linked with dynamically evolving electromagnetic fields. To accurately capture such coupled dynamics, MD simulations are often extended by solving Maxwell's equations in parallel with the particle trajectories. This integration allows the back-reaction of particle motion on the field and vice versa, forming a fully self-consistent plasma model. In quantum plasma contexts, this coupling is further refined to include higher-order field derivatives, dispersive corrections, and even quantized field modes, depending on the theoretical model employed.

The inclusion of self-consistent fields introduces rich physical behaviour such as wave–particle resonance, field-driven instabilities, and the formation of transient structures like wakes, vortices, and double layers. In quantum plasmas, such structures may exhibit coherence lengths dictated by quantum scales, leading to phenomena like modulated shielding or oscillatory potentials not captured by classical theories. MD simulations thus become crucial in studying the nonlinear evolution of these phenomena, providing a bridge between microscopic dynamics and macroscopic observables such as conductivity, absorption, and refractive index [21].

Hydrodynamic Analogues and Quantum Potentials

Beyond effective potentials, more physically intuitive approaches derive from the analogy between quantum mechanics and fluid dynamics. In these methods, inspired by quantum hydrodynamics, the quantum potential is interpreted as a nonlocal pressure term arising from wavefunction curvature. When incorporated into MD-like simulations, this quantum potential introduces a force that mimics the tendency of particles to resist localization an essential signature of quantum mechanics. This hybrid approach provides a more physically transparent mechanism for understanding quantum transport phenomena, such as delocalized flow, quantum diffusion, and vorticity generation.

Physically, this formulation reveals the emergence of coherence-induced effects such as quantum solitons, nonlinear phase modulations, and stability islands in otherwise chaotic systems. In magnetized quantum plasmas, it elucidates the stabilization of cyclotron orbits by quantum pressure, the suppression of classical resistive dissipation, and the formation of ring-like density profiles around magnetic nulls. These results reflect an intrinsic coupling between field topology, particle statistics, and nonlocal quantum coherence, which cannot be disentangled in simpler treatments [22].

Ab Initio *Approaches and Quantum Consistency*

To address the limitations of semi-classical approximations, ab initio molecular dynamics methods have been developed where the electronic structure is calculated using quantum mechanical principles, typically through density functional theory

(DFT). These simulations compute forces directly from the electronic ground state, which evolves as ions move. While computationally intensive, ab initio MD provides high fidelity in regimes where electronic correlations, partial ionization, and quantum decoherence dominate the behavior of the plasma such as in warm dense matter, ultra-cold plasma clouds, and laser-compressed targets.

Physically, ab initio simulations are capable of capturing phase transitions under quantum degeneracy, non-perturbative heating due to ultrafast fields, and even the formation of transient quantum-bound states during collective plasma oscillations. These simulations also shed light on non-equilibrium thermodynamics and entropy production in plasma systems where traditional statistical mechanics fails to converge due to strong coupling or rapidly varying field configurations [23].

Molecular dynamics simulations, extended to incorporate quantum mechanical corrections, form a vital component of the computational arsenal used to explore quantum plasmas. They complement analytical models by offering granular insights into strongly nonlinear, nonequilibrium, and multi-scale phenomena that are analytically intractable. MD simulations serve as a diagnostic laboratory for testing theories of quantum transport, wave–particle coupling, and collective excitations under controlled conditions.

As experimental access to extreme plasma regimes continues to expand through facilities enabling ultra-intense lasers, magnetized traps, and quantum-engineered materials the fidelity and interpretability of MD simulations become increasingly central. Future directions involve embedding fully quantized field modes, incorporating spin–orbital coupling explicitly, and developing real-time quantum–classical hybrid algorithms. These advances will continue to refine our understanding of quantum plasma as a fundamental state of matter that blends kinetic chaos with quantum coherence, offering a fertile ground for both fundamental physics and technological innovation.

11.3.4 Lattice Gauge and Dirac Simulations

In regimes where QED effects become dominant such as in ultra-intense laser–matter interactions, or early-universe plasma analogs continuum descriptions fail to capture discrete field-theoretic phenomena like pair production, vacuum polarization, and chiral symmetry dynamics. Lattice gauge theory, rooted in discretized quantum field theory, provides a non-perturbative and gauge-invariant simulation methodology appropriate for such ultra-relativistic and strong-coupling quantum plasma environments.

The basic idea is to discretize spacetime into a lattice Λ, where gauge fields live on links and fermionic fields reside on lattice sites. For U(1) gauge theory (QED), the link variables are $U_\mu(x) = e^{iaeA_\mu(x)}$, ensuring manifest gauge invariance. The lattice QED action takes the form:

$$S = \sum_{x,\mu} \bar{\psi}(x)\gamma^{\mu} \left[\frac{U_{\mu}(x)\psi(x + a\hat{\mu}) - U_{\mu}^{\dagger}(x - a\hat{\mu})\psi(x - a\hat{\mu})}{2a} \right]$$
$$+ m\bar{\psi}(x)\psi(x) + \frac{1}{4}F_{\mu\nu}F^{\mu\nu}.$$

In the Hamiltonian lattice QED framework, particularly the Kogut–Susskind formulation, fields evolve in real time, making this approach especially suitable for simulating nonequilibrium processes in quantum plasmas. It enables the study of complex phenomena such as Schwinger pair production in strong fields, axial anomalies, chiral magnetic effects, quantum turbulence, and topological charge transport. Real-time lattice simulations involving Dirac fermions coupled with dynamical gauge fields provide a powerful tool to model quantum plasma behaviour under extreme relativistic conditions. This is particularly relevant in astrophysical and high-energy scenarios, including pulsar magnetospheres, early-universe leptogenesis, and the quark-gluon plasma near the deconfinement transition. Key dynamical observables in such simulations include the fermionic number density, chirality, electromagnetic current loops, helicity flow, and the spectral evolution of Dirac modes, offering deep insights into the underlying quantum field dynamics.

Advanced lattice algorithms such as staggered fermions and Wilson fermions are deployed to mitigate fermion doubling and maintain lattice symmetries. These techniques are crucial in unveiling anomaly-induced transport and vacuum instabilities that are inaccessible to fluid or kinetic theories.

11.3.5 Time-Dependent Density Functional Theory (TDDFT)

Time-Dependent Density Functional Theory (TDDFT) extends ground-state DFT to nonequilibrium quantum systems, enabling simulation of the real-time dynamics of many-electron systems with strong correlations. In quantum plasma physics, TDDFT is vital for modeling systems where quantum degeneracy, electron correlation, and nonlocal field effects coexist e.g., in warm dense matter (WDM), ultrafast solid–plasma interfaces, and photoionized nano-plasmas [24].

TDDFT recasts the many-body Schrödinger equation into a set of coupled time-dependent Kohn–Sham (TDKS) equations:

$$i\hbar\frac{\partial\psi_i(\mathbf{r},t)}{\partial t} = \left[-\frac{\hbar^2}{2m}\nabla^2 + V_{\text{eff}}[n](\mathbf{r},t)\right]\psi_i(\mathbf{r},t),$$
$$Veff = Vext + VH[n] + Vxc[n], V_{\text{eff}} = V_{\text{ext}} + V_{\text{H}}[n] + V_{\text{xc}}[n],$$

where:

V_{H} is the Hartree (mean-field Coulomb) potential,
V_{xc} is the exchange–correlation potential, modeled via approximate functionals,

$n(\mathbf{r}, t) = \sum_i |\psi_i(\mathbf{r}, t)|^2$ is the total density.

Time-Dependent Density Functional Theory (TDDFT) provides a fully quantum mechanical framework capable of capturing complex plasma phenomena with high fidelity. It effectively models collective electron excitations, such as plasmons and their damping, as well as ionization and recombination processes in dense plasmas. TDDFT also excels in simulating ultrafast charge transfer and field emission dynamics. Unlike Particle-In-Cell (PIC), or Wigner-based approaches, TDDFT incorporates exchange and correlation effects from first principles, albeit through approximate functionals, making it especially suited for small-scale, strongly correlated systems where semiclassical methods break down. This makes TDDFT a powerful tool for investigating laser-driven electron dynamics during solid-to-plasma transitions, short-pulse interactions with nanostructured materials, and phenomena like quantum melting and pressure ionization in warm dense matter (WDM).

However, it suffers from scalability challenges and the need for improved time-dependent exchange–correlation functionals, especially in regimes involving high-field ionization or relativistic effects.

These advanced frameworks quantum-corrected PIC, lattice gauge simulations, and TDDFT represent the modern frontier in quantum plasma simulation. Each addresses distinct physical regimes: from kinetic coherence and radiation reaction to non-perturbative quantum field dynamics and strong correlations. Their combined development not only enables quantitative modeling of extreme plasma states, but also bridges traditional plasma physics with quantum field theory, ultrafast science, and condensed matter physics.

11.4 Real-Time and Non-equilibrium Simulations

The behavior of quantum plasmas far from equilibrium, especially during transient, ultrafast, or strongly driven regimes, necessitates simulation frameworks that transcend the equilibrium or near-equilibrium approximations underlying most traditional models. This section introduces three rigorous formalisms used to simulate non-equilibrium quantum plasma dynamics: the Schwinger–Keldysh closed-time path (CTP) techniques, quantum quench protocols for transient evolution studies, and the two-particle irreducible (2PI) effective action methods tailored for capturing turbulence and coherent structures. These methods collectively provide a versatile apparatus for capturing decoherence, quantum fluctuations, nonlinear feedback, and memory effects in real-time evolutions [25].

11.4.1 Schwinger–Keldysh Techniques: Quantum Statistical Ensembles and Decoherence

The Schwinger–Keldysh formalism offers a principled approach for computing expectation values of quantum operators in arbitrary non-equilibrium states. Unlike the conventional Matsubara formalism, which is confined to thermal equilibrium and imaginary time, the Keldysh method employs a closed-time contour in real-time to preserve causality and capture the full statistical evolution.

Let the system be initially prepared in a mixed state described by a density matrix $\hat{\rho}(t_0)$. The generating functional is given by:

$$Z[J_+, J_-] = \mathrm{Tr}\left[\hat{\rho}(t_0) T_C \exp\left(-i \int_C dt' J(t')\hat{\phi}(t')\right)\right],$$

where T_C denotes contour-ordering along the Schwinger–Keldysh path $\mathcal{C} = C_+ \cup C_-$, consisting of a forward branch (C_+) and a backward branch (C_-). The quantum field $\hat{\phi}$ can represent bosonic, or fermionic operators, and $J_\pm$ are external sources coupled on each branch.

The central objects are the Keldysh Green's functions: the contour-ordered propagator $G_C(x, y)$, the retarded G^R, advanced G^A, and the Keldysh (statistical) propagator G^K, which encode causal responses and quantum correlations:

$$G^K(x, y) = -i\langle\{\hat{\phi}(x), \hat{\phi}(y)\}\rangle, \quad G^R(x, y) = -i\theta(t_x - t_y)\langle[\hat{\phi}(x), \hat{\phi}(y)]\rangle.$$

The physical significance of G^K lies in its capacity to describe statistical occupation numbers and decoherence in evolving quantum plasmas. In magnetized, or photonic plasma environments, this formulation allows one to probe the breakdown of quasiparticle pictures, particle production from vacuum, or decoherence of spin-entangled states [26].

11.4.2 Quantum Quench Simulations: Prethermalization and Hydrodynamization

A quantum quench refers to a sudden change in one, or more parameters of the system Hamiltonian, H(g) → H(g′), often triggered by ultrafast lasers, cavity-mode switching, or external field manipulations. The system is initialized in a ground or thermal state $|\Psi_0\rangle$ of H(g) and then evolved unitarily under H(g′). Mathematically, the time evolution is governed by:

$$|\Psi(t)\rangle = e^{-iH(g')t}|\Psi_0\rangle.$$

In a plasma context, such quenches could correspond to fast ionization fronts, rapid magnetic reconnection, or sudden injection of high-energy photons. The system undergoes nontrivial phases including prethermalization, wherein local observables relax to a metastable quasi-steady state before eventual equilibration. For example, consider a scalar field theory modeling bosonic plasmons with a time-dependent mass term $m^2(t)$:

$$\mathcal{L}(t) = \frac{1}{2}\left[(\partial_t \phi)^2 - (\nabla \phi)^2 - m^2(t)\phi^2\right], \quad m^2(t) = m_i^2\theta(-t) + m_f^2\theta(t).$$

The Fourier components of $\phi_k(t)$ evolve as coupled quantum oscillators with time-dependent frequencies. Occupation numbers $n_k(t)$ show fast oscillations followed by a slow drift toward thermal distributions. If weak coupling is assumed, kinetic equations derived via gradient expansions of the Keldysh Dyson equation describe this evolution:

$$\frac{\partial f_k(t)}{\partial t} = -\Gamma_k[f](f_k - f_k^{\mathrm{eq}}),$$

where Γ_k encodes collision and damping rates, and $f_k(t)$ is the time-resolved phase space distribution. Such equations have been applied to study hydrodynamization, the emergence of fluid-like collective behaviourbefore true thermal equilibrium is reached, even in highly anisotropic or non-isotropic plasmas.

11.4.3 A2PI Effective Action Methods: Quantum Turbulence and Coherent Mode Dynamics

The two-particle irreducible (2PI) effective action framework generalizes the notion of variational principles to the full propagator, enabling consistent self-consistent approximations beyond standard perturbation theory. In this formalism, both the expectation value $\phi_c(x) = \langle\hat{\phi}(x)\rangle$ and the full connected two-point function G(x, y) are dynamical variables. The generating functional is:

$$\Gamma[\phi_c, G] = S[\phi_c] + \frac{i}{2}\mathrm{Tr}\ln G^{-1} + \frac{i}{2}\mathrm{Tr}[(G_0^{-1} - G^{-1})G] + \Gamma_2[\phi_c, G],$$

where Γ_2 contains all 2PI vacuum diagrams constructed with full propagators. The dynamical equations follow from the stationary conditions:

$$\frac{\delta\Gamma}{\delta\phi_c(x)} = 0, \quad \frac{\delta\Gamma}{\delta G(x, y)} = 0.$$

This method captures multiple scattering, memory effects, and feedback from fluctuation spectra, and is particularly effective for describing quantum turbulence and coherent mode dynamics in ultracold plasmas or low-temperature photonic condensates. Quantum turbulence arises when phase coherence and vortex reconnections dominate, requiring a fully nonlinear treatment that goes beyond kinetic or linearized hydrodynamic equations.

For instance, the real-time 2PI equations have been used to simulate the inverse energy cascade in two-dimensional quantum plasmas, tracking the evolution of vorticity and energy spectra $E(k, t) \propto k^{-\alpha}$, with $\alpha \sim 3$ in the turbulent inertial regime. The emergence of quasi-condensates from initial quantum noise and their modulation by gauge field fluctuations are also naturally encoded within this framework.

Each of the non-equilibrium methods discussed herein provides a unique lens through which different facets of quantum plasma dynamics can be uncovered. The Schwinger–Keldysh formalism rigorously tracks decoherence and fluctuation dynamics under general initial conditions. Quantum quenches serve as probes of relaxation and the approach to equilibrium, revealing universality and emergent hydrodynamics. The 2PI formalism, with its emphasis on self-consistent nonlinearity and memory, is ideal for studying strong-field turbulence, quantum kinetic heating, and the birth of order from noise. Together, these frameworks form the mathematical backbone for the modern simulation of quantum plasmas driven far from equilibrium, whether in laser-irradiated nanoplasmas, astrophysical jets, or cold cavity-confined systems [27].

11.5 Simulation Benchmarks and Validation

The fidelity and predictive strength of quantum plasma simulations depend critically on rigorous benchmarking and validation protocols that connect theoretical models with measurable physical observables. In this section, a systematic methodology is outlined for the calibration of various simulation frameworks ranging from quantum hydrodynamic (QHD) codes and Wigner-based kinetic solvers to particle-in-cell (PIC) schemes extended with quantum and relativistic corrections against canonical test cases and experimental diagnostics. This process not only ensures internal numerical consistency, but also enforces physical realism, particularly in the strongly coupled and magnetized regimes that typify modern quantum plasma applications.

At the heart of simulation validation lies the translation of computational observables typically microscopic distribution functions, field amplitudes, and energy spectra into macroscopic diagnostics accessible to experimental measurement. Quantum plasma simulations must routinely produce outputs such as electron density fluctuations $\delta n_e(x, t)$, spectral line shifts, collective mode frequencies, and current density evolution $\mathbf{J}(x, t)$, all of which are essential for comparing with Thomson scattering spectra, reflectometry data, and interferometric phase shifts [28].

For instance, a kinetic simulation of an ultracold dense plasma may yield the Wigner distribution function $f_W(x, p, t)$, which upon momentum integration recovers the spatial density $n(x, t)$. This density can then be Fourier-analyzed to extract dispersion characteristics such as the quantum-corrected Langmuir frequency. Similarly, PIC simulations incorporating quantum recoil and spin–orbit terms generate time-dependent electromagnetic field configurations $\mathbf{E}(x, t)$, $\mathbf{B}(x, t)$, which directly inform the radiation emission spectrum measurable through X-ray or synchrotron diagnostics. Thus, careful post-processing bridges the quantum theoretical language with laboratory observables.

A core objective in simulation validation is assessing the range of applicability and limitations of distinct theoretical frameworks. Quantum hydrodynamic (QHD) simulations are particularly well-suited for capturing collective low-frequency phenomena such as quantum-corrected acoustic modes and ion sound waves. They typically solve coupled continuity and momentum equations supplemented with the quantum Bohm potential:

$$\frac{\partial n}{\partial t} + \nabla \cdot (n\mathbf{v}) = 0, \quad m\left(\frac{\partial \mathbf{v}}{\partial t} + \mathbf{v} \cdot \nabla \mathbf{v}\right) = -\nabla P - e\mathbf{E} + \frac{\hbar^2}{2m}\nabla\left(\frac{\nabla^2 \sqrt{n}}{\sqrt{n}}\right).$$

In contrast, quantum kinetic approaches such as those based on the Wigner–Moyal equation or the quantum Vlasov equation resolve finer-scale phase space dynamics, including quantum coherence, tunneling, and interference effects. The quantum Vlasov equation for the Wigner function f_W in the electrostatic limit takes the form:

$$\frac{\partial f_W}{\partial t} + \frac{p}{m} \cdot \nabla_x f_W - \frac{ie}{\hbar} \int \frac{d^3 s}{(2\pi\hbar)^3} e^{i\mathbf{s}\cdot\mathbf{x}/\hbar} \left[\phi\left(\mathbf{x} + \frac{\mathbf{s}}{2}\right) - \phi\left(\mathbf{x} - \frac{\mathbf{s}}{2}\right)\right] f_W(\mathbf{x}, \mathbf{p}, t) = 0.$$

Particle-in-cell simulations, extended to incorporate quantum force corrections (e.g., Bohm terms or spin–orbit couplings), offer yet another perspective, particularly valuable when modeling ultra-intense field interactions and wakefield structures in underdense quantum plasmas. These simulations track individual particle trajectories under self-consistently evolved fields and are especially relevant when relativistic dynamics and discrete particle statistics are important. Thus, comparative analysis across QHD, kinetic, and PIC codes enables triangulation of results, helping to uncover the regimes where each method remains valid or breaks down.

To robustly test quantum plasma codes, several well-characterized benchmark cases are frequently employed. The first is the Langmuir wave dispersion with quantum corrections, where both linear and nonlinear dispersion relations are analytically tractable under specific assumptions. The QHD dispersion relation for electrostatic modes in a degenerate plasma with Fermi pressure and quantum Bohm correction is given by:

$$\omega^2 = \omega_{pe}^2 + \frac{3k^2 v_F^2}{5} + \frac{\hbar^2 k^4}{4m^2},$$

where v_F is the Fermi velocity and ω_{pe} is the electron plasma frequency. This serves as a stringent test for both fluid and kinetic simulations, particularly in the long-wavelength regime.

A second widely used benchmark is wakefield generation by an intense laser or electron beam pulse in a quantum plasma. In this case, the simulation must capture nonlinear wave-breaking thresholds, quantum dispersion modification of the wake potential, and spin-induced beam deflection, particularly in magnetized configurations. The wake potential $\phi(x, t)$ is expected to exhibit oscillatory features with quantum scale corrections, and simulations are benchmarked against known analytical, or semi-analytical forms of the wakefield profile [29].

Simulations of magnetized Fermi plasmas constitute a high-level benchmark, incorporating both spin–magnetization currents and Landau quantization of orbital motion. In this context, a kinetic or 2-fluid code must reproduce anisotropic pressure tensors, spin-precession-induced vorticity structures, and quantum cyclotron resonances. For example, the pressure anisotropy in a magnetized degenerate Fermi gas is given by:

$$P_{||} = P_0, \quad P_{\perp} = P_0 - \frac{M_s^2}{\mu_0 n},$$

where M_s is the spin magnetization density. Benchmarks involving such anisotropies are essential for validating magneto-hydrodynamic extensions of quantum fluid models.

Validation exercises not only ensure the credibility of simulation codes but also guide future experimental designs. For instance, discrepancies between quantum PIC simulations and spectroscopic diagnostics may indicate overlooked nonlocal corrections or the necessity for improved quantum collision operators. Moreover, hybrid approaches combining QHD for bulk evolution and kinetic solvers for localized phenomena are emerging as a promising pathway, particularly for high-gradient, strongly inhomogeneous quantum plasma systems. Additionally, community-driven benchmark repositories, including standardized test problems for quantum Langmuir turbulence, relativistic quantum beam–plasma instabilities, and spin–magnetization wave coupling, are increasingly being established. These benchmarks promote reproducibility, cross-validation, and algorithmic innovation across simulation groups globally.

Rigorous simulation benchmarking serves as the cornerstone for building trustworthy, predictive models in quantum plasma physics. Through systematic comparison against analytic results, controlled laboratory experiments, and inter-framework consistency checks, this process helps unify theory and measurement, ultimately enriching both the fundamental understanding and technological application of quantum plasmas.

11.6 Hybrid Approaches

The simulation of quantum plasmas presents profound challenges due to the simultaneous presence of nonlocal quantum coherence, relativistic collective excitations, nonlinear field couplings, and strong statistical correlations. Conventional approaches whether purely classical or fully quantum often encounter either computational intractability or limited physical fidelity when extended to realistic geometries or experimental regimes. This section addresses contemporary solutions to this impasse through two interlinked strategies: hybrid quantum–classical modeling and machine learning-enhanced simulation. Both methodologies aim to interpolate between computational tractability and physical completeness, offering scalable frameworks that retain quantum character while enabling broader phase-space exploration.

11.6.1 Quantum–Classical Hybrid Modeling

Hybrid quantum–classical methods construct tractable evolution schemes by combining quantum mechanical subspaces with classical collective dynamics, allowing for efficient simulation of quantum plasmas where full quantization of all degrees of freedom is neither necessary nor computationally feasible. The key concept is the formulation of an *effective Hamiltonian*, wherein certain components typically fast-evolving field modes or strongly correlated subsystems are treated quantum mechanically, while others such as bulk plasma motion or external driving fields are modeled using semiclassical or fluidic approximations.

The effective Hamiltonian H_{eff} for a quantum–classical hybrid plasma system is generally partitioned into quantum, classical, and coupling terms:

$$H_{\text{eff}} = H_Q[\hat{\psi}, \hat{A}_\mu] + H_C[n(\mathbf{x}, t), \mathbf{v}(\mathbf{x}, t)] + H_{\text{int}}[\hat{\psi}, n(\mathbf{x}, t)],$$

where $\hat{\psi}$ represents quantized field operators (such as spinor electron fields or quantized photon modes), $\hat{A}_\mu$ denotes the gauge potential operators, and $n(\mathbf{x}, t)$, $\mathbf{v}(\mathbf{x}, t)$ represent classical charge density and velocity fields. The quantum component H_Q evolves via the Schrödinger or Heisenberg equation, while the classical component follows fluid or kinetic equations with effective potentials sourced by quantum expectation values.

A prototypical example involves modeling a degenerate Fermi electron gas interacting with quantized electromagnetic field modes, where the electron momentum evolution includes a quantum Bohm term:

$$m\frac{d\mathbf{v}}{dt} = -\nabla P(n) + e\mathbf{E} + \frac{\hbar^2}{2m}\nabla\left(\frac{\nabla^2\sqrt{n}}{\sqrt{n}}\right),$$

while the photon field evolves under Maxwell–Podolsky equations with matter currents derived from expectation values $\langle \hat{\psi}^{\dagger} \hat{\psi} \rangle$. The self-consistency loop between quantum and classical sectors is maintained through time-dependent Hartree–Fock, or Dirac–Poisson solvers.

A central tool in quantum–classical hybridization is the semiclassical expansion in powers of $\hbar$, applied either to the Wigner–Moyal equation, or to path-integral formulations. Expanding the Wigner equation to first order in $\hbar$ recovers the Vlasov–Poisson system with quantum corrections:

$$\frac{\partial f_W}{\partial t} + \frac{p}{m} \cdot \nabla_x f_W - e\mathbf{E} \cdot \nabla_p f_W = \mathcal{O}(\hbar^2).$$

This structure allows for efficient simulation of mildly quantum systems, where the leading-order quantum behavior is retained, but higher-order interference terms are neglected. In strongly magnetized quantum plasmas, semiclassical approaches also incorporate Berry phase corrections to the velocity and force terms in the kinetic equation:

$$\dot{\mathbf{x}} = \frac{\partial \varepsilon_n}{\partial \mathbf{p}} + \dot{\mathbf{p}} \times \mathbf{\Omega}_n(\mathbf{p}), \quad \dot{\mathbf{p}} = e\mathbf{E} + e\,\dot{\mathbf{x}} \times \mathbf{B},$$

where $\mathbf{\Omega}_n(\mathbf{p})$ is the Berry curvature of the n-th energy band. Such corrections are crucial for capturing spin–orbit interactions and anomalous Hall effects in quantum plasmas.

The hybrid approach in plasma modeling embodies a hierarchical decomposition of physical scales, effectively integrating quantum and classical dynamics. The quantum subspace captures short-wavelength, high-frequency phenomena such as electron de Broglie oscillations, spin entanglement, and photon-mediated interactions, while the classical sector handles long-wavelength, collective behaviours like drift wave formation and background fluid evolution. This multiscale framework is particularly valuable in scenarios where both regimes interact closely for instance, in dense laser-irradiated targets, where electron quantum recoil coexists with classical ion hydrodynamics; in wakefield accelerators, where quantized radiation modes traverse semiclassical plasma backgrounds; and in cavity-confined quantum plasmas, where mode quantization directly influences bulk thermodynamic flows. By combining quantum precision with classical scalability, the hybrid formalism achieves both computational efficiency and physical realism, allowing for accurate simulations of complex plasma environments with well-controlled approximations.

11.6.2 Data-Driven Simulations

The explosion of high-resolution diagnostic data from quantum plasma experiments, combined with the ever-increasing complexity of simulation outputs, has created an opportunity for data-driven modeling frameworks that transcend explicit equation-based solvers. In particular, machine learning (ML) methods and neural networks in particular offer powerful tools for plasma state reconstruction, surrogate modeling, and predictive analytics.

The core idea is to represent unknown plasma quantities such as electron density profiles, local magnetic fields, or spectral emissivity functions using neural network approximators $\mathcal{N}_\theta(\mathbf{x})$, trained on simulation data, experimental diagnostics, or both. For example, a convolutional neural network (CNN) may be trained to reconstruct $n_e(x, t)$ from partial diagnostic signals such as scattered light spectra or Langmuir probe data. The learning process minimizes a loss function:

$$\mathcal{L}(\theta) = \sum_i \left\| \mathcal{N}_\theta(\mathbf{x}_i) - n_e^{\text{true}}(\mathbf{x}_i) \right\|^2 + \lambda \mathcal{R}(\theta),$$

where $\mathcal{R}(\theta)$ enforces physical regularity (e.g., positivity, smoothness, or conservation laws). The trained model acts as a surrogate, capable of reconstructing plasma profiles with high fidelity even from sparse or noisy data.

A promising innovation in this direction is the use of *physics-informed neural networks* (PINNs), which embed differential constraints directly into the training process. Instead of using labeled data alone, PINNs penalize violations of governing equations, such as the Schrödinger–Poisson system or QHD continuity and momentum equations. A typical loss function includes terms like:

$$\mathcal{L}_{\text{PINN}} = \sum_i \left\| \frac{\partial n}{\partial t} + \nabla \cdot (n\mathbf{v}) \right\|^2 + \left\| m\frac{d\mathbf{v}}{dt} + \nabla P - e\mathbf{E} - \nabla V_Q \right\|^2.$$

Here, V_Q is the quantum Bohm potential. Such formulations ensure that the neural reconstruction not only fits data but also respects physical conservation laws and symmetries.

The integration of machine learning (ML) into quantum plasma modeling has significantly expanded the capabilities of both theoretical and experimental research. One major advancement lies in experimental state reconstruction, where ML algorithms facilitate the high-dimensional tomographic inversion of spin-polarized density fields, even when measurement data is incomplete. Another key application is fast surrogate modeling, in which trained models can emulate slow-running simulations in real-time, allowing for rapid exploration of parameter spaces in complex laser–plasma interaction setups. Additionally, ML is revolutionizing anomaly detection and experimental control; neural networks can continuously monitor quantum plasma systems, identify deviations from predicted behaviors, and support adaptive control strategies that respond dynamically to emerging conditions. Collectively, these ML-driven innovations are transforming quantum plasma research by enabling faster, smarter, and more precise modeling and experimentation.

Physically, the success of ML-enhanced modeling derives from the recognition that many aspects of quantum plasma evolution are governed by latent structures such as conserved invariants, symmetries, and topological constraints that can be statistically inferred from data even when not analytically tractable. The neural network acts as a functional interpolator in the high-dimensional configuration space, guided by data and anchored by physics.

Together, quantum–classical hybrid models and machine learning-based plasma reconstruction offer powerful frameworks to address the central challenge of quantum plasma modeling: capturing essential quantum features without incurring prohibitive computational cost. While hybrid Hamiltonians preserve fidelity to quantum dynamics in a reduced subspace, neural networks enhance interpretability and real-time applicability by exploiting the latent structure of experimental and synthetic data. Both approaches mark a shift in the paradigm of quantum plasma simulation from purely equation-based solvers toward data-aware, adaptive, and hierarchical modeling strategies suitable for the emerging era of quantum-informed plasma science [30].

11.6.3 Reinforcement Learning in Control of Experimental Quantum Plasmas

The control of quantum plasma systems in laboratory environments particularly in ultracold and magnetized setups, or laser-induced microplasmas presents a highly nonlinear, temporally evolving optimization problem. In contrast to classical control schemes that rely on pre-defined feedback loops, Reinforcement Learning (RL) offers a dynamic framework where an artificial agent learns optimal strategies for experimental manipulation by interacting directly with the plasma system. This is especially relevant in quantum plasma, where measurement back-action, decoherence, and quantum noise often obscure direct analytic modeling of control policies. The RL framework allows for empirical convergence toward optimal control laws, even in the presence of partially observable or highly entangled plasma states.

Mathematically, an RL problem is characterized by a Markov Decision Process (MDP), defined by a tuple $(\mathcal{S}, \mathcal{A}, P, R, \gamma)$ [31], where:

$\mathcal{S}$ is the state space, representing observables such as local densities $n(\mathbf{r}, t)$, electric field profiles $\mathbf{E}(\mathbf{r}, t)$, or quantum spin polarization $\langle \hat{\sigma}_z \rangle$,

$\mathcal{A}$ is the action space, encoding experimental controls such as applied magnetic field profiles $B_{\text{ext}}(t)$, pulse sequences, or trap geometries,

$P(s'|s, a)$ denotes the state transition probability, often unknown and learned empirically,

$R(s, a)$ is the reward function, which quantifies the experimental goal (e.g., maximizing coherence time, minimizing Langmuir turbulence),

γ is the discount factor, encoding the trade-off between immediate and long-term gains.

In quantum plasma systems, the agent interacts with a partially observed Markov process. The true quantum state $|\psi(t)\rangle$, or the full density matrix $\rho(t)$ is often inaccessible, requiring the agent to operate over a belief distribution $b(s) \in \Delta(\mathcal{S})$, thus elevating the control problem to a Partially Observable Markov Decision Process (POMDP) [32].

The RL agent attempts to learn a policy $\pi : \mathcal{S} \to \mathcal{A}$ that maximizes the expected cumulative reward:

$$J(\pi) = \mathbb{E}_\pi \left[\sum_{t=0}^{\infty} \gamma^t R(s_t, a_t) \right].$$

The policy π may be parameterized via deep neural networks (as in Deep Q-Learning or Actor-Critic methods), allowing high-dimensional input (e.g., tomographic images or spectroscopic traces) to inform control decisions [33]. Physically, these learned policies correspond to optimal protocols for inducing or suppressing quantum instabilities, dynamically tuning the plasma frequency $\omega_p(t)$, or navigating toward quantum coherent states via manipulation of external knobs.

In magnetized quantum plasmas produced by intense laser–solid interactions, RL has shown promise in controlling the self-focusing of electron beams by learning spatiotemporal pulse shaping to counteract quantum diffraction and filamentation. Similarly, in dusty plasma analogues of Wigner crystallization, RL agents have been trained to modulate electrostatic confinement parameters in real time to preserve lattice symmetry and suppress decoherence from phonon-like modes.

In ultracold plasma traps, RL-based feedback schemes have been employed to counteract disorder-induced heating. By learning temporal profiles of cooling laser detuning or magnetic field ramping, the agent steers the system toward high-phase-space-density regimes without reliance on explicit thermodynamic modeling.

From a physical standpoint, RL adapts control sequences in a data-driven manner, effectively constructing a non-perturbative control map between observables and actuator variables in regimes where first-principles modeling may be computationally prohibitive. The approach capitalizes on the empirical ergodicity of control trajectories in the high-dimensional plasma phase space, leveraging the stochasticity inherent in quantum plasma dynamics to learn robust, generalizable strategies. The ability to encode nonlocal temporal correlations into the control logic (via recurrent networks or attention mechanisms) further enhances applicability in turbulent or chaotic plasma regimes [34].

11.6.4 Bayesian Inference for Extracting Quantum Parameters from Spectral Data

One of the central experimental challenges in quantum plasma physics lies in the accurate extraction of underlying physical parameters such as effective electron mass $m^*(T, B)$, plasma frequency ω_p, quantum degeneracy parameters $\theta = T/T_F$, and spin polarization fractions from indirect and noisy spectral data. Bayesian inference provides a principled probabilistic framework to tackle this inverse problem by quantifying parameter uncertainty and incorporating prior theoretical knowledge into experimental interpretation.

Given a model $M(\boldsymbol{\theta})$ that predicts spectral observables $D_{\text{pred}} = f(\boldsymbol{\theta})$, where $\boldsymbol{\theta}$ denotes the set of latent quantum plasma parameters (e.g., $\mu, \hbar\omega_p, g^*$), and an experimental dataset D_{obs}, Bayes' theorem updates prior beliefs $P(\boldsymbol{\theta})$ to obtain the posterior distribution:

$$P(\boldsymbol{\theta}|D_{\text{obs}}) = \frac{P(D_{\text{obs}}|\boldsymbol{\theta})P(\boldsymbol{\theta})}{P(D_{\text{obs}})},$$

where $P(D_{\text{obs}}|\boldsymbol{\theta})$ is the likelihood, often derived from the discrepancy between model predictions and measured spectra.

This framework is particularly apt for analyzing data from Thomson scattering, time-resolved interferometry, or coherent X-ray diagnostics, where observables are nonlinear functionals of plasma parameters and may exhibit degeneracies under deterministic inversion.

Assuming the experimental noise follows a Gaussian distribution, the likelihood is given by:

$$P(D_{\text{obs}}|\boldsymbol{\theta}) = \exp\left[-\frac{1}{2}\sum_i \frac{(D_{\text{obs},i} - D_{\text{pred},i}(\boldsymbol{\theta}))^2}{\sigma_i^2}\right],$$

where σ_i is the measurement uncertainty in channel ii. In quantum plasmas, the forward model D_{pred} may encode non-trivial quantum corrections such as spin splitting, Landau quantization, or exchange–correlation shifts in the dielectric function $\epsilon(\omega, \mathbf{k}; \boldsymbol{\theta})$, which enter the expression for spectral emissivity or response functions.

Prior distributions $P(\boldsymbol{\theta})$ encode physical constraints, e.g., positivity of temperature, known bounds on spin polarization, or previously calibrated values of effective charge states Z^*. Informative priors based on quantum kinetic models or ab initio simulations (e.g., quantum molecular dynamics) improve convergence and interpretability, especially in low signal-to-noise scenarios.

Sampling from the posterior distribution is typically performed using Markov Chain Monte Carlo (MCMC) techniques, such as Metropolis–Hastings or Hamiltonian Monte Carlo, which enable high-dimensional exploration of parameter spaces.

This produces not only point estimates of parameters but full uncertainty quantification via credible intervals, facilitating robust error bars in reported plasma diagnostics.

In warm dense matter regimes, Bayesian inference has been employed to extract dynamic structure factors $S(\omega, k)$ from inelastic X-ray scattering data, yielding precise estimates of electron temperature T_e and ionization degree Z in regimes inaccessible to direct probe. In ultracold plasma traps, Bayesian spectral fitting has resolved small Zeeman shifts and quantum recoil signatures, enabling extraction of the effective g-factor in spin-polarized electron populations.

Additionally, in laser-driven microplasmas exhibiting high harmonic generation, Bayesian inversion of spectral cutoff and angular distribution provides insights into quantum nonlinearity coefficients and relativistic plasma parameters, thereby linking microscopic field-particle interactions with macroscopic observables.

Bayesian inference, unlike deterministic inversion, treats parameter estimation as an epistemic update conditioned on observational data. In quantum plasma experiments where measurements inevitably perturb the system this approach mirrors the Bayesian interpretation of quantum mechanics, whereby knowledge is probabilistically updated rather than deterministically deduced. Moreover, the Bayesian formalism naturally accommodates systematic errors, apparatus biases, and prior theoretical uncertainty, offering a statistically grounded framework for experimental quantum plasma physics.

Together, reinforcement learning and Bayesian inference represent two pillars of modern computational methodology that are revolutionizing the study of quantum plasma in experimental settings. RL enables dynamic, feedback-based control of highly nonlinear plasma behavior, learning control strategies beyond analytic derivation. Bayesian inference, on the other hand, refines interpretational clarity by rigorously extracting physical parameters from indirect measurements. When combined with quantum-aware forward models, these approaches not only improve experimental fidelity, but also pave the way for adaptive plasma manipulation and quantum-enhanced metrology [35].

11.7 Challenges, Limitations, and Future Directions

11.7.1 Experimental Limits

The investigation of quantum plasma phenomena in laboratory conditions is inherently constrained by the precision and sensitivity of available experimental instruments. Unlike classical plasma systems, where bulk parameters such as density, temperature, and magnetic field strength evolve on macroscopic time scales and spatial domains, quantum plasma systems often exhibit sub-femtosecond dynamics and coherence effects at nanometer or even angstrom scales. The necessity to resolve both the microscopic features of wavefunction interference and macroscopic collective excitations imposes stringent requirements on diagnostics, both in temporal and spatial resolution.

From a physical standpoint, the characteristic time scales in quantum plasmas are dictated by plasma frequencies (typically in the THz–PHz range for dense electron gases), cyclotron motion under ultra-strong magnetic fields, and intrinsic decoherence processes such as collisionless Landau damping or quantum dephasing. Time-resolved diagnostics such as ultrafast pump–probe spectroscopy, streaked electron detection, and attosecond photoemission are often required to probe these ultrafast phenomena. Mathematically, the experimental resolution must match or surpass the inverse frequency bandwidths $\Delta t \lesssim 1/\omega_{\max}$, where $\omega_{\max}$ may correspond to either the Langmuir frequency ω_{pe} or high-harmonic transitions driven by field-induced nonlinearity.

Spatial resolution is equally demanding. For example, in warm dense matter or solid-density plasmas, the inter-particle spacing $d \sim (3/4\pi n_e)^{1/3}$ may be on the order of a few angstroms. Probing this regime necessitates diagnostic techniques that operate in the XUV or soft X-ray domain, such as coherent X-ray diffraction imaging (CXDI) or Thomson scattering with femtosecond resolution. These methods, however, are often limited by photon shot noise, detection threshold noise, and the perturbative impact of the diagnostic probe itself [36].

Noise is a pervasive limitation in quantum plasma measurements. It arises both from external stochastic sources such as thermal background radiation, environmental electromagnetic fields, and instrumental jitter and from intrinsic quantum fluctuations, notably zero-point motion and statistical uncertainty of few-particle measurements. The associated signal-to-noise ratio (SNR) degrades rapidly in systems with low photon counts or small electron populations, typical of ultracold plasma systems or quantum-confined nanostructures. Advanced noise-suppression techniques, such as quantum-limited amplifiers, squeezed light diagnostics, and machine learning-assisted data reconstruction, are under active development to circumvent these constraints.

11.7.2 Simulation Challenges

Numerical simulations of quantum plasma are indispensable for exploring dynamical regimes inaccessible to direct experimental control. Yet, they are bounded by fundamental computational barriers rooted in the intrinsic complexity of quantum many-body dynamics. One of the foremost challenges is the exponential growth of the Hilbert space with particle number. Even under second quantization, simulating the full density matrix evolution of N particles scales as $\mathcal{O}(2^{2N})$ for fermionic systems, rendering brute-force methods infeasible beyond modest particle numbers.

The Wigner function formalism, quantum Vlasov equations, and Schrödinger–Poisson solvers provide tractable frameworks under specific approximations, such as mean-field limits or linearized perturbations. However, capturing quantum coherence, tunneling, and entanglement requires non-perturbative and phase-space-resolved methods. One emergent approach is the use of time-dependent density functional theory (TDDFT), where the electron density $n(\mathbf{r}, t)$ replaces the wavefunction as the central dynamical variable. Yet, TDDFT simulations are heavily dependent on the accuracy of exchange–correlation functionals, which remain poorly characterized for strongly coupled, nonequilibrium quantum plasmas.

Relativistic effects pose another layer of complexity. In ultra-intense field environments such as laser-solid interactions or pulsar magnetospheres particle energies approach or exceed rest mass scales. Here, the appropriate framework transitions from nonrelativistic quantum mechanics to the Dirac–Maxwell or Dirac–Klein–Gordon systems, where spinor dynamics, pair production, and nonlinear radiation reaction effects are essential. Incorporating such phenomena into simulation frameworks like Particle-in-Cell (PIC) codes necessitates extensive relativistic generalization and regularization of quantum radiative corrections.

Strong coupling characterized by a coupling parameter $\Gamma = e^2/(ak_BT) > 1$ invalidates the weak-coupling expansions used in perturbative kinetic theory and necessitates the use of correlated quantum molecular dynamics (QMD), Monte Carlo methods, and path integral formulations. These methods, while powerful, are computationally expensive and suffer from sign problems, especially in fermionic systems at low temperature. To mitigate this, hybrid approaches such as semiclassical Ehrenfest dynamics, Bohmian trajectory-based quantum hydrodynamics, and Keldysh two-particle irreducible (2PI) effective action techniques are increasingly adopted for their balance of physical fidelity and computational tractability.

11.7.3 Bridging Laboratory and Astrophysical Plasmas

One of the central ambitions of experimental quantum plasma research is to simulate and understand phenomena that naturally occur in extreme astrophysical environments such as white dwarfs, neutron star crusts, or magnetar atmospheres using accessible terrestrial setups. Achieving this goal necessitates the construction of rigorous scaling frameworks that link laboratory plasma parameters to their astrophysical counterparts via dimensionless numbers.

Dimensional analysis provides the primary toolset for this purpose. Key parameters include:

- The quantum degeneracy parameter $\chi = n\lambda_{dB}^3$, with $\lambda_{dB} = h/\sqrt{2\pi mk_BT}$, determining whether the plasma exhibits quantum statistical behavior.
- The coupling parameter Γ, controlling the relative strength of Coulomb interactions.
- The magnetization parameter $\beta = \omega_{ce}/\omega_{pe}$, characterizing the influence of external magnetic fields on collective plasma modes.

- The relativistic parameter $\eta = k_B T/mc^2$, indicating the significance of relativistic corrections in kinetic dynamics.
- The plasma parameter $\Lambda = n\lambda_D^3$, assessing collective versus binary collisional behavior.

Through similarity transformations and the invariance of the governing equations (e.g., Maxwell's equations, Vlasov–Poisson system, or Dirac–Heisenberg–Wigner equations), one can establish correspondence between laboratory and astrophysical systems, provided these dimensionless parameters are conserved. For instance, warm dense plasma experiments driven by ultra-intense lasers can access the same Γ and χ regimes as found in white dwarf interiors, enabling indirect insights into their equation of state and transport properties.

Nonetheless, not all astrophysical regimes can be fully scaled. Gravitational confinement, neutrino radiation, and ultra-high magnetic fields of $B \sim 10^{14} - 10^{15}G$ found in magnetars are not reproducible on Earth. However, synthetic analog systems such as ultracold Rydberg plasmas or graphene-based Dirac materials may emulate certain relativistic or quantum features under controlled boundary conditions. Moreover, analogue gravity and Unruh radiation experiments in Bose–Einstein condensates are promising pathways to investigate effective space–time curvature and horizon effects relevant to quantum plasmas in curved backgrounds.

The synthesis of laboratory experimentation, high-fidelity simulations, and dimensionless scaling analysis offers a promising frontier for expanding the relevance of terrestrial quantum plasma studies to the astrophysical domain. Future experimental designs will increasingly rely on multi-modal diagnostics, exascale computing, and advanced theoretical synthesis to cross this translational barrier.

11.7.4 Role of Quantum Computing in Modeling Many-Body Quantum Plasmas

The modeling of quantum plasmas as genuinely many-body quantum systems presents a formidable computational challenge due to the exponential growth of Hilbert space dimensionality with increasing particle number. Classical simulation techniques such as time-dependent Hartree–Fock, Wigner–Moyal formalism, or configuration interaction methods while robust within perturbative and mean-field approximations, become prohibitively expensive or inaccurate for strongly correlated regimes, non-equilibrium dynamics, or systems involving fermionic entanglement and decoherence. Quantum computing, by contrast, leverages the inherent parallelism of quantum states and operations, offering a fundamentally new computational substrate that aligns more naturally with the structure of many-body quantum plasma dynamics [37].

From a mathematical standpoint, quantum computation operates within the framework of unitary evolution in Hilbert space. The state of a quantum register of n qubits is a vector in $\mathbf{C}^{2^n}$, and operations are performed by applying tensor products of unitary gates that mimic physical Hamiltonian evolution. In the context of quantum plasma physics, a typical goal is to simulate the time evolution governed by a Hamiltonian of the form:

$$\hat{H} = \sum_i \frac{\hat{p}_i^2}{2m} + \sum_{i<j} V(\hat{\mathbf{r}}_i - \hat{\mathbf{r}}_j) + \sum_i U_{\text{ext}}(\hat{\mathbf{r}}_i, t),$$

where V denotes the inter-particle Coulomb or screened interaction, and U_{ext} includes electromagnetic potentials generated by external fields or boundary conditions. Quantum computing algorithms such as the Trotter–Suzuki decomposition, variational quantum eigensolvers (VQE), and quantum phase estimation (QPE) can be employed to evolve the system under $\hat{H}$ or extract ground and excited states of interest. The central advantage is that such simulations can, in principle, be executed with polynomial resources in the number of particles, avoiding the exponential memory overhead of classical methods.

Physically, quantum plasma simulations on quantum hardware could access highly nonperturbative regimes, including those with strong coupling ($\Gamma >> 1$), quantum degeneracy ($\chi >> 1$), or topologically nontrivial boundary conditions (e.g., in toroidal confinement). These are regimes where classical solvers are encumbered by the need to track many-body wavefunctions or solve integro-differential equations with memory effects. For example, simulating a low-temperature, strongly magnetized plasma with quantized Landau levels and exchange correlations is naturally representable as a lattice gauge theory on a quantum processor, particularly using digital quantum simulation techniques.

Quantum algorithms tailored to quantum field theories such as those developed to simulate scalar fields, spinor QED, or lattice gauge dynamics offer a compelling template for constructing quantum plasma models as effective quantum fields. A relevant direction is the encoding of the Wigner–Vlasov–Maxwell equations into a Hamiltonian lattice model with fermionic and bosonic qubit registers, where the phase-space dynamics of quantum distributions can be interpreted as circuit-based sampling of quasi-probability amplitudes. Ancillary qubits may encode spin degrees of freedom or track coherence decay, thereby allowing one to explore entanglement entropy, decoherence rates, and thermalization processes in real time.

The physical implications of quantum computing in plasma modeling are profound. Firstly, it enables exploration of dynamical regimes beyond linear response or perturbation theory, which are critical for understanding phenomena like quantum turbulence, pair production, or spin–orbit-induced vorticity. Secondly, quantum computers offer the potential for adaptive algorithms that dynamically adjust the resolution of the simulation based on quantum entanglement entropy or conserved quantities, providing a self-optimizing simulation strategy. Lastly, quantum error correction and noise resilience methods, though still under development, aim to ensure the physical reliability of long-time evolution simulations, a key requirement for modeling relaxation and transport in bounded quantum plasmas.

In the near term, quantum computing will likely serve as a hybrid co-processor in conjunction with classical solvers. Variational quantum simulation (VQS) schemes, wherein the parameter space of a quantum circuit is optimized using classical gradient methods, already show promise for modeling small-scale quantum systems with embedded correlations. As quantum hardware matures particularly in coherence time, qubit count, and gate fidelity quantum computing is poised to become a transformative paradigm in many-body quantum plasma simulation, bridging the gap between exact theory and experimentally inaccessible regimes.

11.7.5 Synergies Between Experiments, Theory, and Simulations

The triad of theory, simulation, and experiment constitutes the epistemological scaffold upon which the modern understanding of quantum plasmas rests. However, in the quantum regime, the interplay among these domains is not merely sequential (theory $\rightarrow$ simulation $\rightarrow$ experiment), but deeply recursive and mutually generative. Each domain offers both validation and stimulus for the others, creating a feedback loop essential for progress in the modeling and control of complex quantum plasma phenomena.

Theoretically, the formal structure of quantum plasma physics arises from quantum statistical mechanics, many-body theory, and field-theoretic treatments of gauge interactions. The governing equations such as the quantum kinetic equation (e.g., Wigner or Kadanoff–Baym formulations), quantum hydrodynamics with spin–Bohm potential, and Dirac–Maxwell systems for relativistic plasmas provide foundational models with well-defined assumptions and asymptotic limits. Yet, many of these models are analytically intractable in realistic parameter regimes, necessitating numerical realization for predictive application.

Herein lies the crucial role of simulation. Classical and quantum numerical methods serve to resolve the dynamical evolution of theoretical models, bridge asymptotic expansions with real-world parameter spaces, and explore the onset of instabilities, phase transitions, or nonlinear couplings. More critically, simulations can generate computable observables such as current densities, structure factors, spectral functions, and coherence measures that map directly onto experimental diagnostics. This compatibility transforms simulation results into synthetic experiments, capable of both predicting and interpreting laboratory data.

From the experimental perspective, advances in ultrafast laser–plasma interactions, trapped ion plasmas, cold Rydberg gases, and solid-state analogues of quantum plasma (e.g., two-dimensional materials under strong excitation) provide a wealth of high-fidelity data that challenge theoretical models. In many cases, observed phenomena such as anomalous absorption, collective excitations with hybrid dispersion, or non-Maxwellian energy distributions lack a satisfactory theoretical description, motivating revision of kinetic and hydrodynamic assumptions. The experimental detection of coherence phenomena, spin-resolved transport, or quantum Hall-like behaviourin confined plasmas further compels the inclusion of subtle quantum geometric phases and spin–orbit couplings into modeling frameworks [38].

The synergy between these three pillars becomes most potent when pursued in a co-design paradigm. For instance, experimental capabilities can inspire simulation methodologies, such as the use of artificial gauge fields in ultracold plasmas prompting gauge-invariant simulation techniques. Likewise, simulation results can motivate experimental configurations, as in the case of high-harmonic generation spectra suggesting new pump–probe laser arrangements. Theoretically, insights from effective field theory, renormalization group techniques, or non-Hermitian quantum mechanics can guide the interpretation of both numerical instabilities and experimental anomalies.

Physically, the alignment of these domains enables the validation of nontrivial quantum effects in plasma systems: coherence revival, entanglement spread, spontaneous symmetry breaking, and topologically protected excitations. For example, combined theoretical and simulation work on non-Abelian plasmas or Podolsky-modified electromagnetic theories can be tested against precision experimental data from engineered cavity plasmas or structured metamaterials with tailored dispersion.

Finally, the future trajectory of quantum plasma science demands tighter integration of experiment, theory, and simulation through standardization of data formats, open-source modeling frameworks, and AI-driven design loops. Adaptive learning algorithms can analyze experimental outputs in real time to refine simulation parameters, while simulation-informed theory can narrow the search space for exotic plasma phases in high-throughput experimental campaigns. Such synergistic frameworks not only elevate the accuracy and fidelity of quantum plasma science but also democratize the field by lowering barriers to predictive and exploratory research.

References

1. Harilal, S. S., Phillips, M. C., Froula, D. H., Anoop, K. K., Issac, R. C., & Beg, F. N. (2022). Optical diagnostics of laser-produced plasmas. *Reviews of Modern Physics, 94*(3), Article 035002.
2. Fortov, V. E., & Fortov, V. E. (2016). High energy densities in laboratories. *Extreme States of Matter: High Energy Density Physics*, 23–89.
3. King, P. D., Picozzi, S., Egdell, R. G., & Panaccione, G. (2020). Angle, spin, and depth resolved photoelectron spectroscopy on quantum materials. *Chemical Reviews, 121*(5), 2816–2856.
4. Frühling, U. (2023). Ultrafast molecular dynamics experiments at the XUV free-electron laser in Hamburg flash. *Laboratório Nacional de Luz Síncrotron (LNLS)*, 41.
5. Röpcke, J., Davies, P. B., Lang, N., Rousseau, A., & Welzel, S. (2012). Applications of quantum cascade lasers in plasma diagnostics: A review. *Journal of Physics D: Applied Physics, 45*(42), Article 423001.

6. Van Bokhoven, J. A., & Lamberti, C. (2016). *X-ray absorption and X-ray emission spectroscopy, 2 volume set: Theory and applications* (vol. 1). Wiley.
7. Stransky, M. (2016). Monte Carlo simulations of ionization potential depression in dense plasmas. *Physics of Plasmas, 23*(1).
8. Zeng, J., Li, Y., Hou, Y., Gao, C., & Yuan, J. (2020). Ionization potential depression and ionization balance in dense carbon plasma under solar and stellar interior conditions. *Astronomy and Astrophysics, 644*, A92.
9. Janik, M., Myśliwiec, A. K., Koba, M., Celebańska, A., Bock, W. J., & Śmietana, M. (2017). Sensitivity pattern of femtosecond laser micromachined and plasma-processed in-fiber Mach-Zehnder interferometers, as applied to small-scale refractive index sensing. *IEEE Sensors Journal, 17*(11), 3316–3322.
10. Colliex, C., Kociak, M., & Stéphan, O. (2016). Electron energy loss spectroscopy imaging of surface plasmons at the nanometer scale. *Ultramicroscopy, 162*, A1–A24.
11. Arber, T. D., Bennett, K., Brady, C. S., Lawrence-Douglas, A., Ramsay, M. G., Sircombe, N. J., Gillies, P., Evans, R. G., Schmitz, H., Bell, A. R., & Ridgers, C. P. (2015). Contemporary particle-in-cell approach to laser-plasma modelling. *Plasma Physics and Controlled Fusion, 57*(11), 113001.
12. Wang, Y., Shukla, P. K., & Eliasson, B. (2013). Instability and dynamics of two nonlinearly coupled intense laser beams in a quantum plasma. *Physics of Plasmas, 20*(1).
13. Filinov, V. S., Bonitz, M., Fortov, V. E., Ebeling, W., Fehske, H., Kremp, D., Kraeft, W. D., Bezkrovniy, V., & Levashov, P. (2006). Monte Carlo simulations of dense quantum plasmas. *Journal of Physics A: Mathematical and General, 39*(17), 4421.
14. Filinov, V. S., Ivanov, Y. B., Fortov, V. E., Bonitz, M., & Levashov, P. R. (2013). Color path-integral Monte-Carlo simulations of quark-gluon plasma: Thermodynamic and transport properties. *Physical Review C—Nuclear Physics, 87*(3), 035207.
15. Shevkunov, S. V. (2002). A dense hydrogen plasma modeled by the path integral-Monte Carlo method. *Journal of Experimental and Theoretical Physics, 94*, 943–965.
16. Li, Z. X., & Yao, H. (2019). Sign-problem-free fermionic quantum Monte Carlo: Developments and applications. *Annual Review of Condensed Matter Physics, 10*(1), 337–356.
17. Longo, S. (2000). Monte Carlo models of electron and ion transport in non-equilibriumplasmas. *Plasma Sources Science and Technology, 9*(4), 468.
18. Trieschmann, J., Vialetto, L., & Gergs, T. (2023). Machine learning for advancing low-temperature plasma modeling and simulation. *Journal of Micro/Nanopatterning, Materials, and Metrology, 22*(4), 041504–041504.
19. Hau-Riege, S. P., Weisheit, J., Castor, J. I., London, R. A., Scott, H., & Richards, D. F. (2013). Modeling quantum processes in classical molecular dynamics simulations of dense plasmas. *New Journal of Physics, 15*(1), Article 015011.
20. Bonitz, M., Filinov, A., Abraham, J. W., & Loffhagen, D. (2018). Extending first principle plasma-surface simulations to experimentally relevant scales. *Plasma Sources Science and Technology, 27*(6), Article 064005.
21. Donkó, Z. (2009). Molecular dynamics simulations of strongly coupled plasmas. *Journal of Physics A: Mathematical and Theoretical, 42*(21), Article 214029.
22. Holm, D. D., Rawlinson, J. I., & Tronci, C. (2021). The bohmion method in nonadiabatic quantum hydrodynamics. *Journal of Physics A: Mathematical and Theoretical, 54*(49), Article 495201.
23. Curchod, B. F., & Martínez, T. J. (2018). Ab initio nonadiabatic quantum molecular dynamics. *Chemical reviews, 118*(7), 3305–3336.
24. Rosa, A., & Ricciardi, G. (2009). A time-dependent density functional theory (TDDFT) interpretation of the optical spectra of zinc phthalocyanine π-cation and π-anion radicals. *Canadian Journal of Chemistry, 87*(7), 994–1005.
25. Zurek, W. H. (2003). Decoherence, einselection, and the quantum origins of the classical. *Reviews of Modern Physics, 75*(3), 715.
26. Hu, B. L. (2014). Gravitational decoherence, alternative quantum theories and semiclassical gravity. *Journal of Physics Conference Series, 504*(1), 012021. IOP Publishing.

27. Le, Y., Zhang, Y., Gopalakrishnan, S., Rigol, M., & Weiss, D. S. (2023). Observation of hydrodynamization and local prethermalization in 1D Bose gases. *Nature, 618*(7965), 494–499.
28. Mirnov, V. V., Brower, D. L., Den Hartog, D. J., Ding, W. X., Duff, J., & Parke, E. (2014). Electron kinetic effects on interferometry, polarimetry and Thomson scattering measurements in burning plasmas. *Review of Scientific Instruments, 85*(11).
29. Muggli, P., & Hogan, M. J. (2009). Review of high-energy plasma wakefield experiments. *Comptes Rendus Physique, 10*(2–3), 116–129.
30. Döpp, A., Eberle, C., Howard, S., Irshad, F., Lin, J., & Streeter, M. (2023). Data-driven science and machine learning methods in laser–plasma physics. *High Power Laser Science and Engineering, 11*, Article e55.
31. Degrave, J., Felici, F., Buchli, J., Neunert, M., Tracey, B., Carpanese, F., Ewalds, T., Hafner, R., Abdolmaleki, A., de Las Casas, D., Donner, C., & Riedmiller, M. (2022). Magnetic control of tokamak plasmas through deep reinforcement learning. *Nature, 602*(7897), 414–419.
32. Bukov, M., Day, A. G., Sels, D., Weinberg, P., Polkovnikov, A., & Mehta, P. (2018). Reinforcement learning in different phases of quantum control. *Physical Review X, 8*(3), Article 031086.
33. Niu, M. Y., Boixo, S., Smelyanskiy, V. N., & Neven, H. (2019). Universal quantum control through deep reinforcement learning. *NPJ Quantum Information, 5*(1), 33.
34. Wakatsuki, T., Suzuki, T., Oyama, N., & Hayashi, N. (2021). Ion temperature gradient control using reinforcement learning technique. *Nuclear Fusion, 61*(4), Article 046036.
35. Pavone, A., Merlo, A., Kwak, S., & Svensson, J. (2023). Machine learning and Bayesian inference in nuclear fusion research: An overview. *Plasma Physics and Controlled Fusion, 65*(5), Article 053001.
36. Godyak, V. A., & Alexandrovich, B. M. (2015). Comparative analyses of plasma probe diagnostics techniques. *Journal of Applied Physics, 118*(23).
37. Dodin, I. Y., & Startsev, E. A. (2021). On applications of quantum computing to plasma simulations. *Physics of Plasmas, 28*(9).
38. Hsieh, D., Qian, D., Wray, L., Xia, Y., Hor, Y. S., Cava, R. J., & Hasan, M. Z. (2008). A topological Dirac insulator in a quantum spin hall phase. *Nature, 452*(7190), 970–974.

Chapter 12
Future Directions and Challenges

Quantum plasma physics has matured into a robust interdisciplinary domain at the intersection of quantum field theory, condensed matter physics, plasma dynamics, and modern high-energy phenomenology. With the preceding chapters of this volume having established the theoretical underpinnings, practical formulations, specialized domains of applicability, and experimental methodologies relevant to quantum plasmas, this final chapter transitions the reader toward the forward frontier charting the conceptual trajectories, technological ambitions, and foundational hurdles that define the emerging research landscape.

12.1 Open Theoretical Problems in Quantum Plasma Physics

As quantum plasma theory extends its reach toward more extreme physical regimes and increasingly intricate systems, its internal structure reveals a number of outstanding theoretical problems that challenge conventional techniques.

12.1.1 Nonperturbative and Strong Coupling Regimes

At high densities, low temperatures, or ultra-intense field environments conditions routinely encountered in high-energy laser–plasma setups, neutron star crusts, or early-universe relic plasmas the coupling strength between particles can grow so large that perturbative expansions, such as those employed in quantum Vlasov, or Wigner–Poisson approaches, become quantitatively unreliable and qualitatively incorrect. These situations give rise to nonperturbative dynamics, where collective excitations,

P. Kumar, *Quantum Plasma*, https://doi.org/10.1007/978-981-95-0481-7_12

vacuum instabilities, and coherent long-range correlations must be treated on equal footing with single-particle behavior.

The inadequacy of perturbative tools in these domains invites the development of nonperturbative frameworks, such as lattice formulations of quantum electrodynamics adapted for finite-temperature and finite-density plasma conditions. By discretizing spacetime and working within Euclideanized path integrals, lattice QED can, in principle, probe correlation functions, confinement-deconfinement-like transitions in plasma states, and nontrivial topological phenomena. However, applying these techniques to real-time dynamics in non-equilibrium quantum plasma remains an open technical frontier due to the infamous sign problem and the difficulty of implementing real-time evolution on Euclidean lattices.

Parallel to this, the formal apparatus of gauge/gravity dualities, particularly the AdS/CFT correspondence, has been deployed in related domains such as the quark-gluon plasma and could be adapted to the quantum plasma context. In particular, strongly coupled, magnetized plasmas with anisotropic stress-energy tensors may admit holographic duals where transport coefficients like shear viscosity, charge diffusion, and Hall conductivities can be computed geometrically. The challenge lies in constructing a precise dictionary between the bulk (gravitational) and boundary (plasma) theories, especially for systems not conformally invariant or possessing complex spinor-gauge field interactions.

Beyond formal apparatus, the breakdown of perturbative methods also implies the failure of kinetic assumptions commonly used in quantum plasma modeling. The Markovian approximation which underlies most Boltzmann-like treatments becomes suspect in strong coupling due to memory effects, long-range entanglement, and nonlocal collisions. Here, generalized quantum kinetic theories incorporating nonlocal self-energies, dynamical screening, and time-retarded interactions such as those built using the 2PI (two-particle irreducible) effective action formalism offer a promising route forward. These formulations, while computationally intensive, enable the inclusion of resummed propagators and dynamically evolving vertex corrections, capturing the essence of quantum correlations in nontrivial interaction regimes. In all these cases, the central theme is the failure of linear response and mean-field assumptions. The nonlinear, nonperturbative structure of quantum plasma at strong coupling requires a shift from particle-based pictures to field-theoretic, collective mode-centric descriptions. It is expected that the full realization of such a transition will demand new numerical techniques, possibly integrating quantum simulation platforms and tensor network methods to handle the exponential state space entanglement growth intrinsic to these systems [1].

12.1.2 Gauge Symmetry, Renormalization, and Higher-Order QED Corrections

A second class of foundational issues concerns the consistency and completeness of the field-theoretic formulations of quantum plasma models. Quantum plasma theory, particularly in the relativistic and magnetized domains, must rigorously preserve gauge symmetry, ensure renormalizability, and account for higher-order quantum electrodynamic corrections, especially when embedded in anisotropic or inhomogeneous backgrounds.

Traditional Maxwellian electrodynamics, when embedded within quantum kinetic, or fluidic models, often fails to account for the full suite of quantum field theoretical constraints. Podolsky's generalization of electrodynamics, which introduces a higher-order derivative term to regularize the ultraviolet structure of the electromagnetic field, offers a promising alternative by enabling a ghost-free theory with natural cutoff behavior. However, its consistent coupling to dynamic charged plasma environments is nontrivial. The presence of second-order derivatives in the field equations introduces additional propagating modes whose physical interpretation (e.g., longitudinal photons or plasma quasi-modes) is not yet fully settled, especially in the presence of dynamically varying charge densities.

The renormalization group (RG) flow of such models in magnetized or spatially inhomogeneous plasmas presents another unresolved issue. Most RG analyses assume Lorentz invariance, isotropy, and homogeneity, assumptions explicitly broken in magnetized plasma environments. The running of the fine-structure constant, dielectric response functions, and photon self-energies under RG flow becomes dependent on directional scales (set by the magnetic field) and local gradients (induced by inhomogeneity). Moreover, the presence of Landau quantization, Zeeman splitting, and cyclotron motion modifies the infrared structure of the plasma, which in turn feeds back into its effective QED behavior.

From a practical perspective, this leads to ambiguities in defining the effective coupling constants and determining the scale at which perturbative expansion may remain valid. The inclusion of higher-loop diagrams, vacuum polarization effects, and vertex corrections remains an open task, especially for in-medium QED where the fermion propagators are modified by the presence of collective excitations and the photon propagator acquires an effective mass [2].

12.1.3 Quantum Turbulence and Topological Effects

In contrast to the classical turbulence encountered in magnetohydrodynamics (MHD), quantum turbulence in plasmas emerges from the inherently quantized and topologically constrained nature of quantum fields. While classical turbulence is characterized by continuous vorticity fields and a cascade of eddies across scales, quantum turbulence arises from the discrete structure of vorticity, where circulation is quantized in units determined by Planck's constant and the elementary charge. In quantum plasmas especially in ultracold, strongly magnetized, or astrophysically extreme environments such turbulence involves rich topological phenomena including vortex quantization, helicity anomalies, and knotted field line configurations that are inaccessible within purely classical models.

The quantization of vorticity in such systems stems from the single-valuedness of the quantum wavefunction, leading to circulation integrals around closed loops being quantized via the Bohr–Sommerfeld condition. In systems involving coherent charged fluids such as superfluid components of neutron star interiors, or Bose–Einstein-condensed quasiparticles in photonic plasmas this gives rise to quantized vortex filaments. These filaments can entangle, reconnect, and support topologically protected structures such as Hopfions and torus knots. The reconnection dynamics of these filaments in a quantum plasma setting exhibit nonlocal correlations and anomalous energy transfer pathways that do not conform to the Kolmogorov cascade observed in classical turbulence.

One critical topological invariant in magnetized quantum plasmas is magnetic helicity, defined as $\mathcal{H} = \int \mathbf{A} \cdot \mathbf{B} d^3x$, where $\mathbf{A}$ is the vector potential and $\mathbf{B} = \nabla \times \mathbf{A}$ is the magnetic field. In ideal MHD, this quantity is conserved, providing a robust diagnostic for the topology of magnetic field lines. However, in quantum plasmas, particularly under non-equilibrium or dissipative conditions, quantum helicity anomalies may arise. These anomalies occur when quantum effects, such as chiral asymmetry or spin–magnetic coupling, lead to non-conservation of helicity even in the absence of classical resistivity. Notably, such effects can manifest through the axial anomaly in chiral quantum electrodynamics, where the divergence of the axial current is linked to $\mathbf{E} \cdot \mathbf{B}$, enabling a dynamic helicity-violating channel.

Magnetic reconnection a fundamental process that enables rapid topological rearrangement of field lines acquires new significance in quantum plasma. Classical models based on resistive diffusion and tearing modes fail to account for coherence and tunneling effects at microscopic scales. In quantum plasma, quantum-assisted reconnection mechanisms become relevant. These include tunneling of wavepackets across classically forbidden regions, non-Abelian phase entanglements in multi-component plasmas, and quantized circulation exchange during vortex filament annihilation. These phenomena contribute to a modified reconnection rate, potentially allowing for fast reconnection without invoking anomalous resistivity, and are deeply linked to the spectral properties of the field operators and quantum entanglement entropy of the modes involved.

A promising direction lies in the field-theoretic classification of topological states of quantum plasma, using tools from knot theory, Chern–Simons theory, and topological quantum field theory (TQFT). These approaches could yield a taxonomy of plasma states labeled by topological invariants, with implications for the stability and transition dynamics of magnetized plasma configurations. For example, knotted magnetic fields may provide metastable configurations with nontrivial topological protection, relevant for confinement in fusion devices or the longevity of magnetic structures in astrophysical bodies [3].

12.1.4 Unified Frameworks Across Scales

The diversity of physical conditions under which quantum plasma arises from ultra-cold condensed matter systems to relativistic astrophysical jets demands a theoretical architecture capable of bridging disparate energy, length, and interaction scales. Present-day models often treat quantum plasma within highly specialized frameworks: quantum hydrodynamics at low energies, kinetic field theory for moderately energetic systems, and perturbative quantum field theory for high-energy plasmas. However, such fragmentation leads to inconsistent scaling behaviors, incomplete transport coefficients, and incompatible treatments of gauge fields and correlation functions. This motivates the development of unified frameworks that retain fidelity across regimes and interpolate seamlessly between them.

One strategic approach involves mapping analogies between low-energy condensed matter systems and high-energy plasma models, particularly through emergent gauge symmetries, topological states, and quasiparticle behavior. Systems such as graphene, topological insulators, and Weyl semimetals exhibit chiral transport phenomena, axial anomalies, and Berry curvature-induced dynamics features that parallel those in chiral quantum plasmas and magnetized pair plasmas. Establishing formal correspondences between such systems allows controlled laboratory emulation of otherwise inaccessible plasma regimes, particularly those relevant to early-universe cosmology or black hole magnetospheres.

Another avenue is the application of effective field theory (EFT) methods that preserve symmetry principles while incorporating relevant degrees of freedom at each scale. By successively integrating out high-energy modes and encoding their influence via higher-dimensional operators, EFTs can link Planck-scale physics with the phenomenology of quantum plasma. For example, integrating quantum electrodynamics with nonlinear scalar fields representing order parameters for condensation or symmetry breaking could yield effective actions relevant for strongly correlated quantum plasma states with emergent collective excitations [4].

12.2 Frontier Computational and Experimental Challenges

12.2.1 Quantum Simulation and Emulation

The pursuit of an accurate understanding of quantum plasma dynamics spanning coherent many-body quantum effects, non-equilibrium field interactions, and quantized electromagnetic phenomena has necessitated the exploration of emulation techniques that replicate quantum plasma behavior in controllable laboratory environments. In this context, quantum simulators based on ultracold atomic gases, trapped ions, and photonic lattices have emerged as indispensable tools, not merely for qualitative analogy, but as platforms for quantitatively resolving parameter regimes currently inaccessible to direct measurement or classical computation.

Ultracold atomic systems offer unique advantages in this arena. By confining bosonic, or fermionic atoms in optical lattices with tunable interaction strengths, external potentials, and synthetic gauge fields, one can reproduce effective Hamiltonians mimicking those in low-energy plasma configurations. Notably, when atoms interact via long-range dipolar, or Rydberg-mediated forces, they emulate collective behaviors characteristic of Coulombic plasmas with quantum coherence. These systems permit the resolution of phase transitions, such as those associated with quantum criticality, or coherent mode condensation, under externally imposed analogs of magnetic confinement or radiation pressure.

Trapped ion arrays, due to their precision control and high-fidelity state preparation, further enable the real-time simulation of spin–boson couplings and non-trivial gauge interactions, thereby mirroring aspects of quantum plasmas governed by field-theoretic interactions. Notably, models simulating effective Abelian, or non-Abelian gauge theories such as emergent photon-like excitations or Schwinger pair production have been realized in these systems, offering a concrete experimental probe into regimes where gauge fields acquire dynamical character and couple nonlinearly to charged matter.

However, fundamental limitations persist in these platforms. Chief among them is the challenge of scaling such emulators to represent realistic, high-dimensional plasma systems with relativistic particle content, strong-field nonlinearity, or dissipative and decohering environments. The emergence of relativistic effects, essential for describing quantum plasma in intense laser fields or astrophysical jets, requires simulators capable of incorporating Dirac-type dispersion and Lorentz-invariant interactions, an area still under rapid theoretical development. Further, many-body decoherence due to system-environment entanglement remains a key constraint, especially for long-time evolutions required to model relaxation, transport, or non-thermal fixed points. Finally, while emulation of Hamiltonian dynamics has seen great progress, the incorporation of dissipative, open-system features (e.g., Landau damping, collisional relaxation) necessitates reservoir engineering that is not yet universally achievable across platforms.

Thus, while quantum simulation holds profound promise for exploring complex quantum plasma regimes, its current applicability remains most effective in modeling idealized or reduced-dimensionality systems. Bridging the gap toward fully emulating quantum plasma phenomena particularly those involving real-time relativistic and gauge-theoretic dynamics demands continued innovation in both experimental capability and theoretical embedding of plasma-relevant interactions [5].

12.2.2 High-performance and Quantum Computing for Plasma Simulation

The theoretical description of quantum plasma, especially in strongly correlated or non-equilibrium scenarios, routinely demands a computational framework capable of capturing nonperturbative dynamics, multi-scale interactions, and emergent phenomena far from equilibrium. This necessity has driven the development of advanced numerical techniques grounded in quantum field theory, such as the Two-Particle Irreducible (2PI) effective action formalism and the Schwinger–Keldysh contour method. These methods enable the computation of real-time evolution, memory effects, and collective excitations in a fully quantum setting. However, their implementation is computationally intensive, necessitating state-of-the-art high-performance computing (HPC) infrastructures.

Exascale computing platforms are thus not merely desirable, but essential for resolving quantum kinetic equations derived from 2PI or Kadanoff–Baym formalisms. These approaches, which self-consistently propagate the two-point correlators and effective propagators, scale unfavorably with system size, interaction complexity, and dimensionality. For instance, the evaluation of self-energy insertions and vertex corrections in a gauge-invariant manner leads to a hierarchy of coupled integro-differential equations whose solution requires adaptive grid techniques, parallelized iterative solvers, and careful renormalization procedures. The inclusion of spinor fields, gauge constraints, and fluctuating background geometries further exacerbates the computational demand.

To alleviate this bottleneck, hybrid computational schemes combining HPC with variational tensor networks, stochastic sampling, and machine-learned effective models have begun to play a complementary role. In particular, surrogate models trained on small-scale exact solutions can provide rapid predictions in parameter regimes where direct simulations are intractable. Nevertheless, caution must be exercised to ensure that such interpolative techniques retain fidelity to gauge invariance and conservation laws, which are non-trivially encoded in quantum plasma dynamics.

In parallel, quantum computing has emerged as a conceptual frontier for simulating quantum plasma systems. The fundamental premise is that a quantum processor comprising a network of qubits and entangling gates can natively represent the Hilbert space of a quantum field, thereby avoiding the exponential overhead of classical memory requirements. Quantum algorithms designed for lattice gauge theories, such as the digital quantum simulation of U(1) or SU(2) gauge fields, hold potential for modeling gauge-mediated interactions in quantum plasmas, including photon propagation, screening, and confinement-like phenomena.

Nevertheless, practical deployment of quantum algorithms remains constrained by current quantum hardware limitations, including decoherence, gate infidelity, and the relatively shallow circuit depth achievable before noise renders outputs untrustworthy. Moreover, while variational quantum eigensolvers and quantum phase estimation algorithms have demonstrated initial success in quantum chemistry and condensed matter systems, their extension to fully dynamical plasma evolution, particularly incorporating fermionic fields, time-dependent backgrounds, and open-system dynamics remains a formidable challenge. One promising approach involves hybrid quantum–classical schemes, where quantum processors compute non-trivial correlation functions or spectral properties, which are then fed into classical solvers for dynamical evolution [6].

12.2.3 Diagnostic Gaps and Instrumentation

The experimental validation of quantum plasma theories hinges critically on the development of diagnostic tools capable of probing the defining characteristics of these systems namely, quantum coherence, spin polarization, and microscopic electromagnetic field fluctuations on ultrafast temporal and nanoscopic spatial scales. Unlike classical plasmas, where macroscopic field variables and thermodynamic observables often suffice to characterize behavior, quantum plasmas demand access to phase-sensitive quantities, correlation functions, and internal degrees of freedom that are inherently nonlocal and often hidden from conventional diagnostics. This necessitates a paradigm shift in diagnostic science, rooted in the quantum nature of matter and light.

One of the foremost challenges is the noninvasive measurement of quantum coherence within the plasma ensemble. Quantum coherence manifests in the off-diagonal components of the reduced density matrix, encoding phase relationships between superposed quantum states. In many-body systems, coherence is central to collective excitations, entangled quasiparticle dynamics, and the emergence of macroscopic quantum phenomena such as condensates and coherent photon emission. Traditional electromagnetic diagnostics such as Langmuir probes or interferometric scattering inevitably disturb the plasma and fail to resolve coherent phase information at the single-particle, or field-theoretic level. Consequently, new methodologies based on quantum nondemolition measurements, homodyne tomography, or quantum back-action evasion are under active exploration. These techniques aim to extract observables such as the Wigner function, or Glauber coherence functions without collapsing the system's wavefunction, thereby enabling the real-time tracking of phase-space distributions and coherence decay.

Closely related to coherence diagnostics is the need to measure spin polarization and spin–current correlations within quantum plasma environments. In magnetized quantum plasmas, spin degrees of freedom couple non-trivially to magnetic fields, flow vorticity, and quantum fluctuations, giving rise to phenomena such as spin–vorticity coupling, magnetization waves, and anomalous transport. Measuring such effects requires spin-resolved detection with high spatial and temporal resolution. Recent advances in spin-resolved electron energy loss spectroscopy (SREELS), Mott scattering, and nitrogen-vacancy (NV) center magnetometry offer potential platforms for localized spin diagnostics. For instance, NV-based quantum sensors owing to their intrinsic atomic-scale sensitivity and tunable spectral response can serve as embedded probes within plasma-facing components, measuring local magnetic fields, spin fluctuations, and even current vortices with sub-nanometer precision. However, their deployment in high-temperature or dense plasma environments remains limited by thermal fragility and electromagnetic noise.

Beyond coherence and spin, an equally critical diagnostic frontier involves microfield fluctuations the rapidly varying local electric and magnetic fields arising from the stochastic motion of charged carriers and field quanta in quantum plasmas. These microfields influence energy level broadening, spectral line shifts (e.g., via the Stark effect), and the onset of collective instabilities. From a theoretical standpoint, the microfield is encoded in higher-order correlation functions such as $\langle E(x,t)E(x\prime,t\prime)\rangle$ or in the noise kernel of the electromagnetic field's Schwinger–Keldysh propagator. Experimentally, resolving such quantities demands ultrafast, high-sensitivity techniques that can temporally resolve field fluctuations on femtosecond scales and spatially on nanometer scales.

In this domain, attosecond streaking and pump–probe spectroscopy have shown great promise. Attosecond pulses generated via high-harmonic generation can interrogate electronic motion with unprecedented temporal resolution, revealing transient field strengths, phase shifts, and decoherence rates. Similarly, Thomson scattering using x-ray free-electron lasers (XFELs) offers a path to access dynamic structure factors and collective mode spectra, from which microfield distributions can be inferred. These techniques, however, require significant infrastructure and synchronization, limiting their accessibility.

Realizing real-time diagnostics at the necessary sensitivity and resolution thus demands both instrumentation innovation and theoretical support. Detector architectures must evolve to accommodate quantum-limited noise floors, single-photon resolution, and multi-dimensional data acquisition. Concurrently, theoretical models must translate quantum observables such as quasiprobability distributions, spectral functions, or non-Markovian memory kernels into experimentally measurable signatures. This translation is nontrivial, especially in nonequilibrium regimes where standard fluctuation–dissipation relations do not hold [7].

12.3 Cross-Disciplinary and Technological Applications

12.3.1 Quantum Plasma in Quantum Information Science

As the quantum technological revolution advances, the interface between quantum plasma physics and quantum information science (QIS) is emerging as a fertile ground for innovation. While traditionally these domains have evolved independently plasma physics rooted in many-body classical and semiclassical kinetics, and QIS built upon isolated qubits, entanglement, and coherence the discovery of quantum plasma phenomena exhibiting long-range coherence, collective quantum states, and photon-mediated coupling has opened novel conceptual and practical bridges. These include emerging fields such as quantum plasmonics, spintronic information transport via plasma modes, and decoherence-resilient signal propagation through structured plasma media.

A major cross-disciplinary frontier arises from quantum plasmonics, where collective charge oscillations plasmons at nanostructured interfaces exhibit quantum behavior that can be exploited in quantum networks. In metallic, or doped semiconductor nanostructures, surface plasmon polaritons (SPPs) support subwavelength confinement of electromagnetic energy, offering ultrafast and highly localized information carriers. When these structures are embedded within, or interfaced with quantum plasmas e.g., degenerate electron systems at nanoscales, or photonic cavities with quantized plasma oscillations the resulting plasmonic modes acquire quantum coherence and entanglement characteristics. These quantum plasmonic states can be engineered to act as flying qubits, or quantum buses, transmitting entanglement across nodes in a network. Further, the strong field localization inherent to plasmonic modes enables non-Markovian coupling between quantum emitters, thereby providing enhanced control over decoherence processes.

The core mathematical framework for analyzing such systems involves the coupling of quantum hydrodynamic, or quantum kinetic models (e.g., the Wigner–Poisson system) with quantized electromagnetic field modes under boundary confinement. In this context, quantum plasma oscillations modify the local density of photonic states, enabling Purcell-like enhancements, or suppressions of spontaneous emission, and also influence two-photon interference statistics. A quantum field theoretical description utilizing the Keldysh contour or input–output formalism becomes essential to capture the interplay between plasmonic coherence and dissipation.

Parallelly, spin-polarized quantum plasmas are increasingly considered as functional components in spintronic and photonic information transport platforms. In environments where spin-dependent interactions dominate such as magnetized quantum wells, low-dimensional heterostructures, or optically driven plasma states the coupling of spin to charge and current densities leads to novel modes such as spin–plasmons and magnetoplasmons. These collective excitations carry spin information over mesoscopic distances and may serve as robust channels for spin-based logic gates, or memory elements. Importantly, such spin–plasma modes can be engineered to be less susceptible to environmental decoherence compared to conventional spintronic channels, owing to their collective and topologically protected nature.

The theoretical modeling of such systems requires a simultaneous treatment of the spin–orbit coupling, Berry curvature effects, and the evolution of spinor-valued distribution functions. Moreover, gauge-theoretic approaches incorporating non-Abelian potentials are often invoked to describe coherent spinor transport in plasma media. These formalisms are essential to accurately capture the spin Hall-like effects and quantum anomalous transport that emerge in magnetized and strongly coupled quantum plasmas [5].

12.3.2 Nanoplasma and Quantum Photonics

A second rapidly evolving axis of cross-disciplinary relevance is the application of quantum plasma physics to ultrafast laser–matter interactions at nanoscopic scales. At these scales, the plasma is neither entirely classical nor strictly quantum, but occupies a regime where quantum degeneracy, strong coupling, and field quantization co-exist a regime particularly relevant in nanoplasma formation induced by femtosecond or attosecond laser pulses. In such scenarios, the ionization of nanoclusters or thin films by ultrashort pulses generates transient, high-density electron populations confined within spatial domains on the order of tens of nanometers and evolving over timescales comparable to the electron plasma period. These systems exhibit quantum collisionality, nonlocal dielectric responses, and coherent field feedback, placing them outside the scope of standard plasma models.

A quantum kinetic description of nanoplasmas involves the use of Wigner–Moyal formalisms, where the evolution of phase-space quasidistributions is driven by both classical forces and quantum interference terms. The response of the plasma to the laser field is mediated by nonlocal permittivity tensors, and the onset of self-organized structures (such as coherent current sheets or localized fields) is governed by the interplay of quantum diffraction and collective Coulomb effects. These effects are critical for controlling laser–matter interactions in nanofabrication, high-harmonic generation, and laser-driven particle acceleration.

Of particular importance are hollow-core plasmas and waveguide-confined environments, which constitute engineered geometries where plasma and light are co-confined in tight spatial domains. In such structures, plasma waves can be guided alongside photons, leading to coupled plasma–photon polariton modes. Quantum electrodynamical (QED) effects in these confined systems, including vacuum polarization and radiative corrections to dispersion relations, become non-negligible. For instance, in ultra-intense fields, the effective plasma permittivity can exhibit nonlinearities induced by the Heisenberg–Euler action, while quantum recoil and tunneling effects modulate the propagation of plasma modes in the longitudinal direction.

Moreover, strong-field QED processes such as Breit–Wheeler pair creation, photon splitting, or vacuum birefringence may be realized in hollow-core plasmas where field strengths approach the Schwinger limit in a confined geometry. Here, the plasma serves not only as a medium for light–matter interaction, but also as an active participant in quantum field processes. The confinement enhances the photon–photon and photon–plasma coupling, creating a setting where nonperturbative quantum field effects are both theoretically tractable and experimentally observable.

To model these interactions, one must employ field-theoretic treatments that combine quantum plasma response functions with quantized electromagnetic field operators in curved or bounded spacetime manifolds. The eigenmodes of such systems are solutions to modified Maxwell–Dirac–Klein–Gordon systems, subject to boundary and gauge constraints that reflect the nontrivial topology of the confinement geometry.

The mutual fertilization between quantum plasma physics and frontier domains such as quantum information science and nanophotonic technology is catalyzing a new era of both fundamental and applied research. Quantum plasmas, with their unique ability to mediate collective quantum interactions, support spin–charge coupling, and manipulate photonic modes at ultrafast timescales and nanoscales, are positioned to contribute significantly to future quantum technologies. Theoretical and experimental efforts aimed at harnessing these properties not only expand the scope of quantum plasma physics but also illuminate pathways toward next-generation quantum devices where plasma phenomena are no longer peripheral, but central to functionality [8].

12.3.3 Biomedical and Environmental Prospects

The emergence of quantum plasma research in biomedical and environmental sciences represents a transformative shift in how matter–field interactions are conceptualized and applied at bio-relevant and ecological scales. Plasma medicine, a field that has matured significantly over the past two decades through low-temperature, partially ionized plasmas for sterilization and tissue interaction, is now poised to undergo a second revolution by incorporating quantum plasma principles into diagnostic, therapeutic, and atmospheric control modalities. This integration requires a comprehensive reevaluation of plasma–biological matter interactions, especially when the characteristic dimensions, energy scales, or timescales of the interaction approach quantum limits be it through coherence, tunneling, or quantized energy transfer mechanisms.

In the domain of plasma medicine at quantum precision, quantum plasmas comprising degenerate electrons, coherent photons, and spin-polarized carriers enable unprecedented control over energy delivery, momentum transfer, and biochemical pathway activation. For example, in quantum-enhanced sterilization, the ability to tune plasmonic and polaritonic modes within structured plasma sources allows localized energy deposition at resonance frequencies corresponding to molecular bond energies (e.g., C–H, N–O, or S–S bonds), resulting in targeted microbial inactivation without bulk heating. This precise deposition is a result of resonance conditions that arise when quantized plasma oscillations couple with vibrational modes in the molecular spectra of pathogens.

Mathematically, these interactions can be described through a non-equilibrium density matrix formalism, where the coherent interaction Hamiltonian includes time-dependent plasma fields, spin–orbit interactions, and quantized electromagnetic field operators. The decoherence dynamics and population inversion of biological macromolecules under plasma exposure are then governed by a master equation of Lindblad, or Redfield type, parameterized by field–molecule coupling constants and plasma-mode lifetimes. The critical control mechanism lies in the manipulation of quantum coherence lifetimes (T_2 processes) and energy relaxation times (T_1

processes), which can be dynamically tuned through plasma parameters such as carrier density, confinement geometry, and external magnetic field configuration.

Another profound frontier is targeted cancer therapy via quantum plasmas, where ultrashort plasma pulses generated by tightly focused lasers or microwave cavities produce quantum nanoplasma states in situ within biological tissue. The resultant local heating, ionization, and reactive species generation is confined within nanometer to micrometer scales, ensuring minimal collateral damage to healthy cells. When embedded with engineered nanoparticles (e.g., gold nanorods or quantum dots), the system enables resonant quantum plasmon excitation, which enhances selectivity and cytotoxicity at the tumor site through thermoplasmonic and non-thermal pathways, including singlet oxygen production and oxidative stress induction.

From an environmental perspective, quantum plasmas are instrumental in atmospheric interaction control, especially for high-altitude applications and planetary boundary layer modulation. In rarefied environments where quantum degeneracy of charged particles is non-negligible, such as upper ionospheric layers, or cryogenic atmospheric interfaces, the collective modes of the plasma interact with trace gases in ways that defy classical predictions. These include quantum-modified ion–neutral reaction cross-sections, nonclassical electron attachment processes, and coherence-assisted radical generation. The generation of quantum atmospheric plasmoids coherently bound charge–field configurations stabilized by quantum pressure and field quantization presents a novel pathway for pollutant degradation, cloud seeding, or artificial ionospheric layer construction.

These effects are theoretically captured through a hybrid quantum–classical treatment of the Boltzmann or BGK-type kinetic equations, where quantum corrections such as the Bohm potential, spin–vorticity coupling, and coherence flux terms are retained. The reactive pathways are further modulated by photon statistics and field quantization, necessitating a fully time-resolved, operator-based formulation of atmospheric plasma dynamics. This suggests a new paradigm of "quantum geoengineering" where atmospheric behavior is controlled at the level of elementary quantum interactions.

12.3.4 Compact Plasma-Based Accelerators and Fusion Reactors

One of the most promising technological frontiers of quantum plasma physics lies in the design and optimization of next-generation plasma-based accelerators and fusion reactor systems, especially under regimes where quantum corrections to classical plasma models become significant. These quantum corrections become essential either due to the high densities and low temperatures leading to Fermi degeneracy and quantum diffraction or due to extreme field conditions that induce quantum recoil and radiation reaction effects on the accelerated particles.

In the context of compact plasma-based accelerators, particularly those driven by ultraintense laser pulses (laser wakefield acceleration, LWFA) or high-gradient microstructured beamlines, quantum plasma physics introduces nontrivial corrections to the wakefield structure, phase velocity, and energy transfer dynamics. Quantum degeneracy pressure, relativistic mass corrections, and spin–orbit coupling alter the dispersion relation of the Langmuir and transverse electromagnetic waves that mediate the acceleration process. These corrections become crucial in regimes approaching the quantum critical density, where the de Broglie wavelength of electrons becomes comparable to the plasma skin depth.

A rigorous theoretical treatment of these systems involves extending the classical Vlasov–Maxwell system to incorporate quantum kinetic effects via the Wigner–Maxwell, or Dirac–Maxwell equations. This includes the self-consistent evolution of the Wigner function with full inclusion of quantum interference terms, nonlocal field operators, and radiation damping. For instance, the quantum radiation reaction force becomes relevant at field strengths near or beyond the Schwinger limit, modifying the energy gain and spectral broadening of accelerated particle beams. These quantum corrections can also help suppress classical beam instabilities through modified Landau damping rates or coherent field backreaction, thereby improving beam quality and luminosity.

In the domain of fusion energy, particularly inertial confinement fusion (ICF) and magneto-inertial fusion (MIF), quantum plasmas play a dual role both in the core fuel dynamics and in the diagnostic and control architectures. In ICF, during peak compression, the fuel pellet reaches conditions where the plasma becomes partially degenerate and strongly coupled, invalidating classical assumptions of ion–electron equilibration, opacity, and heat transport. The quantum-statistical effects, including Fermi–Dirac pressure, quantum tunneling in fusion cross-section enhancement, and bound-state continuum transitions, must be incorporated to accurately model ignition thresholds and energy gain.

The development and deployment of such diagnostics require a convergence of quantum optics, non-equilibrium field theory, and high-performance plasma simulation. The theoretical models must incorporate both the backaction of the probe fields on the plasma and the quantum measurement-induced collapse of the system's wavefunction a fundamentally new approach compared to classical probe diagnostics.

12.4 Interfacing Quantum Plasma with Emerging Paradigms

12.4.1 *Quantum Thermodynamics and Entropy Production*

As the domain of quantum plasma physics matures, there is a compelling imperative to forge robust connections with the broader theoretical infrastructure of quantum thermodynamics. The interplay between non-equilibrium statistical mechanics, quantum coherence, and collective plasma behavior has begun to illuminate a new theoretical frontier wherein quantum plasmas are reconceived not merely as statistical ensembles of quantum-charged particles, but as intrinsically open quantum systems governed by irreversibility, information flow, and energy-resource constraints.

In conventional plasma theory, entropy evolution is commonly approached through macroscopic quantities defined via Boltzmann, or Gibbs distributions. However, in quantum plasmas, particularly those embedded in electromagnetic cavities, astrophysical structures, or controlled laboratory environments where quantum coherence and entanglement persist over macroscopically relevant timescales, this perspective must be fundamentally revised. Quantum plasmas in such settings are best modeled as open quantum systems subsystems that interact with an environment (e.g., radiation baths, external fields, boundary layers) necessitating a non-unitary evolution described via master equations or path integral approaches on the Schwinger–Keldysh contour.

In this non-equilibrium framework, entropy production arises not solely from dissipative particle interactions, but from quantum decoherence, information leakage to the environment, and internal redistribution of quantum correlations. The von Neumann entropy $S = -\mathrm{Tr}[\rho \ln \rho]$, where ρ is the reduced density matrix of the plasma subsystem, does not remain invariant under non-unitary dynamics. The entropy production rate $\sigma(t) = \frac{dS}{dt} + \Phi(t)$, where $\Phi(t)$ represents the entropy flux to the environment, encapsulates how quantum correlations decay and how microscopic reversibility is broken.

Recent developments show that in strongly coupled quantum plasmas where the interaction energy per particle becomes comparable to thermal, or quantum kinetic energy the entropy landscape is shaped not only by collisional or radiative processes, but also by the structure of the plasma's entangled eigenstates. This imposes a new form of 'entropic topology' in phase space, where the configurational manifold becomes stratified by coherence-driven entropy barriers. These effects are particularly pronounced in photonic–plasma systems with cavity confinement, wherein photon number statistics and mode entanglement introduce non-trivial entropy growth mechanisms that defy classical thermodynamic intuition.

A profound development in nonequilibrium quantum thermodynamics is the emergence of fluctuation theorems, such as the Jarzynski equality and Crooks relation, which provide exact identities for the probability distributions of work and entropy production, irrespective of system size or proximity to equilibrium. These theorems have been rigorously derived for open quantum systems undergoing non-Markovian dynamics, thus offering a natural extension to quantum plasmas subjected to time-dependent fields, laser pulses, or turbulent fluctuations.

The incorporation of fluctuation theorems into quantum plasma research reveals key insights into time-reversal symmetry breaking, irreversibility, and energetics of transient processes such as quantum shock waves, decoherence bursts, and local heating events. For instance, in magnetized quantum plasmas under time-varying field configurations, the work distribution becomes asymmetric, reflecting the quantum hysteresis and memory effects embedded in the plasma response. The Jarzynski equality provides a route to compute free energy differences between plasma states from non-equilibrium measurements thereby enabling diagnostics of plasma reconfigurations that do not rely on equilibrium assumptions.

Complementary to these thermodynamic principles is the conceptual framework of quantum resource theories, which treat coherence, entanglement, and athermality as operational resources. Within this paradigm, a quantum plasma can be analyzed in terms of its ability to perform useful work, generate coherence, or enable communication protocols. The degree of non-passivity in the plasma's state i.e., its deviation from thermal equilibrium quantifies the available free energy for performing unitary or dissipative operations.

The path forward involves a dual imperative: formalizing the theoretical linkages between open quantum systems theory and collective plasma dynamics, and developing experimental tools to measure entropy production, coherence loss, and fluctuation statistics in plasma systems. One promising direction involves embedding quantum plasma simulations within tensor network frameworks or quantum computing architectures, where entropic and thermodynamic constraints can be explicitly encoded and monitored. Another frontier is the design of quantum control protocols to harness non-equilibrium plasma states as working mediums in nanoscale engines or information processors.

However, several fundamental challenges remain. The definition of local versus global entropy in spatially extended, magnetically structured plasmas is still under debate, particularly in strongly nonlocal or topologically constrained regimes. Similarly, the construction of effective field theories that embed thermodynamic irreversibility, yet remain compatible with gauge and Lorentz symmetries, is incomplete. A quantum plasma's entropy is not merely a measure of disorder it is a dynamic quantity reflecting entanglement, coherence, and information flow in a non-trivial many-body system. The full exploration of this landscape will require new mathematical tools, interdisciplinary synthesis, and persistent theoretical innovation.

12.4.2 *Quantum Gravity, Cosmology, and Early Universe Plasmas*

Quantum plasma physics, while rooted in electromagnetic and many-body interactions of charged particles, finds an increasingly rich interface with high-curvature and extreme energy density environments. The nascent overlap between quantum plasma theory and quantum gravity, as well as its relevance to early-universe cosmology, introduces a new class of conceptual and formal challenges. These extend the scope of quantum plasma far beyond traditional laboratory and astrophysical settings into regimes where spacetime itself is a dynamic participant in plasma evolution.

The early universe, during its first microseconds, underwent a sequence of phase transitions wherein the coupling constants of the fundamental interactions evolved and symmetry breaking dynamically restructured the vacuum. During the electroweak epoch and prior to hadronization, the primordial plasma was a strongly interacting mixture of photons, leptons, gluons, and quarks a QED-QCD hybrid plasma governed by ultra-relativistic field dynamics. The classical notion of Debye screening, familiar in ordinary plasmas, here must be generalized to accommodate chromodynamic screening, thermal gluon masses, and quark–antiquark polarization clouds.

Quantum plasmas in this cosmological regime are inherently relativistic and non-Abelian, and thus the statistical structure of their field fluctuations, transport coefficients, and instabilities cannot be derived from linearized Vlasov–Poisson systems. Instead, the Hard Thermal Loop (HTL) framework provides the effective propagators and vertices required to describe gauge-invariant excitations in thermal QCD plasmas. Importantly, the interplay between color conductivity and thermal fluctuations leads to transient magnetic instabilities (e.g., Weibel-like) that could seed primordial magnetic fields.

The analogies with electromagnetic quantum plasmas become tractable in certain Abelian sectors, or under symmetry-breaking reductions. For example, lepton–photon plasmas post-electroweak symmetry breaking retain a resemblance to relativistic QED plasmas under extreme temperatures, where radiative corrections significantly modify dispersion relations and introduce non-local dielectric responses. These early-universe plasma conditions can be approximated in high-energy collider environments, where strongly-coupled QCD plasmas (e.g., in RHIC and LHC) provide experimental insights into the dynamics of collective flow, anisotropy, and viscosity in relativistic quantum matter.

An important direction of research involves constructing quantum plasma analogs of cosmological epochs using quantum simulation platforms, including ultracold atoms, trapped ions, and strongly coupled photonic structures. These analogs offer an emergent 'tabletop cosmology', where horizon-like structures, field fluctuations, and thermal gradients can mimic conditions near the Big Bang or during inflationary reheating. By embedding quantum plasma interactions within curved metric analogs such as in Bose–Einstein condensates exhibiting acoustic horizons insights into early-universe entropy production and field amplification become experimentally tractable.

The confluence of quantum field theory, curved spacetime, and plasma physics culminates in the exploration of effective field excitations in non-inertial frames or near analog event horizons. The Unruh effect, where an observer undergoing constant acceleration perceives a thermal bath in Minkowski vacuum, can be formally derived using the Bogoliubov transformation of field modes. In a quantum plasma, if collective excitations such as Langmuir or Alfvén waves propagate through a spatially accelerated background or gradient-index medium, one may engineer analog thermal spectra that mirror the Unruh effect. The effective temperature $T_U = \frac{\hbar a}{2\pi c k_B}$, where a is the acceleration, governs the spectral character of these analog emissions.

Likewise, Hawking radiation analogs can be realized in dispersive media where horizon conditions are simulated by a variable refractive index or moving plasma fronts, leading to pair creation of field quanta analogous to photon–plasma polaritons. In such setups, the plasma dielectric function plays the role of an effective metric, and event-horizon-like conditions emerge when the phase velocity of the wave matches the flow velocity of the plasma medium. These conditions are being actively pursued in optical plasmas, photonic crystals, and strongly confined nanoplasma structures.

Vacuum polarization, a hallmark of quantum electrodynamics in strong fields also acquires plasma analogs. In ultra-strong magnetic fields approaching the Schwinger limit, virtual electron–positron pairs contribute to the effective permittivity and permeability of the vacuum. When immersed in a quantum plasma, the interplay between the field-induced vacuum birefringence and plasma-induced dispersion results in mixed-mode propagation effects, where collective plasma modes are dressed by quantum vacuum fluctuations. These effects are not only theoretically rich but also experimentally observable in high-field laser–plasma interactions and astrophysical magnetospheres.

12.4.3 Machine Learning and Data-Driven Models

The complexity of quantum plasma dynamics, especially in high-dimensional phase spaces and turbulent regimes, increasingly demands computational strategies that extend beyond first-principles simulations. Machine learning (ML) and data-driven models, with their capacity to interpolate high-dimensional nonlinear manifolds and extract low-dimensional representations of complex behavior, have emerged as indispensable tools in modern quantum plasma research. However, their adoption is accompanied by a suite of theoretical, interpretive, and epistemological challenges that must be rigorously addressed to ensure scientific robustness.

Quantum plasmas exhibit a rich spectrum of collective behaviors: from nonlinear turbulence and quantum vortices to mode-coupling and phase transitions. Capturing the full phenomenology of these effects in either analytic formalisms or direct numerical simulations is computationally demanding and often intractable beyond moderate scales. ML architectures particularly convolutional neural networks (CNNs), recurrent neural networks (RNNs), and variational autoencoders (VAEs) provide powerful alternatives to uncover underlying patterns, detect emergent structures, and classify dynamic regimes.

For turbulence in magnetized quantum plasmas, unsupervised learning techniques such as principal component analysis (PCA) and t-distributed stochastic neighbor embedding (t-SNE) allow for the dimensionality reduction of complex vorticity fields, enabling the identification of coherent structures, intermittency bursts, and phase-slip events. Spectral methods combined with ML classifiers can distinguish between linear and nonlinear mode spectra, helping to isolate unstable modes and prethermalization pathways.

Wave-mode detection, especially in quantum kinetic simulations, benefits from ML-enhanced spectral diagnostics, where neural networks trained on synthetic datasets can rapidly identify the onset of Langmuir, Bernstein, or spin-polarized Alfvén modes. Bayesian inference methods further permit the extraction of uncertainty bounds, ensuring that mode classification is not merely heuristic, but statistically meaningful.

In quantum plasmas exhibiting phase transitions, clustering algorithms and supervised classification can be deployed to detect order parameters, domain wall formation, or condensate fragmentation in photonic–plasma states. More ambitiously, generative adversarial networks (GANs) have been used to synthesize entire plasma evolution sequences that preserve statistical symmetries and conservation laws, offering surrogate models for real-time simulation and control.

Despite their apparent success, ML models face significant limitations in quantum plasma applications, particularly with respect to interpretability, physical consistency, and generalizability. Many ML approaches function as black boxes, obscuring the physical mechanisms responsible for the patterns they detect. This opacity is incompatible with the foundational goals of plasma physics, which seek causal explanations grounded in field theory, conservation laws, and symmetry principles.

To bridge this gap, physics-informed machine learning (PIML) has emerged as a critical paradigm. In PIML, governing equations such as the quantum Vlasov–Maxwell system, Wigner–Moyal dynamics, or generalized Ohm's law are explicitly encoded into the loss functions of neural networks. This ensures that the learned representations conform to physical constraints such as charge conservation, gauge invariance, or Lorentz symmetry. However, embedding such constraints at scale requires sophisticated architectural design and often leads to trade-offs between flexibility and interpretability.

Another challenge lies in dataset generation. Quantum plasma simulations capable of providing high-fidelity training data are computationally intensive, and often limited by resolution, noise, or model assumptions. Experimental datasets, where available, are typically sparse and contaminated by measurement noise, requiring hybrid approaches that blend theory-driven simulations with experimental priors through Bayesian fusion, or transfer learning.

Moreover, quantum plasmas often inhabit regimes where classical assumptions about smoothness, locality, or ergodicity break down. This complicates the statistical learning process, particularly when dealing with non-Gaussian distributions, anomalous transport, or topological defects. Thus, a central task remains: to develop

ML models that do not merely replicate observed patterns, but reveal new structures, suggest physical mechanisms, and propose falsifiable hypotheses.

The synthesis of machine learning and quantum plasma physics represents not merely a computational convenience, but a new epistemic modality. It reorients scientific inquiry from equation-solving to structure-discovery, while demanding a rigorous balance between predictive power and physical insight. The challenge ahead is to train models not just to fit data, but to discover physics thereby advancing quantum plasma theory into an era of hybrid intelligence.

12.5 Synthesis and Future Vision

12.5.1 Recapitulation of Theoretical, Computational, and Experimental Themes

From a theoretical standpoint, quantum plasma research has matured through layers of abstraction: starting from mean-field quantum hydrodynamics, progressing through quantum kinetic and Wigner–Moyal formulations, and culminating in field-theoretic approaches such as Schwinger–Keldysh formalism and 2PI (two-particle-irreducible) effective actions. These layers reveal the increasing necessity of capturing quantum coherence, decoherence, entanglement, and strong coupling effects across multiple scales.

Computationally, quantum plasma simulations now extend beyond traditional particle-in-cell (PIC) frameworks to include quantum lattice algorithms, quantum Vlasov–Poisson solvers, and real-time quantum field theory simulations under non-equilibrium conditions. These developments have enabled the resolution of phenomena such as pair production, wave-function bifurcations, quantum turbulence, and quantum Landau damping with unprecedented fidelity.

Experimentally, while direct probing of fully quantum plasma remains challenging, there is notable progress in generating quantum-degenerate electron beams, plasmonic crystals with quantum coherent excitations, and ultra-intense laser–plasma interactions that mimic early-universe or neutron star conditions. Techniques such as attosecond laser interferometry, quantum dot diagnostics, and cavity QED setups have opened portals to probe these systems with high spatial and temporal resolution.

12.5.2 Defining Central Questions for the Next Decades

Despite its growing sophistication, quantum plasma physics is far from complete. Several deep and unresolved questions will shape the discipline's trajectory over the coming decades:

1. Unified Effective Actions: Can a generalized effective field theory (EFT) emerge that unifies low-energy quantum plasma phenomena with high-energy QED and QCD plasma regimes? What role will non-Abelian gauge fields, topological terms, and higher-derivative corrections play in this synthesis?
2. Non-Perturbative Quantum Dynamics: What are the universal principles governing quantum turbulence, phase transitions, and transport in regimes where traditional perturbative expansions fail? Can insights from holographic duality or lattice quantum chromodynamics (lattice QCD) be translated to plasma dynamics?
3. Quantum–Classical Transition and Irreversibility: How does a quantum plasma evolve toward classical behavior under extreme decoherence? What governs the emergence of entropy, irreversible heating, and memory loss in out-of-equilibrium evolution?
4. Extreme Magnetic Topologies and Spin Dynamics: How does spin–vorticity coupling, spin-Hall currents, and anomalous magnetic moment renormalization affect the global dynamics of magnetized quantum plasmas in neutron stars, white dwarfs, and magnetars?
5. Quantum Vacuum Engineering: To what extent can artificial quantum vacuum structures such as those created in cavity QED or metamaterials be used to simulate, probe, or even manipulate plasma states that are otherwise inaccessible in nature?

These questions are not merely academic; they lie at the heart of humanity's quest to understand matter and energy at their most fundamental levels.

12.5.3 Toward Unified Frameworks: QED, QCD, Optics, and Plasma

A future vision for quantum plasma science cannot afford to remain compartmentalized. Theoretical and phenomenological intersections must be fostered between:

i. Quantum Electrodynamics and Quantum Plasma: QED processes such as vacuum polarization, radiation reaction, and pair production must be understood not as isolated events but as constituents of collective plasma behavior in ultra-intense fields. Real-time QED corrections must be systematically embedded within quantum plasma Hamiltonians.
ii. QCD Plasmas and Dense Matter: Insights from quark-gluon plasma studies especially regarding deconfinement transitions, color screening, and chiral anomalies must be extrapolated to explore potential "hadronic quantum plasma" regimes in heavy-ion collisions, early-universe epochs, and compact astrophysical objects.
iii. Quantum Optics and Photonic Plasmas: Strong light–matter coupling regimes, especially in cavity-confined or photonic crystal environments, offer a fertile ground for engineering designer plasmas with tunable quantum coherence.

Quantum optomechanics and squeezed-light injection can be harnessed to manipulate plasma fluctuations and mode spectra.

iv. Non-Abelian and Topological Effects: Future theoretical frameworks must integrate topological solitons, Chern–Simons terms, Berry curvatures, and emergent gauge fields. These features will play a pivotal role in understanding topologically protected plasma modes, quantum anomalies, and spin transport in inhomogeneous environments.

In essence, the unified vision is not simply the amalgamation of these subfields, but the emergence of a transdisciplinary science in which plasmas are not just ionized gases, but laboratories for exploring the quantum structure of space, time, and matter.

12.5.4 Quantum Matter Under Extreme Conditions: Toward an Integrated Theory

Ultimately, the aspiration of quantum plasma research is to become a cornerstone of a broader framework for quantum matter under extreme conditions. Such a theory must accommodate the following:

(i) Strong Coupling and Relativity: It must describe systems far from equilibrium, strongly coupled, and relativistic, in which lightcone fluctuations, Schwinger mechanisms, and Hawking–Unruh radiation become dynamically relevant.

(ii) Multiscale and Multiphysics Interactions: It must incorporate interactions across energy scales from electron volts to tera-electron volts, alongside multiphysics couplings between fluid dynamics, electromagnetism, quantum mechanics, and statistical physics.

(iii) Quantum Information and Entanglement: It must integrate the language of quantum information, using entanglement entropy, decoherence functionals, and quantum Fisher information to understand coherence loss and information flow in plasma dynamics.

(iv) Thermodynamic Consistency and Renormalization: It must obey rigorous thermodynamic and renormalization constraints, ensuring consistent entropy production, gauge invariance, and ultraviolet completeness.

Thus, the synthesis of quantum plasma physics into a grander, integrated framework is not merely a theoretical ambition but an imperative dictated by the physical phenomena themselves. The behaviors of quantum matter under extreme conditions ranging from laboratory-generated pair plasmas to the early universe demand a new physics of cohesion, where particles, fields, and information are treated on equal footing.

References

1. Hu, G., Huang, W. W., Cai, R., Wang, L., Yang, C. H., Cao, G., Xue, X., Huang, P., & He, Y. (2025). Single-electron spin qubits in silicon for quantum computing. *Intelligent Computing, 4*, 0115.
2. Gupta, S., Zhang, J. J., Lei, J., Yu, H., Liu, M., Zou, X., & Yakobson, B. I. (2025). Two-dimensional transition metal dichalcogenides: A theory and simulation perspective. *Chemical Reviews, 125*(2), 786–834.
3. Yang, Y. (2025). The role of quantum computing in advancing plasma physics simulations for fusion energy and high-energy. *Frontiers in Physics, 13*, 1551209.
4. Garcia-Gasulla, M., & Mantsinen, M. J. (2025). Challenges and opportunities in exascale fusion simulations. *Nature Reviews Physics*, 1–10.
5. Tennie, F., Laizet, S., Lloyd, S., & Magri, L. (2025). Quantum computing for nonlinear differential equations and turbulence. *Nature Reviews Physics*, 1–11.
6. Cao, C., & An, Q. (2025). Elucidating thin film growth mechanisms for high-performance II–VI photovoltaic semiconductors: Simulation-driven insights and challenges. *CrystEngComm.*
7. Caldwell, A., Torre, S. D., Ent, R., Levy, A., Newman, P., Olness, F., & Rojo, J. (2025). Future opportunities with Lepton-Hadron collisions. arXiv preprint arXiv:2503.18208.
8. Hu, M., Fan, J., Tong, Y., Sun, Z., & Jiang, H. (2025). Deep learning for ultrafast X-ray scattering and imaging with intense X-ray FEL pulses. *Advanced Optical Technologies, 14*, 1546386.

Index

A
Active Galactic Nuclei (AGN), 12
Alfvén eigenmodes, 261
Alfvénoid oscillations, 153
Alfvén waves, 36, 254
Ampère–Maxwell law, 119
Ampère's law, 37, 41, 179
Angle-Resolved Photoemission Spectroscopy (ARPES), 330
Attosecond pulse generation, 315

B
Ballooning instability, 24
Bargmann–Michel–Telegdi (BMT) equation, 187
Barnett effect, 258
Bayesian fusion, 383
Bayesian inference, 355
BCS-type energy gap, 223
Beam–plasma instabilities, 190
Bernstein modes, 162, 302
Berry curvature, 257
Berry-phase curvature, 171
Bethe–Salpeter equations, 214
Bhatnagar–Gross–Krook (BGK), 266
Biermann battery effects, 154
Bloch-type equation, 118
Bogoliubov modes, 80
Bogoliubov transformation, 382
Bohm potential, 72, 73, 110, 113–116, 178, 179
Bohm potential gradient, 180
Bohr magneton, 77, 166, 176
Bohr–van Leeuwen theorem, 157, 158
Boltzmann distribution, 15, 63
Boltzmann equation, 26, 110
Boltzmann relation, 15
Boltzmann-weighted population, 197
Bose–Einstein, 154
Braginskii's formulation, 249
Breit–Wheeler pair, 307, 375
Breit–Wheeler pair production, 291
Bremsstrahlung, 99
Brown dwarfs, 225

C
Canonical quantum vorticity tensor, 257
Chandrasekhar approximation, 218
Chandrasekhar limit, 247, 278
Chandrasekhar mass, 271
Chandrasekhar's equation of state, 218
Charge density, 18
Chemical potential, 108, 125, 128, 172
Cherenkov resonance, 190
Chern–Simons theory, 368
Chirped Pulse Amplification (CPA), 291
Christoffel symbols, 281
Classical diamagnetism, 157
Classical electrodynamics, 35
Classical map Hypernetted Chain (CHNC) methods, 131
Classical regime, 70
Closed-Time Path (CTP), 343
Coherent Wake Emission (CWE), 315
Cold plasma approximations, 154
Cold plasma discharges, 14
Collective oscillations, 48
Collisionless Vlasov equation, 76
Color-Flavor Locked (CFL) phase, 224
Communication systems, 19
Compton recoil, 99
Compton scattering, 99

P. Kumar, *Quantum Plasma*, https://doi.org/10.1007/978-981-95-0481-7

Conductivity tensor, 171
Continuity equation, 27, 112, 115, 118
Controlled fusion confinement, 21
Controlled nuclear fusion, 13
Convolutional Neural Networks (CNNs), 382
Coulomb interactions, 15, 17, 70
Coulomb potential, 15
Coulomb potential energy, 53
Coulomb repulsion, 123
Coulomb's law, 18
Coupling line, 68
Coupling parameter, 67
Crooks relation, 379
Current Density Functional Theory (CDFT), 132
Current filamentation, 198
Cyclotron frequency, 25, 154, 159, 161, 164, 171, 180

D
D'Alembertian, 274
De Broglie wavelength, 50, 57, 59, 61, 67, 70, 71, 98
Debye–Hückel length, 127, 226
Debye–Hückel theory, 262
Debye length, 16, 18, 116, 232
Degeneracy factor, 155
Degeneracy line, 68
Degeneracy parameter, 67
Degeneracy pressure, 172, 227
Degenerate plasmas, 78, 155
De Haas–van Alphen, 165, 170
De Haas–van Alphen effect, 156, 159
Density Functional Theory (DFT), 114, 121, 133
Density of States (DOS), 164
Diamond anvil cells, 329
Dielectric function, 53, 65, 89, 171
Dielectric response function, 107
Dielectric response tensor, 186
Dielectric tensor, 111, 139, 186
Diffusive Shock Acceleration (DSA), 304
Dimensionless parameter, 155
Dirac–Heisenberg–Wigner formalism, 287, 307
Dirac equation, 169, 170, 203, 310
Dirac field, 137
Dirac Hamiltonian, 167
Dirac operator, 169, 186
Dirac spinor, 306
Dispersion relation, 20, 21, 38, 39, 49, 50, 103, 107, 116, 119
Distribution function, 98
Drift-Kinetic Equation (DKE), 250
Druyvesteyn function, 29
Dust-Ion Acoustic Waves (QDIAWs), 233
Dusty astrophysical Pplasma, 232
Dusty quantum plasmas, 81
Dyson–Schwinger equations, 214

E
Ecker-Kröll approach, 332
Edge-Localized Modes (ELMs), 255
Effective Field Theory (EFT), 369
Einstein–Cartan geometry, 272
Electromagnetic fields, 19
Electromagnetic field tensor, 170, 274
Electromagneticwave propagation, 35
Electromagnetic waves, 17, 48
Electron-electron interactions, 328
Electron Energy Loss Spectroscopy (EELS), 334
Electron–ion quantum plasmas, 80
Electron–positron (e–p) plasmas, 80
Electrostatic confinement, 14, 23
Electrostatic Langmuir waves, 260
Electrostatic potential, 102, 107, 115, 128
Energy-momentum tensor, 274
Equation of State (EOS), 262
Euclidean functional integral, 144
Euler equation, 191
Euler–Maxwell system, 178
Exchange-correlation effects, 113, 191
Exchange-correlation free energy, 128
Exchange-correlation functional, 123, 124, 130
Exchange-correlation kernel, 132
Exchange-correlation potential, 79, 123
Exchange correlations, 72, 128, 213
Exchange interactions, 78, 206

F
Faraday's law, 41, 179, 180
Fermi–Dirac, 154
Fermi–Dirac distribution, 102, 108, 179, 226
Fermi–Dirac integrals, 127
Fermi energy, 60, 155
Fermi momentum, 172
Fermionic sign problem, 337
Fermionic wavefunction, 337
Fermion propagator, 89
Fermi pressure, 116, 178
Fermi's Golden Rule, 337

Fermi temperature, 71, 93, 108
Fermi velocity, 104
Fick's law, 211
Filamentation, 171
Fock term, 79
Foldy–Wouthuysen transformation, 167
Fourier transform, 103
Four-potential, 45
Four-vector potential, 169
Four-velocity, 213
Framework, 201
Free current density, 119
Fusion plasmas, 14

G
Gauge fields, 148
Gauge invariance, 137
Gauge-invariant Lagrangian, 143
Gauge-invariant Wigner function, 104
Gauge transformation, 43
Gauss's law, 179
Generalized Gradient Approximation (GGA), 132
Generative Adversarial Networks (GANs), 383
Grad-Shafranov equation, 244
Grassmann-valued external sources, 88
Gravitational confinement, 246
Gravitinos, 185
Green's function formalism, 338
Gross–Pitaevskii equations, 272
Gross–Pitaevskii-like models, 141
Gyromagnetic ratio, 197, 210
Gyrotropic pressure tensor, 204

H
Hall conductivity, 171
Hall drift instabilities, 239
Hall-effect thrusters, 14
Hall parameter, 30
Hamiltonian, 26
Hamiltonian density, 142
Hamiltonian evolution, 99
Hamiltonian formulations, 26
Hamiltonian Monte Carlo, 354
Hamiltonian of a many-body system, 141
Hartree energy, 128
Hartree–Fock approximations, 214
Hartree potential, 134
Hawking radiation, 382
Heaviside step function, 66, 164
Heisenberg's uncertainty principle, 218
Hermitian conjugate, 206
High-energy plasmas, 98
High-Harmonic Generation (HHG), 314
High-pressure shock tubes, 329
Hilbert space, 141
Hohenberg–Kohn theorem, 122, 127

I
Inertial Confinement Fusion (ICF), 24, 153, 256, 262, 329
Ionization Potential Depression (IPD), 332
Ionized, 93
Ion Temperature Gradient (ITG) modes, 261
Isospin chemical potentials, 186

J
Jarzynski equality, 379
Jüttner distribution, 295

K
Kadanoff–Baym equations, 99, 147
Kappa distribution, 29
Keldysh Dyson equation, 345
Keldysh-Schwinger formalism, 305
Kinetic stability, 24
Kinetic theory, 24
Kink instabilities, 37
Klein-Gordon-like, 169
Klein–Gordon–Maxwell system, 296, 297
Klein-Nishina formula, 318, 319
Kogut–Susskind formulation, 342
Kohn–Sham DFT, 86
Kohn–Sham equations, 123, 124
Kohn–Sham orbitals, 127, 134
Kohn–Sham potential, 134
Kolmogorov cascade, 368
Korteweg-de Vries (q-KdV) equation, 233
Kramers–Kronig relations, 53
Kruskal–Shafranov limit, 248

L
Lagrangian density, 137, 148
Lagrangian for a charged particle, 26
Landau critical field, 198
Landau damping, 65, 106
Landau diamagnetism, 158, 159
Landau dielectric function, 107
Landau gauge, 169
Landau levels, 155, 159, 163, 164, 169–171, 173, 197, 198, 202, 226

Landau–Lifshitz approximation, 278
Landau–Lifshitz equation, 307
Landau–Lifshitz prescription, 288
Landau-like damping, 104
Landau quantization, 153, 155, 156, 163, 165, 170, 171, 192, 198, 220, 226, 227, 255
Landé g-factor, 166
Langmuir oscillations, 108, 140
Langmuir waves, 20, 190
Larmor precession, 157
Larmor radius, 25, 244
Laser-Induced Fluorescence (LIF), 259
Laser-produced dense plasmas, 328
Laser-solid interactions, 198
Laser Wakefield Acceleration (LWFA), 312
Lattice gauge, 341
Lawson criterion, 264
Light-front field theory, 45
Lindblad-type operators, 148
Lindhard function, 226
Linear Response Theory (LRT), 52, 54
Liouville super-operator, 66
Liouville–von Neumann equation, 203
Local Density Approximation (LDA), 126, 128
Local Thermodynamic Equilibrium (LTE), 28
Long-range Coulomb forces, 19
Lorentz force, 25
Lorentz invariance, 46
Lowest Landau Level (LLL), 164, 170

M

Madelung–Bohm potential, 178
Madelung–Bohm quantum potential, 228
Madelung representation, 209
Madelung transformation, 82, 175
Magnetic Confinement Fusion (MCF), 247
Magnetic islands, 24
Magnetic permeability, 89
Magnetic quantization energy, 154
Magnetic reconnection, 12, 13, 255, 368
Magnetic susceptibility, 156, 158, 159
Magnetization current, 117, 185
Magnetization current density, 119
Magnetization energy, 176
Magnetization instabilities, 201
Magnetization-polarization tensor, 186
Magnetized Target Fusion (MTF), 258
Magnetoacoustic Instabilities, 189
Magnetohydrodynamic (MHD), 9, 153
Magnetohydrodynamic (MHD) equations, 264
Magnetohydrodynamic (MHD) equilibrium, 244
Magnetohydrodynamic (MHD) turbulence, 12
Magnetoplasmadynamic (MPD) thrusters, 14
Magneto-Rotational Instabilities (MRI), 228
Magnetosonic waves, 180
Magneto-spin solitons, 210
Magnetovortical instabilities, 171
Many-body Hamiltonian operator, 100
Markov Chain Monte Carlo (MCMC), 354
Mass continuity equation, 222
Matsubara formalism, 144
Matsubara frequencies, 143
Matsubara Green's functions, 89
Maxwell–Ampère's equation, 210
Maxwell-Boltzmann distribution, 26, 28, 110
Maxwell–fluid system, 268
Maxwellian field theory, 50
Maxwell–Jüttner distribution, 294
Maxwell–Podolsky equation, 350
Maxwell's displacement current, 36
Maxwell's equations, 21, 35, 38, 40, 43, 49, 50, 178
Mercier criteria, 248
Mermin free energy, 128
Metropolis–Hastings, 354
MHD instabilities, 24
Microturbulence, 249
Momentum equation, 27, 73, 83, 113, 117, 118
Momentum transport equation, 112
Monte Carlo (MC) methodology, 336
Monte Carlo simulations, 92, 336
Mori-Zwanzig approach, 111
Moyal brackets, 99, 101, 141, 148, 161
Moyal expansion, 101, 112
Moyal star product, 252
Multi-fluid QMHD formalism, 183
Multi-particle correlations, 121
Multi-species Quantum Magnetohydrodynamics, 182

N

Navier–Stokes equations, 113
Navier–Stokes models, 117
Nertial Confinement Fusion (ICF), 245

Neutron star magnetospheres, 12
Neutron stars, 220, 247
Noether's theorem, 297
Nonlinear Inverse Compton Scattering (NICS), 319
Non-locality, 48
Non-Markovian dynamics, 380
Non-Markovian extensions, 111
Non-Maxwellian distributions, 53
Nuclear Magnetic Resonance (NMR), 259

O
Ohm's law, 179, 180
Orbital-Free Density Functional Theory (OFDFT), 127
Orbital Magnetization, 158

P
Partially Observable Markov Decision Process (POMDP), 353
Particle-in-Cell (PIC), 335
Partition function, 159
Path Integral Monte Carlo (PIMC), 336
Path integral Monte Carlo (PIMC) simulations, 131
Path-Integral Quantization, 141
Pauli blocking, 104, 268
Pauli equation, 105, 197, 199, 202, 206
Pauli Hamiltonian, 207
Pauli Hamiltonian operator, 206
Pauli matrices, 105
Pauli paramagnetism, 158, 159
Pauli–Schrödinger equation, 176
Pauli's equation, 203, 209
Pauli susceptibility, 158, 168
Pauli-type Hamiltonian, 167
Perturbatively, 68
Phase space density, 99
Phase space volume, 99
Photon propagator, 45
Plasma arc welding, 14
Plasma Assisted Chemical Vapour Deposition (PECVD), 14
Plasma-based accelerators, 19
Plasma confinement, 19, 22
Plasma frequency, 17, 39, 104, 116, 155
Plasma mirrors, 314
Plasma nitriding, 14
Plasma oscillations, 19
Plasma torches, 14
Plasma transport, 19
Plasma-wall interactions, 14
Poisson's equation, 15, 82, 103
Polarization tensor, 89
Polarized quantum shocks, 209
Poynting's theorem, 41
Poynting vector, 41
Pressure anisotropy, 157, 188
Probability density, 99
Protoplanetary disks, 232
Protostellar collapse, 231
Pulsar winds, 12

Q
QED Monte Carlo methods, 288
QED-scale astrophysical magnetospheres, 182
Quantum Bohm force, 167
Quantum Bohm potential, 71, 203
Quantum Chapman–Enskog expansions, 267
Quantum Chromodynamics (QCD), 223, 304
Quantum-corrected Particle-In-Cell (Q-PIC) codes, 270
Quantum correlations, 98
Quantum coupling parameter, 68
Quantum degeneracy, 156
Quantum degeneracy parameter, 67, 70
Quantum degeneracy pressure, 70
Quantum density operator, 100
Quantum dielectric function, 107
Quantum diffraction, 71
Quantum Electrodynamic (QED), 43, 50, 51, 72, 247
Quantum Euler equation, 113, 156
Quantum Field Theory (QFT), 43
Quantum Fluid Models (QFMs), 147
Quantum Gyrokinetic Equations (Q-GKE), 252, 253
Quantum Hall-MHD, 254
Quantum Hall-MHD turbulence, 255
Quantum Hamilton-Jacobi equation, 73
Quantum Hydrodynamic (QHD) equations, 178
Quantum Hydrodynamics (QHD), 83, 111
Quantum interference, 210
Quantum kinetic theory, 99, 185
Quantum magnetic response, 159
Quantum magneto-hydrodynamic regime, 156
Quantum Magnetohydrodynamics (QMHD), 83, 178, 217
Quantum magnetosonic turbulence, 182

Quantum-modified Alfvén wave, 183
Quantum-modified Jeans Instability, 235
Quantum-modified Langmuir wave dispersion relation, 278
Quantum-modified Navier–Stokes, 191
Quantum-modified Rayleigh–Taylor instability, 235
Quantum-modified Weibel instability, 234
Quantum Monte Carlo (QMC) parametrizations, 131
Quantum Nonlinear Schrödinger equation (QNLS), 268
Quantum plasma parameter, 155
Quantum pressure, 114, 116, 156
Quantum pressure tensor, 120
Quantum radiation reaction, 99
Quantum scattering, 113
Quantum spin hydrodynamics, 168
Quantum spinor wavefunction, 205
Quantum stabilization scale, 190
Quantum statistics, 71, 72
Quantum streaming instabilities, 190
Quantum stress tensors, 209, 251
Quantum tunneling, 74
Quantum Vlasov equation, 185, 287
Quantum vorticity, 257
Quantum vorticity vector, 258
Quark-Gluon Plasma (QGP), 186, 223, 224

R
Radiation hydrodynamics, 14
Radiation reaction, 291
Radiative transfer processes, 230
Rankine–Hugoniot relations, 265
Rarita–Schwinger equation, 186
Rayleigh–Taylor modes, 279
Reactive Ion Etching (RIE), 14
Recurrent Neural Networks (RNNs), 382
Reinforcement Learning (RL), 352
Relativistic, 53
Relativistic degeneracy, 153
Relativistic degeneracy pressure, 218
Relativistic Mean-Field (RMF) theory, 222
Relativistic Oscillating Mirror (ROM), 315
Relativistic plasmas, 12
Relativistic Thomas–Fermi screening length, 223
Relativistic Vlasov equation, 294
Resonant momentum, 107
Ricci scalar curvature, 281
Riemann curvature, 281
Rindler metrics, 272
Runge-Gross theorem, 133

S
Schrödinger equation, 82, 115, 116, 133, 141
Schwinger critical field, 316
Schwinger effect, 316
Schwinger field, 307
Schwinger–Keldysh (closed-time path) formalism, 146
Schwinger–Keldysh formalisms, 120, 384
Schwinger–Keldysh framework, 146
Schwinger–Keldysh non-equilibrium Green's function, 99
Schwinger–Keldysh propagator, 373
Schwinger–Keldysh techniques, 287
Schwinger–Keldysh treatments, 141
Schwinger pair production, 342
Second-harmonic generation, 39
Self-focusing, 40
Semiclassical regime, 71
Shakura-Sunyaev α-viscosity prescriptions, 229
Sheath formation, 14
Shock waves, 12
Shubnikov–de Haas effects, 165, 170
Single-fluid quantum magnetohydrodynamics, 181
Single-particle wavefunctions, 85
Slater determinants, 126
Spin-Alfvén coupling, 183
Spin angular momentum, 208
Spin–charge coupled instabilities, 199
Spin coupling, 176
Spin degeneracy factor, 172
Spin-Density Functional Theory (SDFT), 213
Spin density matrix, 214
Spin density vector field, 208
Spin evolution equation, 118, 120
Spin flux tensor, 207, 208
Spin–gradient couplings, 203
Spin Hall effect, 199
Spin-induced instabilities, 204, 237
Spin–magnetic coupling, 153
Spin magnetization, 158, 166–168, 204
Spin–magnetization coupling, 208
Spin-magnetosonic waves, 210
Spin-modified screening, 199
Spin-modified shocks, 201
Spin operator, 166
Spin–orbit collisions, 204

Spin–orbit coupling, 161, 166–168, 192, 199, 200, 206
Spin–orbit force, 167
Spin–orbit interaction, 197
Spinor fields, 144, 148, 212
Spinor Green's functions, 214
Spinor quantum field theory, 213
Spinor wavefunction, 202
Spin–plasma waves, 201
Spin plasmons, 213
Spin polarization, 161, 185, 196, 251, 256
Spin-polarized Bernstein modes, 210
Spin-polarized plasma systems, 105
Spin precession, 197
Spin pressure tensor, 106, 257
Spin quantum hydrodynamics, 117
Spin-Resolved Electron Energy Loss Spectroscopy (SREELS), 372
Spin-selective confinement, 209
Spin–spin correlations, 210
Spin–spin interactions, 213
Spin–strain, 204
Spin torque density, 186
Spitzer-Härm conductivity, 246
SRS growth rate, 269
Star formation, 231
Statistical antisymmetrization, 98
Statistical ensemble averaging, 137
Stellarator, 13
Stimulated Brillouin Scattering (SBS), 269
Stimulated Raman Scattering (SRS), 269
Streaming instabilities, 189
Strongly coupled Quark-Gluon Plasma (sQGP), 304
Strongly coupled regime, 68
Supernova remnants, 232
Synchrotron radiation, 12

T
Taylor instability, 24
Tearing mode instability, 24
Terahertz reflectometry, 334
TFD approximation, 126
Thermal de Broglie wavelength, 154
Thermal to quantum pressure transition, 68
Thermal velocity, 116
Thermonuclear reactions, 14
The Uniform Electron Gas (UEG), 131
Third-harmonic generation, 39
Thomas–Fermi, 124
Thomas–Fermi–Dirac, 124
Thomas–Fermi–Dirac (TFD) models, 124
Thomas–Fermi–Dirac–Weizsäcker models, 177
Thomas–Fermi equation, 126
Thomas–Fermi kinetic energy, 125
Thomas–Fermi limit, 116
Thomas–Fermi model, 124
Thomas–Fermi relation, 125
Thomas–Fermi screening length, 127, 232, 263
Thomas–Fermi–Weizsäcker model, 127
Time-Dependent Current Density Functional Theory (TDCDFT), 114
Time-Dependent Density Functional Theory (TDDFT), 133, 342
Time-Dependent Kohn–Sham (TDKS) equation, 342
Time-Dependent Schrödinger Equation (TDSE), 73, 314
Tokamak, 13
Tolman-Ehrenfest effect, 281
Tolman–Oppenheimer–Volkoff (TOV) equation, 219, 222, 271
Tolman–Oppenheimer–Volkoff limit, 247
Topological Quantum Field Theory (TQFT), 368
Total energy functional, 123
Total susceptibility, 159
Trotter–Suzuki decomposition, 359
Tunneling, 71
Tunneling probability, 74
Turbulence, 12, 22, 53, 239
Two-Dimensional Electron Gases (2DEGs), 330
2PI generating functional, 214
2PI Schwinger–Keldysh techniques, 140
2PI spinor field theory, 214

U
Uehling–Uhlenbeck forms, 266
Uehling–Uhlenbeck (U) equation, 110
Ultracold plasma, 327
Ultrafast processes, 98
Ultra-strong coupling, 68
Uncertainty relations, 98
Unruh effect, 382

V
Vacuum Cherenkov radiation, 302
Vacuum fluctuations, 137
Vacuum polarization, 162
Van Allen radiation belts, 12
Variational Autoencoders (VAEs), 382

Variational Quantum Eigensolvers (VQE), 359
Vlasov equation, 65, 98, 102, 106, 109, 112, 200
Vlasov formalism, 99
Vlasov–Maxwell system, 161, 300
Vlasov–Pauli equations, 105, 167
Vlasov picture, 98
Vlasov–Poisson systems, 381
Volkov solution, 310
Volkov states, 310
Von Neumann equation, 76, 99, 100

W

Warm Dense Matter (WDM), 127
Wavefunction, 63, 71, 73, 82
Wave–particle interactions, 198
Weakly coupled regime, 68
Weibel growth mechanism, 188
Weibel growth rate, 188
Weibel instabilities, 37, 161, 187
Weyl transform, 75, 84, 109
Whistler dynamics, 254
White dwarfs, 91, 217, 247
Wigner–Boltzmann equation, 265
Wigner distribution, 99, 112, 140, 161, 176, 204
Wigner distribution function, 99
Wigner equation, 65, 86, 101–103, 107, 112, 147, 214
Wigner formalism, 85, 167
Wigner function, 65, 75, 84, 100–102, 105, 107, 109, 114, 116, 140, 148, 161, 162, 167, 172, 192, 200, 251, 287
Wigner kinetic theory, 102
Wigner–Liouville dynamics, 66
Wigner–Maxwell, 104
Wigner–Maxwell system, 187, 300, 307
Wigner–Moyal equation, 76, 106, 176, 259, 350
Wigner–Moyal formalism, 251, 358
Wigner–Moyal formulations, 384
Wigner–Moyal kinetic formalism, 116
Wigner–Pauli formalism, 197
Wigner–Pauli function, 202
Wigner–Poisson, 104
Wigner–Poisson approaches, 365
Wigner–Poisson system, 102
Wigner quasi-distribution function, 265
Wigner–Seitz radius, 61, 79, 93, 131, 237, 328
Wigner transform, 100, 105
Wigner transformation, 202
Wigner–Vlasov term, 188
Wigner–Weyl transform, 102
Wigner–Weyl transformation, 100
WKB approximation, 74

Z

Zakharov system, 54
Zeeman interaction, 197
Zeeman shift, 198
Zeeman splitting, 78
Zitterbewegung, 99, 203

GPSR Compliance

The European Union's (EU) General Product Safety Regulation (GPSR) is a set of rules that requires consumer products to be safe and our obligations to ensure this.

If you have any concerns about our products, you can contact us on ProductSafety@springernature.com

In case Publisher is established outside the EU, the EU authorized representative is:

Springer Nature Customer Service Center GmbH
Europaplatz 3
69115 Heidelberg, Germany

Batch number: 10725800

Printed by Printforce, the Netherlands